MYCELIUM
RUNNING

MYCELIUM RUNNING

How Mushrooms Can Help Save the World

PAUL STAMETS

TEN SPEED PRESS
Berkeley

All rights reserved. Published in the United States by Ten Speed Press, an imprint of the Crown Publishing
Group, a division of Random House, Inc., New York.
www.crownpublishing.com
www.tenspeed.com

Ten Speed Press and the Ten Speed Press colophon are registered trademarks of Random House, Inc.

Library of Congress Cataloging-in-Publication Data
Stamets, Paul.
  Mycelium running : how mushrooms can help save the world / Paul Stamets.
      p. cm.
  Includes bibliographical references and index.
  1. Mycelium. 2. Mushroom culture. 3. Fungi--Ecology. I. Title.
  QK601.S73 2005
  579.5163—dc22   2005015898

ISBN 978-1-58008-579-3

Printed in China through Colorcraft Ltd., Hong Kong

Cover design by Betsy Stromberg and Andrew Lenzer
Text design by Betsy Stromberg

15 14 13 12 11

First Edition

Dedicated to Dusty

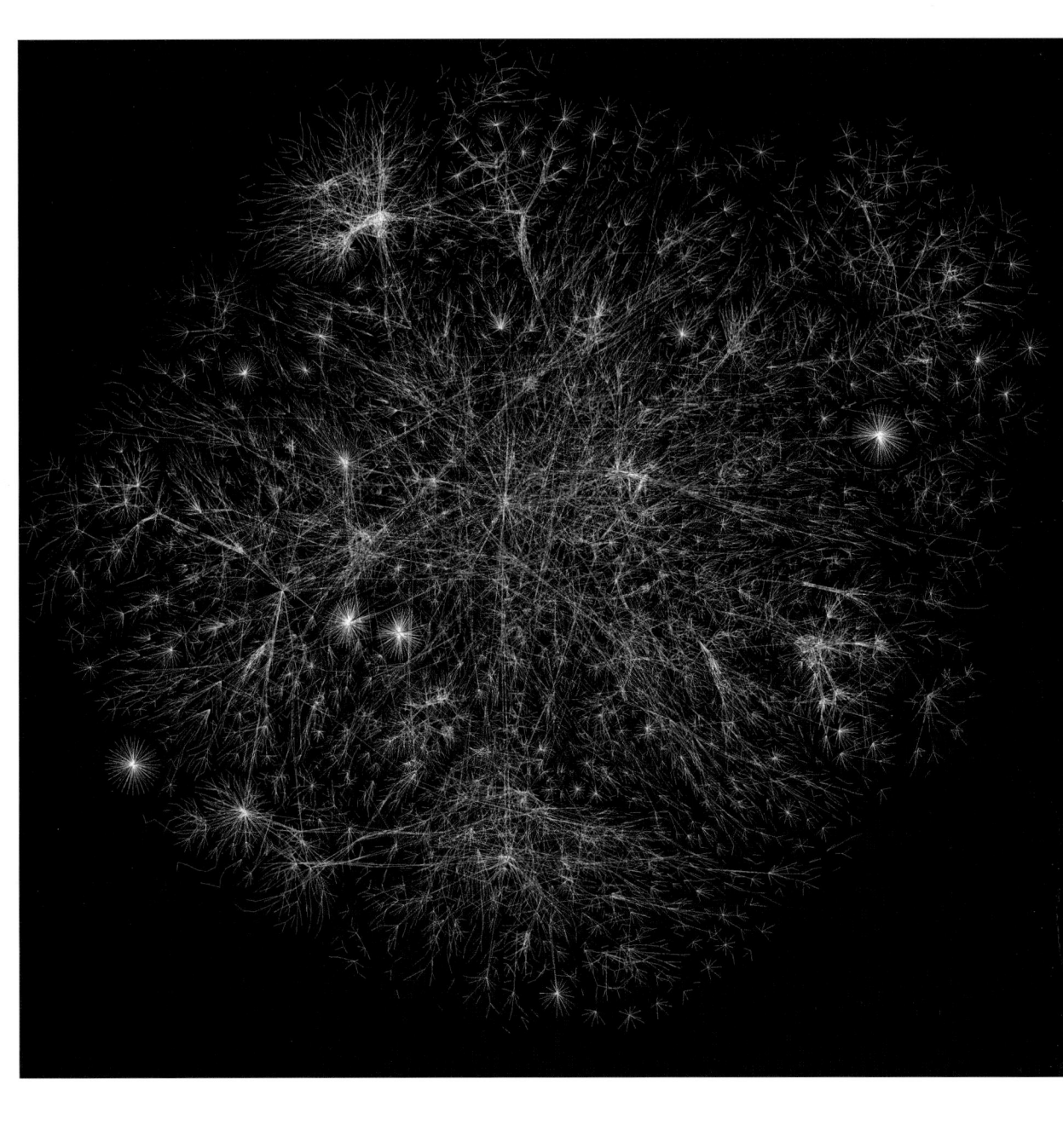

CONTENTS

FOREWORD

Mushrooms—ignored by many, reviled by some—may turn out to be important keys to both human health and planetary health. Their indispensable role in recycling organic matter, especially in forests, has long been known. But how many people realize that trees and other green plants could not grow and reach maturity without symbiotic associations with mushrooms, at least with mycelium, the network of fungal threads in soil that act as interfaces between plant roots and nutrients?

A mushroom is the reproductive structure or fruiting body of mycelium. Mycelium runs through our world, performing many other feats as well, but it is hidden and inconspicuous-a strange life form that has not attracted the same scientific attention as microorganisms or plants or animals. Even conventional mycologists hardly recognize its larger implications and possibilities.

Paul Stamets has never been a conventional thinker. I have known him for 25 years, and during that time, I have been repeatedly impressed by his insights into the interdependence of human beings and nature, his enthusiasm for harnessing and directing biological energies toward higher purposes, and his talent for thinking in novel and inventive ways. He has always looked at mushrooms from unique perspectives and as a result has made remarkable discoveries about them.

When we first met, I was questioning why Western medicine had never looked to mushrooms as sources of new therapeutic agents, given their prominence in the traditional pharmacopeias of China, Japan, and Korea. Paul took that question and ran with it, focusing on the natural competition that exists in soil between mycelium and bacteria. Fungi have evolved novel chemical defenses, a range of antibiotics that are often active against not only bacteria but also viruses and other infectious agents that cause disease in humans. One of the Big Ideas in this book is that fungi, especially fungi from old-growth forests, may be sources of new medicines that are active against a range of germs, including HIV/AIDS and the causative agents of smallpox and anthrax, potential bioterrorist threats.

Another of Paul's Big Ideas is that mycelium can be selected and trained to break down toxic waste, reducing it to harmless metabolites. He calls this strategy *mycoremediation* and has demonstrated its practicality in cleaning up oil spills. He suggests that our mushroom allies may even be able to detoxify chemical warfare agents.

This is one facet of a larger strategy that Paul calls *mycorestoration*, the use of fungi to improve the health of the environment: by filtering water, helping trees to grow in forests and plants to grow in gardens, and by controlling insect pests. The last possibility is especially noteworthy because it has the potential to neutralize pests like termites and fire ants by means that are completely nontoxic to human beings. Paul Stamets holds a number of patents in these areas, and I look forward to seeing his inventions put to use.

As a physician and practitioner of integrative medicine, I find this book exciting and optimistic because it suggests new, nonharmful possibilities for solving serious problems that affect our health and the health of our environment. Paul Stamets has come up with those possibilities by observing an area of the nat-ural world most of us have ignored. He has directed his attention to mushrooms and mycelium and has used his unique intelligence and intuition to make dis-coveries of great practical import. I think you will find it hard not to share the enthusiasm and passion he brings to these pages.

Cortes Island, British Columbia
June 2005

ANDREW WEIL, MD

PREFACE

For 30 years, I have engaged fungi, or perhaps they have engaged me, in a mission to promote the benefits of mushrooms. My previous books *Growing Gourmet and Medicinal Mushrooms* (2000a) and *The Mushroom Cultivator*, coauthored with Jeff Chilton (1983), delve into the methods of cultivating mushrooms. This new book is designed to show readers how to grow mushrooms in gardens, yards, and woods for the purpose of reaping both personal and planetary rewards. As you will discover, mushrooms help us reconnect to nature in profound ways. Mushrooms, mysterious and once feared, can be powerful allies for protecting the planet from the ecological injury we inflict.

More specifically, this book focuses on healing the planet using mycelial membranes, also known as *mycelium*, a fungal network of threadlike cells; it is a mycological manual for rescuing ecosystems. Engaging mycelium for healing habitats is what I call "mycorestoration." The umbrella concept of mycorestoration includes the selective use of fungi for mycofiltration, mycoforestry, mycoremediation, and mycopesticides. Mycofiltration uses mycelium to catch and reduce silt and catch upstream contaminants. Mycoforestry uses mycelium and mushrooms to enhance forest health. Mycoremediation neutralizes toxins. Mycopesticides refers to the use of fungi to help influence and control pest populations. This quartet of strategies can be used to improve soil health, support diverse food chains, and increase sustainability in the biosphere.

This book is written for a readership as diverse as the fungal community. For readers devoted to recycling, organic cultivation, habitat restoration, or applied mycology, I hope this book will be as useful as it is revolutionary. If you are a landscaper, bioremediator, ecoforester, sustainable-village planner, physician, scientist, futurist, or anyone who is passionately bemushroomed, I hope this book enriches your life and that you will pass on the love of mushrooms to future generations. And even if you have never walked through an old-growth forest, cultured fungus in a petri dish, relished a fresh-picked matsutake grilled over an open fire, or taken a mushroom-based medicine, I hope you will find this book—and my pragmatic environmental philosophy described herein—informative and inspiring. I contend that the planet's health actually depends on our respect for fungi. This book will show how you can help save the world using mushrooms.

ACKNOWLEDGMENTS

Writing this book has been an adventure of a lifetime, for which I am indebted to many people. First, to my wife, Dusty, I thank you for your love, companionship, humor, heart, and honor. Many thanks to Azureus and La Dena for all your help with my field work and special projects. To my brother Bill, I thank you for your skill in editing and challenges to my ideas that helped focus my vision. I am also grateful to Meghan Keeffe, Karen O'Donnell Stein, Jasmine Star, Laura Tennen, and Betsy Stromberg for their helpful editorial comments and stewardship as we navigated through the production of this book. To my family, especially my mother and my father, I am grateful for how you supported me with your love and for nurturing my scientific curiosity. To Phil Wood and Jo Ann Deck, thank you for placing your faith in me. To David Sumerlin, Steve Cividanes, Jimmy Gouin, David Brigham, Andrew Lenzer, Noelle Machnicki, Damein Pack, Natalie Parks, Kevin Schoenacker, Bulmaro Solano, George Osgood, Alex Winstead, and the other employees at Fungi Perfecti, I thank you for helping me more times than I can recount. To my mentors, Dr. Alexander Smith, Dr. Daniel Stuntz, and Dr. Michael Beug, who first encouraged me on this path, and to my ally and friend Dr. Andrew Weil, you hold special places in my heart.

Battelle Laboratories, and their mycoremediation team, including Jack Word, Susan Thomas, Ann Drum, Meg Pinza, Pete Becker, and others are acknowledged for their contributions. Roger Gold and Grady Glenn of Texas A&M University are thanked for their work on my mycopesticide projects. David Arora, Kenny Ausubel, William Hyde, Omon Isik-huemhen, Taylor Lockwood, Tom Newmark, Bill Nicholson, John Norris, David Price, Ethan Schaffer, Nina Simons, Phil Stern, and Solomon Wasser also helped in their special ways.

I also want to thank my critics: you have made me stronger, and no doubt you will continue to do so. I thank the thousands of mycologists, from shamans to scientists, whose collective experiences created the body-intellect that has become the springboard for the mycorestoration revolution. Last, I am humbled by the psilocybes who have been my mushroom spirit teachers. May future generations continue to build upon this foundation of knowledge to help the health of people and our planet.

Part I

THE MYCELIAL MIND

There are more species of fungi, bacteria, and protozoa in a single scoop of soil than there are species of plants and vertebrate animals in all of North America. And of these, fungi are the grand recyclers of our planet, the mycomagicians disassembling large organic molecules into simpler forms, which in turn nourish other members of the ecological community. Fungi are the interface organisms between life and death.

Look under any log lying on the ground and you will see fuzzy, cobweblike growths called *mycelium*, a fine web of cells which, in one phase of its life cycle, fruits mushrooms. This fine web of cells courses through virtually all habitats—like mycelial tsunamis—unlocking nutrient sources stored in plants and other organisms, building soils. The activities of mycelium help heal and steer ecosystems on their evolutionary path, cycling nutrients through the food chain. As land masses and mountain ranges form, successive generations of plants and animals are born, live, and die. Fungi are keystone species that create ever-thickening layers of soil, which allow future plant and animal generations to flourish. Without fungi, all ecosystems would fail.

With each footstep on a lawn, field, or forest floor, we walk upon these vast sentient cellular membranes. Fine cottony tufts of mycelium channel nutrients from great distances to form fast-growing mushrooms. Mycelium, constantly on the move, can travel across landscapes up to several inches a day to weave a living network over the land. But mycelium benefits our environment far beyond simply producing mushrooms for our consumption.

Humans collaborate with these cellular networks, using fungi, specifically using mushroom mycelium as spawn, for both short- and long-term benefits. Mushroom spawn lets us recycle garden waste, wood, and yard debris, thereby creating mycological membranes that heal habitats suffering from poor nutrition, stress, and toxic waste. In this sense, mushrooms emerge as environmental guardians in a time critical to our mutual evolutionary survival.

I believe random selection is no longer the dominant force of human evolution. Our political, economic, and biotechnological policies may determine our future, for better or worse. Some forecasts claim that half of the current species could disappear in the next hundred years if current trends continue. A "what-if" Pentagon report issued in October 2003, *An Abrupt Climate Change Scenario and Its Implications for United States National Security* (Schwartz and Randall 2003), hypothesizes that a more dire and imminent collapse of our biosphere may occur as climates radically destabilize as a result of pollution and global warming.

I wonder what would happen if there were a United Organization of Organisms (UOO, pronounced "uh-oh"), where each species gets one vote. Would we be voted off the planet? The answer is pretty clear. When we irresponsibly exploit the Earth, disease, famine, and ecological collapse result. We face the possibility of being rejected by the biosphere as a virulent organism. But if we act as a responsible species, nature will not evict us. Our fungal friends equip us with tools to act responsibly and repair our shared environment, leading the way to habitat recovery. So knowing how to work with fungi—by custom pairing fungal species with plant communities—is critical for our survival. The twenty-first century may be remembered as the Biotech Age, when these kinds of mycotechnologies play a prominent and increasing role in strengthening habitat health.

CHAPTER 1

Mycelium as Nature's Internet

I believe that mycelium is the neurological network of nature. Interlacing mosaics of mycelium infuse habitats with information-sharing membranes. These membranes are aware, react to change, and collectively have the long-term health of the host environment in mind. The mycelium stays in constant molecular communication with its environment, devising diverse enzymatic and chemical responses to complex challenges. These networks not only survive, but sometimes expand to thousands of acres in size, achieving the greatest mass of any individual organism on this planet. That mycelia can spread enormous cellular mats across thousands of acres is a testimonial to a successful and versatile evolutionary strategy.

The History of Fungal Networks

Animals are more closely related to fungi than to any other kingdom. More than 600 million years ago we shared a common ancestry. Fungi evolved a means of externally digesting food by secreting acids and enzymes into their immediate environs and then absorbing nutrients using netlike cell chains. Fungi marched onto land more than a billion years ago. Many fungi partnered with plants, which largely lacked these digestive juices. Mycologists believe that this alliance allowed plants to inhabit land around 700 million years ago. Many millions of years later, one evolutionary branch of fungi led to the develop-

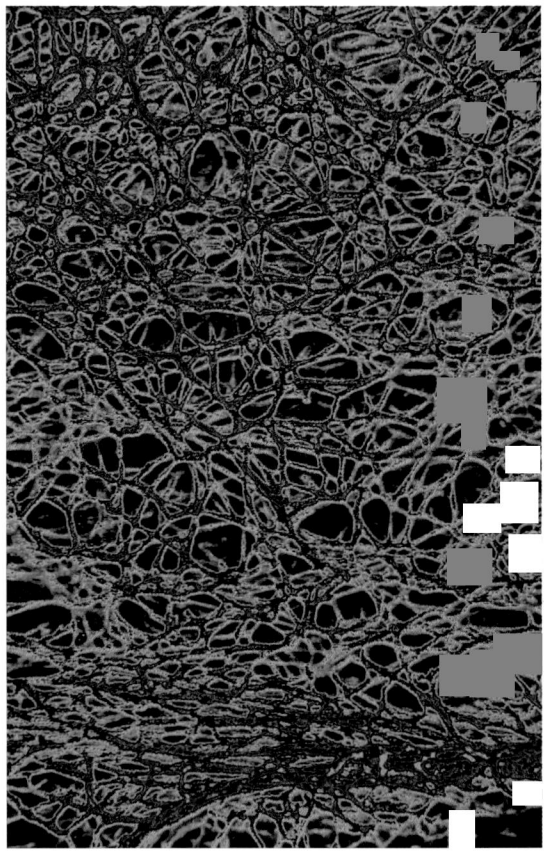

▲ **FIGURE 1**

The mycelial network is composed of a membrane of interweaving, continuously branching cell chains, only one cell wall thick.

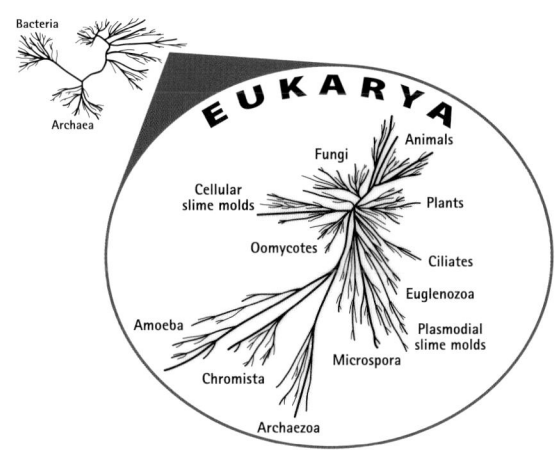

> **FIGURE A**

Evolutionary Branches of Life. Animals have a more common ancestry with fungi than with any other kingdom, diverging about 650 million years ago. A new super-kingdom, Opisthokonta, has been erected to encompass the kingdoms Fungi and Animalia under this one taxonomic concept (Sina et al. 2005).

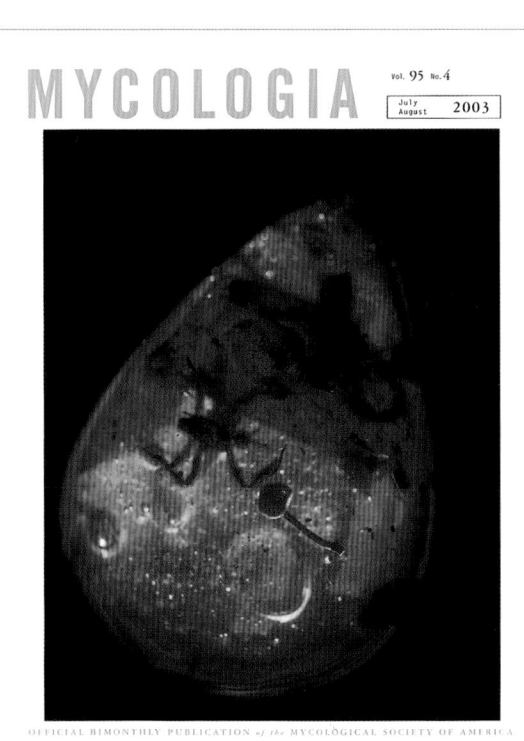

▲ **FIGURE 2**

The journal *Mycologia* featured this 15- to 20-million-year old amber with a mushroom embedded, now called *Aureofungus yaniguaensis,* dating from Miocene time and collected in the Dominican Republic. The oldest mushrooms in amber are estimated at 90 to 94 million years old.

ment of animals. The branch of fungi leading to animals evolved to capture nutrients by surrounding their food with cellular sacs, essentially primitive stomachs. As species emerged from aquatic habitats, organisms adapted means to prevent moisture loss. In terrestrial creatures, skin composed of many layers of cells emerged as a barrier against infection. Taking a different evolutionary path, the mycelium retained its net-like form of interweaving chains of cells and went underground, forming a vast food web upon which life flourished.

About 250 million years ago, at the boundary of the Permian and Triassic periods, a catastrophe wiped out 90 percent of the Earth's species when, according to some scientists, a meteorite struck. Tidal waves, lava flows, hot gases, and winds of more than a thousand miles per hour scourged the planet. The Earth darkened under a dust cloud of airborne debris, causing massive extinctions of plants and animals. Fungi inherited the Earth, surging to recycle the postcataclysmic debris fields. The era of dinosaurs began and then ended 185 million years later when another meteorite hit, causing a second massive extinction. Once again, fungi surged and many symbiotically partnered with plants for survival. The classic cap and stem mushrooms, so common today, are the descendants of varieties that predated this second catastrophic event. (The oldest known mushroom—encased in amber and

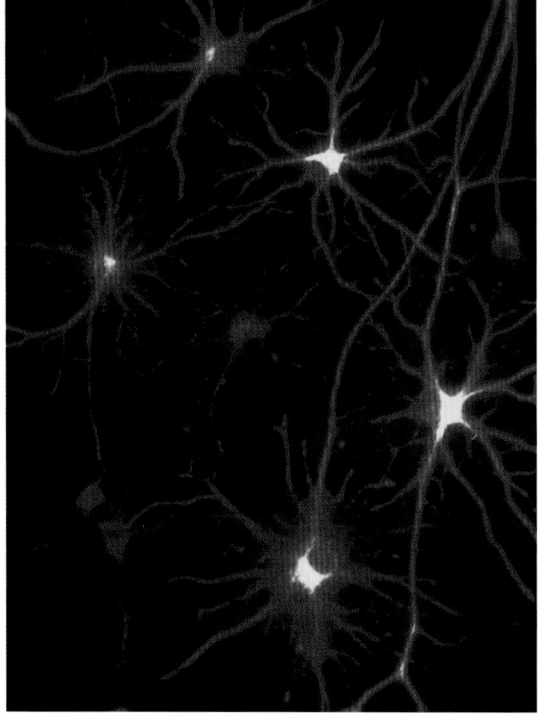

▲ **FIGURE 3**

Micrograph of astrocytic brain cells. Networking of neurons creates pathways for distributing information. Mycelial nets share this same architecture.

collected in New Jersey—dates from Cretaceous time, 92 to 94 million years ago. Mushrooms evolved their basic forms well before the most distant mammal ancestors of humans.) Mycelium steers the course of ecosystems by favoring successions of species. Ultimately, mycelium prepares its immediate environment for its benefit by growing ecosystems that fuel its food chains.

Ecotheorist James Lovelock, together with Lynn Margulis, came up with the Gaia hypothesis, which postulated that the planet's biosphere intelligently piloted its course to sustain and breed new life. I see mycelium as the living network that manifests the natural intelligence imagined by Gaia theorists. The mycelium is an exposed sentient membrane, aware and responsive to changes in its environment. As hikers, deer, or insects walk across these sensitive filamentous nets, they leave impressions, and mycelia sense and respond to these movements. A complex and resourceful structure for sharing information, mycelium can adapt and evolve through the ever-changing forces of

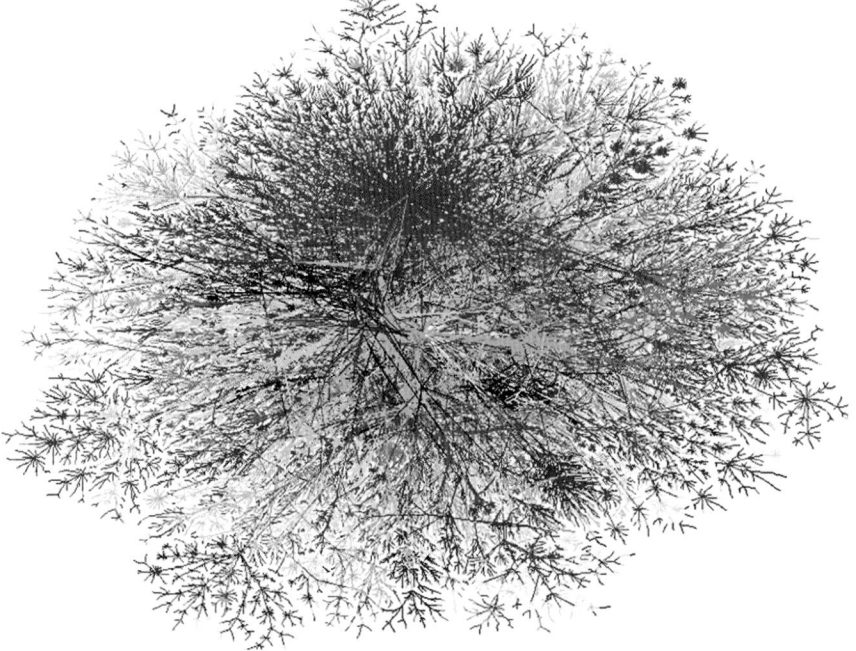

➤ **FIGURE 4**

A diagram of the overlapping information-sharing systems that comprise the Internet.

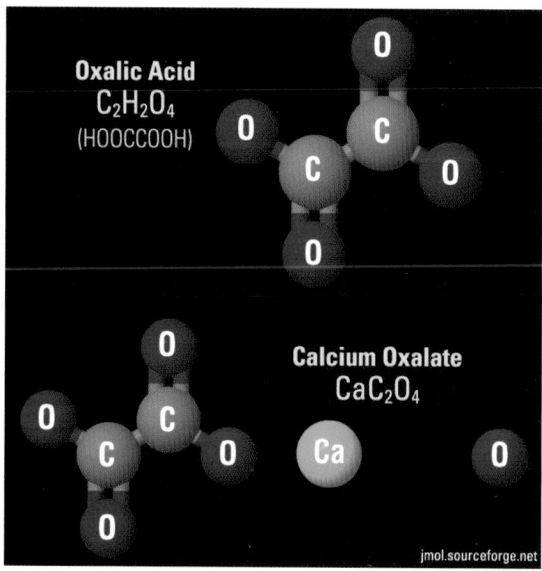

◄ ▼ **FIGURES B AND C**

Oxalic acid and calcium oxalate. Oxalic acid crystals are formed by the mycelia of many fungi. Oxalic acid mineralizes rock by combining with calcium and many other minerals to form oxalates, in this case calcium oxalate. Calcium oxalate sequesters two carbon dioxide molecules. Carbon-rich mushroom mycelia unfold into complex food webs, crumbling rocks as they grow, creating dynamic soils that support diverse populations of organisms. Below: Scanning electron micrograph of calcium oxalate crystals forming upon mycelium.

▲ FIGURE D

Prototaxites was the name given to this fossil—a remnant of a life form approximately 420 million years old, existing at the end of the late Silurian and through the beginning of the Devonian periods. Found in Canada and Saudi Arabia, this organism was widespread across the landscapes of the Paleozoic era. First described in 1859, this fossil remained a mystery until C. Kevin Boyce and others proved that it was a giant fungus in 2007.

◄ FIGURE E

Artist depiction of Prototaxites, which was the tallest known organism on land in its time, laying down or standing upright. The tallest plants, featured next to Prototaxites, were less than a meter high.

Parrish

nature. I especially feel that this is true upon entering a forest after a rainfall when, I believe, interlacing mycelial membranes awaken. These sensitive mycelial membranes act as a collective fungal consciousness. As mycelia's metabolisms surge, they emit attractants, imparting sweet fragrances to the forest and connecting ecosystems and their species with scent trails. Like a matrix, a biomolecular superhighway, the mycelium is in constant dialogue with its environment, reacting to and governing the flow of essential nutrients cycling through the food chain.

I believe that the mycelium operates at a level of complexity that exceeds the computational powers of our most advanced supercomputers. I see the mycelium as the Earth's natural Internet, a consciousness with which we might be able to communicate. Through cross-species interfacing, we may one day exchange information with these sentient cellular networks. Because these externalized neurological nets sense any impression upon them, from footsteps to falling tree branches, they could relay enormous amounts of data regarding the movements of all

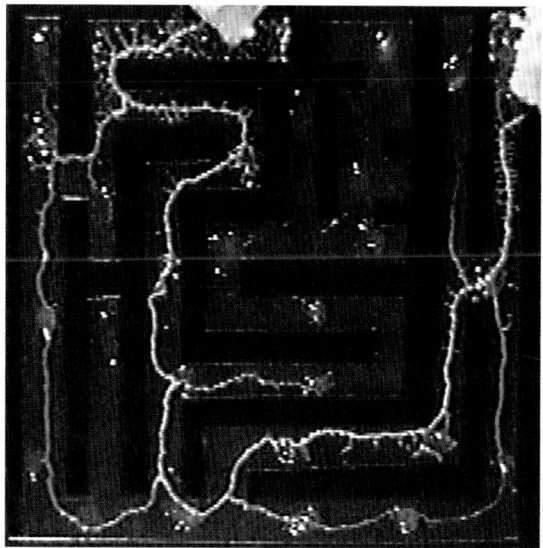

▲ **FIGURE 5**

A slime mold, *Physarum polycephalum,* chooses the shortest route between 2 food sources in a maze, disregarding dead ends. In a controversial article, Toshuyiki Nakagaki proposes that this represents a form of cellular intelligence.

◄ **FIGURE 6**

Computer model of the early universe. These primeval filaments in space resemble the mycelial archetype.

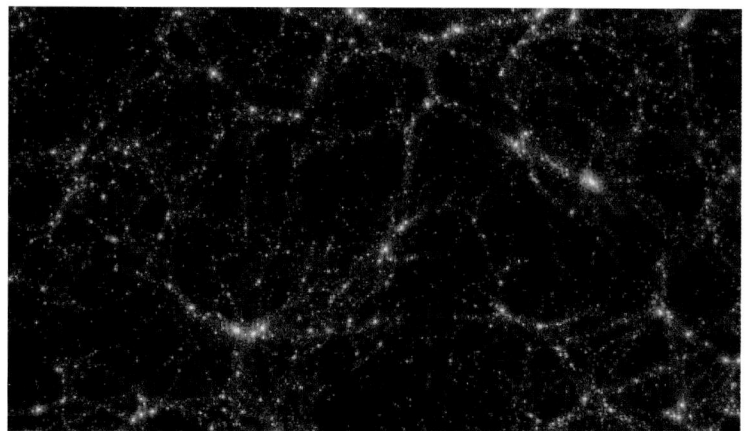

➤ **FIGURE 7**

Computer model of dark matter in universe. In a conjunct of string theory, more than 96 percent of the mass of the universe is theorized to be composed of these molecular threads. Note the galaxies interspersed throughout the myceliumlike matrix.

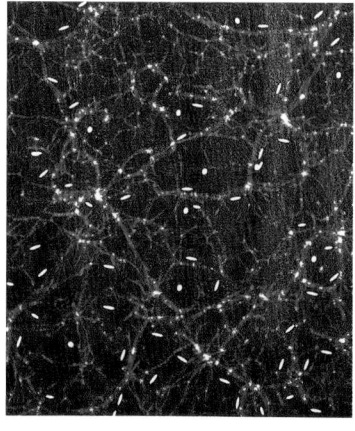

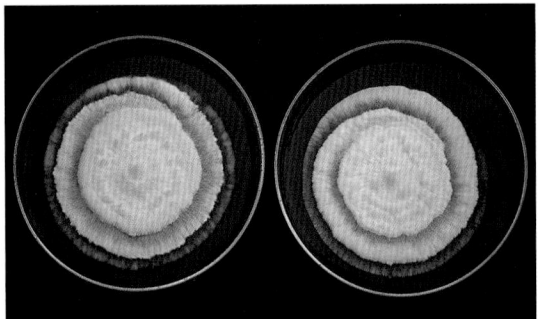

▲ **FIGURE 8**

Cultures of this yet-to-be-named Californian *Psilocybe* mushroom swirl like a cyclone as they grow outward; the rate of growth increases with time.

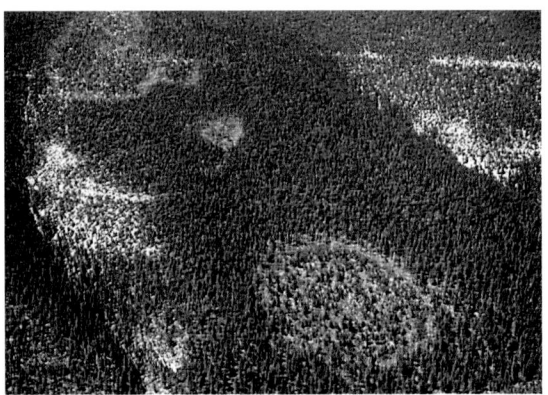

▲ **FIGURE 9**

Several mycelial mats of the root-rot *Armillaria* mushroom spiral outward, killing a forest in Montana. Once these trees die, they become highly flammable. (See also figure 60 for a larger patch of *Armillaria*, the largest organism in the world.)

organisms through the landscape. A new bioneering science could be born, dedicated to programming myconeurological networks to monitor and respond to threats to environments. Mycelial webs could be used as information platforms for mycoengineered ecosystems.

The idea that a cellular organism can demonstrate intelligence might seem radical if not for work by researchers like Toshuyiki Nakagaki (2000). He placed a maze over a petri dish filled with the nutrient agar and introduced nutritious oat flakes at an entrance and exit. He then inoculated the entrance with a culture of the slime mold *Physarum polycephalum* under sterile conditions. As it grew through the maze it consistently chose the shortest route to the oat flakes at the end, rejecting dead ends and empty exits, demonstrating a form of intelligence, according to Nakagami and his fellow researchers. If this is true, then the neural nets of microbes and mycelia may be deeply intelligent.

A few recent studies support this novel perspective—that fungi can be intelligent and may have potential as our allies, perhaps being programmed to collect environmental data, as suggested above, or to communicate with silicon chips in a computer interface. Envisioning fungi as nanoconductors in mycocomputers, Gorman (2003) and his fellow researchers at Northwestern Uni-

versity have manipulated mycelia of *Aspergillus niger* to organize gold into its DNA, in effect creating mycelial conductors of electrical potentials. NASA reports that microbiologists at the University of Tennessee, led by Gary Sayler, have developed a rugged biological computer chip housing bacteria that glow upon sensing pollutants, from heavy metals to PCBs (Miller 2004). Such innovations hint at new microbiotechnologies on the near horizon. Working together, fungal networks and environmentally responsive bacteria could provide us with data about pH, detect nutrients and toxic waste, and even measure biological populations.

Fungi in Outer Space?

Fungi may not be unique to Earth. Scientists theorize that life is spread throughout the cosmos, and that it is likely to exist wherever water is found in a liquid state. Recently, scientists detected a distant planet 5,600 light-years away, which formed 13 billion years ago, old enough that life could have evolved there and become extinct several times over (Savage et al. 2003). (It took 4 billion years for life to evolve on Earth.) Thus far 120 planets outside our solar system have been discovered, and more are being discovered every few months. Astrobiologists believe that the precursors of DNA, prenucleic acids, are forming throughout the cosmos as an inevitable consequence of matter as it organizes, and I have little doubt that we will eventually survey planets for mycological communities. The fact that NASA has established the Astrobiology Institute and that Cambridge University Press has established *The International Journal for Astrobiology* is strong support for the theory that life springs from matter and is likely widely distributed throughout the galaxies. I predict an *Interplanetary Journal of Astromycology* will emerge as fungi are discovered on other planets. It is possible that protogermplasm could travel throughout the galactic expanses riding upon comets or carried by stellar winds. This form of interstellar protobiological migration, known as *panspermia*, does not sound as far-fetched today as it did when first proposed by Sir Fred Doyle and Chandra Wickramasinghe in the early 1970s. NASA considered the possibility of using fungi for interplanetary colonization. Now that we have landed rovers on Mars, NASA takes seriously the unknown consequences that our microbes will have on seeding other planets. Spores have no borders.

The Mycelial Archetype

Nature tends to build upon its successes. The mycelial archetype can be seen throughout the universe: in the patterns of hurricanes, dark matter, and the Internet. The similarity in form to mycelium may not be merely a coincidence. Biological systems are influenced by the laws of physics, and it may be that mycelium exploits the natural momentum of matter, just like salmon take advantage of the tides. The architecture of mycelium resembles patterns predicted in string theory, and astrophysicists theorize that the most energy-conserving forms in the universe will be organized as threads of matterenergy. The arrangement of these strings resembles the architecture of mycelium.

▲ **FIGURE 10**

Hurricane Isabella approaches North America in October 2003.

▲ **FIGURE 11**

Spiral galaxies conform to the same archetypal pattern as hurricanes and mycelium.

When the Internet was designed, its weblike structure maximized the pooling of data and computational power while minimizing critical points upon which the system is dependent. I believe that the structure of the Internet is simply an archetypal form, the inevitable consequence of a previously proven evolutionary model, which is also seen in the human brain; diagrams of computer networks bear resemblance to both mycelium and neurological arrays in the mammalian brain (see figures 3 and 4). Our understanding of information networks in their many forms will lead to a quantum leap in human computational power (Bebber et al. 2007).

Mycelium in the Web of Life

As an evolutionary strategy, mycelial architecture is amazing: one cell wall thick, in direct contact with myriad hostile organisms, and yet so pervasive that a single cubic inch of topsoil contains enough fungal cells to stretch more than 8 miles if placed end to end. I calculate that every footstep I take impacts more than 300 miles of mycelium. These fungal fabrics run through the top few inches of virtually all landmasses that support life, sharing the soil with legions of other organisms. If you were a tiny organism in a forest's soil, you would be enmeshed in a carnival of activity, with mycelium constantly moving through subterranean landscapes like cellular waves, through dancing bacteria and swimming protozoa with nematodes racing like whales through a microcosmic sea of life.

Year-round, fungi decompose and recycle plant debris, filter microbes and sediments from runoff, and restore soil. In the end, life-sustaining soil is created from debris, particularly dead wood. We are now entering a time when mycofilters of select mushroom species can be constructed to destroy toxic waste and prevent disease, such as infection from coliform or staph bacteria and protozoa and plagues caused by disease-carrying organisms. In the near future, we can orchestrate selected mushroom species to manage species successions. While mycelium nourishes plants,

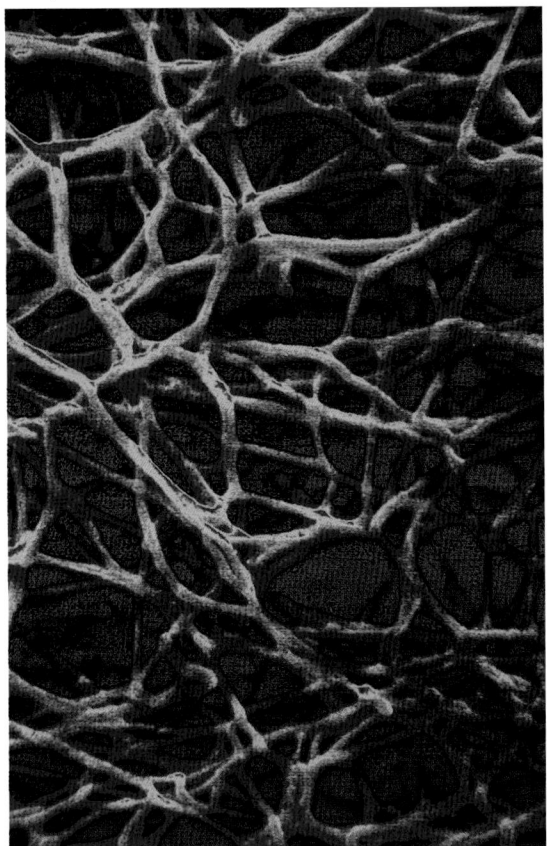

▲ **FIGURE 12**

Close-up of mycelium.

mushrooms themselves are nourishment for worms, insects, mammals, bacteria, and other, parasitic fungi. I believe that the occurrence and decomposition of a mushroom pre-determines the nature and composition of down-stream populations in its habitat niche.

Wherever a catastrophe creates a field of debris—whether from downed trees or an oil spill—many fungi respond with waves of mycelium. This adaptive ability reflects the deep-rooted ancestry and diversity of fungi—resulting in the evolution of a whole kingdom populated with between 1 and 2 million species. Fungi outnumber plants at a ratio of at least 6 to 1. About 10 percent of fungi are what we call mushrooms (Hawksworth 2001), and only about 10 percent of the mushroom species have been identified, meaning that our taxonomic knowledge of mushrooms is exceeded by our ignorance by at least one order of magnitude. The surprising diversity of fungi speaks to the complexity needed for a healthy environment. What has been become increasingly clear to mycologists is that protecting the health of the environment is directly related to our understanding of the roles of its complex fungal populations. Our bodies and our environs are habitats with immune systems; fungi are a common bridge between the two.

All habitats depend directly on these fungal allies, without which the life-support system of the Earth would soon collapse. Mycelial networks hold soils together and aerate them. Fungal enzymes, acids, and antibiotics dramatically affect the condition and structure of soils (see page 18). In the wake of catastrophes, fungal diversity helps restore devastated habitats. Evolutionary trends generally lead to increased biodiversity. However, due to human activities we are losing many species before we can even identify them. In effect, as we lose species, we are experiencing devolution—turning back the clock on biodiversity, which is a slippery slope toward massive ecological collapse. The interconnectedness of life is an obvious truth that we ignore at our peril.

In the 1960s, the concept of "better living through chemistry" became the ideal as plastics, alloys, pesticides, fungicides, and petrochemicals were born in the laboratory. When these synthetics were released into nature, they often had a dramatic and initially desirable effect on their targets. However, events in the past few decades have shown that many of these inventions were in fact bitter fruits of science, levying a heavy toll on the biosphere. We have now learned that we must tread softly on the web of life, or else it will unravel beneath us.

Toxic fungicides like methyl bromide, once touted, not only harm targeted species but also non-targeted organisms and their food chains and threaten the ozone layer. Toxic insecticides often confer a temporary solution until tolerance is achieved. When the natural benefits of fungi have been repressed, the perceived need for artificial fertilizers increases, creating a cycle of chemical dependence, ultimately eroding sustainability. However, we can create mycologically sustainable environments by introducing plant-partnering fungi (mycorrhizal and endophytic) in combination with mulching with saprophytic mushroom mycelia. The results of these fungal activities include healthy soil, biodynamic communities, and endless cycles of renewal. With every cycle, soil depth increases and the capacity for biodiversity is enhanced.

Living in harmony with our natural environment is key to our health as individuals and as a species. We are a reflection of the environment that has given us birth. Wantonly destroying our life-support ecosystems is tantamount to suicide. Enlisting fungi as allies, we can offset the environmental damage inflicted by humans by accelerating organic decomposition of the massive fields of debris we create—through everything from clear-cutting forests to constructing cities. Our relatively sudden rise as a destructive species is stressing the fungal recycling systems of nature. The cascade of toxins and debris generated by humans destabilizes nutrient return cycles, causing crop failure, global warming, climate change and, in a worst-case scenario, quickening the pace towards eco-catastrophes of our own making. As ecological disrupters, humans challenge the immune systems of our environment beyond their limits. The rule of nature is that when a species exceeds the carrying capacity of its host environment, its food chains collapse and diseases emerge to devastate the population of the threatening organism. I believe we can come into balance with nature using mycelium to regulate the flow of nutrients. The age of mycological medicine is upon us. Now is the time to ensure the future of our planet and our species by partnering, or running, with mycelium.

CHAPTER 2

The Mushroom Life Cycle

For you to use mycelia as healing membranes, a basic understanding of the mushroom life cycle is helpful. Although we notice mushrooms when they pop up, their sudden appearance is the completion of cellular events largely hidden from view—until the inquisitive mycophile digs deeper. Although mycologists have a basic understanding of the mushroom life cycle, we are clueless how mushroom species interact with most other organisms coexisting in the same habitat. With each nuance revealed, the body-intellect of mycology expands, and our knowledge slowly inches forward. What is so exciting about mycology is that the depth of undiscovered knowledge laying before us is more vast than our minds can imagine.

Mushrooms reproduce through microscopic spores, visible as dust when they collect en masse. When the moisture, temperature, and nutrients are right, spores freed from a mushroom (essentially mushroom seeds) germinate into threads of cells called *hyphae*. As each hypha grows and branches, it forms connections with other hyphae from compatible spores to create a mycelial mat, which matures, gathering nutrients and moisture. From mycelium, cells aggregate to form a primordium—called "pinheads" or baby mushrooms by growers. Under optimal conditions, the transformation from spores to mycelium to mushroom can take just a few days.

Mushrooms can be divided into 2 basic categories depending upon how they form: predeterminant or

➤ **FIGURE 13**

Depiction of the mushroom life cycle.

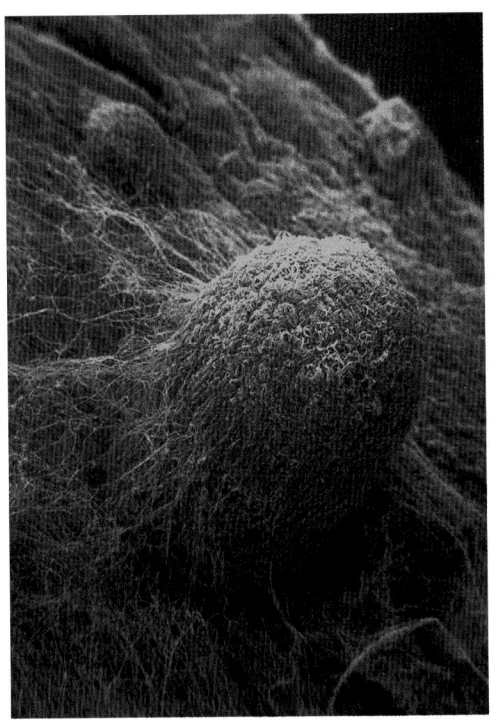

▲ **FIGURE 14**

Scanning electron micrograph of primordium forming from a mycelial mat.

▲ FIGURE 15

A baby mushroom is called a *primordium,* a stage between mycelium and mature mushroom.

▲ FIGURE 16

An example of an indeterminant mushroom species, a *Ganoderma,* perhaps *Ganoderma curtisii,* a sister species to reishi (*Ganoderma lucidum).* The mushrooms formed and grew around twigs and grass—the latter of which remains green, vibrant, and healthy, despite being surrounded by fungal tissue, a phenomenom I find peculiar, and biologically interesting.

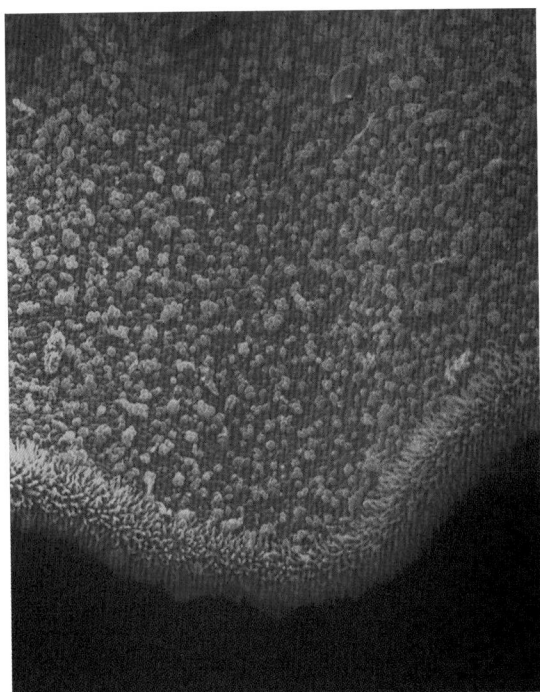

▲ FIGURE 17

Low magnification of a mushroom gill plate showing the gill edge and surface plane populated with spore-producing basidia.

indeterminant. Most mushrooms are predeterminant, meaning the stem, cap, and gills preform in the primordial state. If the young primordia are damaged, deformities appear in adulthood.

Less common are the indeterminant mushrooms, including many *Ganodermas, Phaeolus schweinitzii,* and the rare *Bridgeoporus nobilissimus.* Their mycelia form primordia that envelop sticks and twigs as they grow. If these young mushrooms are damaged at this stage and go on to recover, they mature with little trace of wounds.

Mushrooms display many artful forms, adapted for the purpose of dispersing spores: classic button mushroom, hoof-shaped conk (which has many pores, and hence is called a *polypore*), ridge-forming chanterelle, toothed *Hericium,* coral-like *Ramaria,* leafy *Sparassis,* and cup-forming *Auricularia.* These mushrooms, so diverse in shape, produce spores from similar clublike structures called *basidia,* which arise from a specialized layer of cells called the *hymenium*. In oyster and button mushrooms, the hymenial layer covers the surfaces of the gills. Despite their anatomical differences, these mushrooms produce microscopic spores in a similar way.

Many mushrooms launch spores from basidia, which populate the gills on oyster mushrooms, for instance, and emerge in increasing quantities as the mushroom body matures. The vast majority of species produce 4-spored basidia, which are jettisoned in pairs with enough force to throw them inches away from the mushroom (see figures 18 and 19). Nicholas Money (1998) measured this force as 25,000 g's, approximately 10,000 times the forces experienced by the space shuttle astronauts escaping the gravitational pull of the Earth to obtain orbit.

Although spores tend to fall near their parent mushroom, trails of spores can sometimes be seen wafting in the air. Correspondingly, spores tend to be most concentrated closest to the ripening mushroom, with the concentration decreasing exponentially with distance. However, many insects and mammals also participate in distribution. Drawn by the mushrooms' scent, insects use them as a home for their larvae, which then grow up and carry spores with them when they leave the nest. Mammals eat mushrooms for nourishment, and many spores survive digestion and are dispersed through the animals' waste. Mycologist James Trappe of Oregon State University showed that voles and flying squirrels ate subterranean truffles in old-growth forests, and in turn, spotted owls ate the flying squirrels and voles. (However, scientists do not yet know whether the scat from these

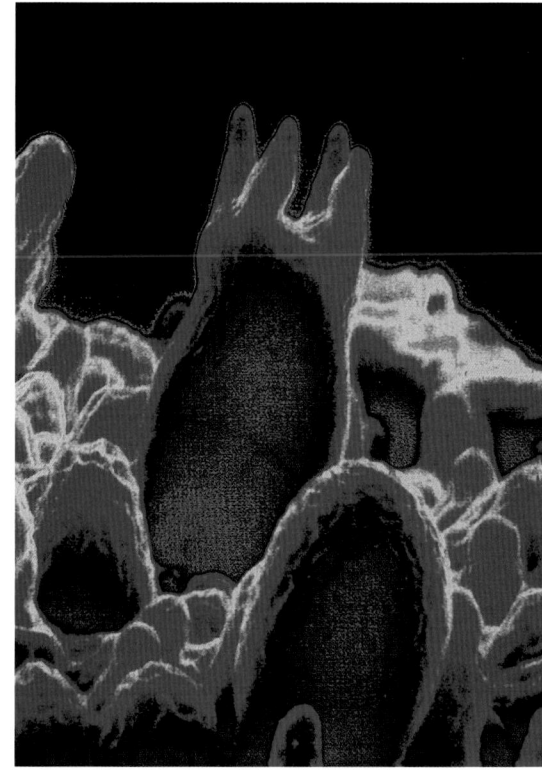

▲ **FIGURE 18**

Emerging young basidium.

➤ **FIGURE 19**

Mature basidium just before spore release.

▲ FIGURE 20

Cedar Cividanes reaches upward to touch the underside of a large specimen of the artist conk *(Ganoderma applanatum)* in the old-growth forest of the Duckabush River basin. In the Pacific Northwest, this mushroom produces prodigious quantities of spores from late spring through early fall.

spotted owls harbors viable truffle spores.) He discovered that these mammals' diets are dependent on truffle mushrooms, and that from the animals' fertile fecal droppings, the subterranean truffle mushroom is assured wider dispersal of its spores through the forest. This interdependency between animals and fungi is only one example of many in nature. That so many mushrooms compete for distribution and safe harbors for their spores may be one reason why so many spores are necessary.

David Arora reports in *Mushrooms Demystified* (1986) that a large *Ganoderma applanatum* is estimated to liberate up to 30 billion spores a day, and more than 5 trillion a year! (See figure 20.) This prodigious output of spores is necessary for fungi to find new habitats in which to thrive. Species like chanterelles are slow to release spores, typically producing mushrooms that persist and continue to release spores for many weeks, in contrast to fast-collapsing inky caps, which sporulate and liquefy within hours. Species vary in the timing and duration of spore release, depending on temperature, moisture, habitat, their animal partners, and their own constitution.

Within a species, younger, thicker-fleshed mushrooms are typically more succulent than older ones and correspondingly have fewer spores. With oysters and buttons, for instance, the flesh above the gills, thick when young, thins as each wave of spores is released by successions of basidia. Generally, when a mature mushroom stops producing spores, it becomes

◄ FIGURE 21

From the artist conk featured in the previous image, we took a thumbnail-size slice of tissue back to the laboratory, where we broke it in half, cut out a tiny fragment, and transferred it to a nutrient-filled petri dish to start a culture. The resulting mushroom that grew is genetically identical to the wild artist conk from which it came. The original mushroom, whose small wound soon healed over, still survives in the old-growth forest. I encourage such low-impact practices for collecting cultures without removing the mushrooms from their ecosystem.

an essential food source for people, deer, bears, squirrels, voles, and insects from gnats to arthropods, and no doubt influences legions of other organisms in the food chain.

Once spores are produced, most are quick to germinate. The spores of some mushrooms, like oysters, can germinate as soon as they leave the basidia and find a hospitable niche, whereas others, like shiitake, germinate more readily after drying out and then rehydrating. With many mushroom species, germination begins in the dimpled depression on the spore. In the first minutes, this process looks like that of a seed sprouting. The sproutlike hypha mitotically divides. Next comes the mating of hyphae from 2 compatible spores, each of which is mononucleate, having half of the code necessary for producing fertile offspring. After their mating, when the hyphae fuse to form one mycelium, the resulting cellular network, called a *dikaryon*, is invigorated, binucleate, and capable of producing descendant fertile mushrooms with spore-bearing ability. In the laboratory and in nature, cultures from mated spores grow far faster than mycelium originating from a single spore.

You can grow mushrooms from spores or tissue. If you are creating your own cultures, it is essential that you use mushrooms that are fresh. If fresh mushrooms are not available, you can purchase cultures (spawn) or spores from commercial sources. What are

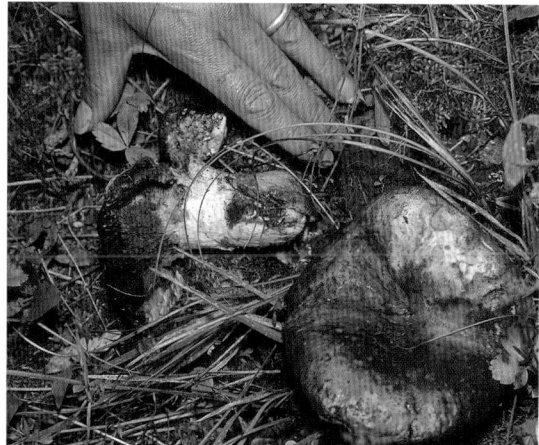

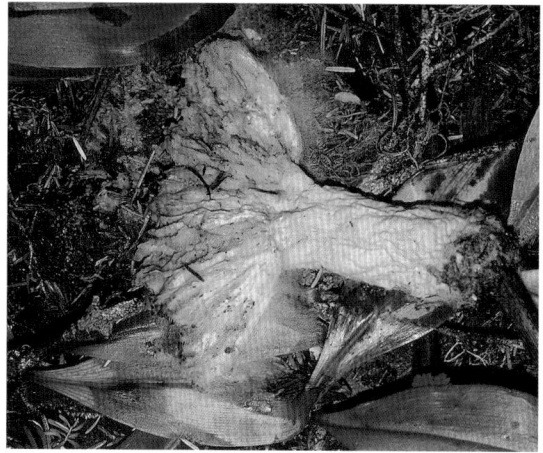

➤ FIGURES 22, 23, AND 24

After a *Russula* mushroom climaxes and disintegrates, its spores germinate into a mycelial matrix. Days later, the mycelium spreads from the disintegrated parent mushroom's corpse, forming a mycelial network. Such surface mycelia soon submerge into the duff or soil, disappearing from view. Mycelia can be found under practically any log, stick, bale of straw, cardboard, or other organic material on the ground. In a gram of this myceliated soil, more than 1 mile of cells form; in a cubic inch more than 8 miles. In this photo, my hiking boot covers approximately 300 miles of mycelium. Hence from a mycelium's point of reference, a journey of 10,000 miles is only 33 plus footsteps!

the differences between cultures created from spores and those created from tissue? Each mating of 2 spores expresses but one of several possible phenotypes from the genome of the contributing mushrooms. In contrast, using a piece of living tissue from the mushroom—cloning—captures the exact genetic composition of the contributing mushroom. Cloning usually requires knowledge of sterile tissue culture technique and a clean room laboratory. (For more information on these techniques, refer to the books listed in the paragraph below.) Many mushrooms can also be propagated naturally from broken stem butts, which is another, although low-tech, form of cloning (see chapter 9). When stem butts regrow, or if you clone a mushroom by taking a piece of internal flesh and placing it on a petri dish filled with sterilized media, you are capturing the exact individual mushroom in hand. This book reveals easy-to-use techniques using spores, spawn, and stem butts for getting mushrooms into culture without needing a laboratory.

For more detailed descriptions of mushroom life cycles, see my book *Growing Gourmet and Medicinal Mushrooms* (2000a). I also highly recommend *The Fungi* by Carlile, Watkinson, and Gooday (2001), and *Fungal Morphogenesis* by David Moore (1998), both of which are available through www.fungi.com.

◀ **FIGURES 25, 26, AND 27**

The path of decomposition: wood chips; wood chips colonized by mycelium; myceliated wood chips after digestion by worms and other organisms.

CHAPTER 3

Mushrooms in Their Natural Habitats

Mushrooms can be placed in 4 basic categories: *saprophytic, parasitic, mycorrhizal,* and *endophytic,* depending upon how they nourish themselves. However, exceptions abound, since some species employ more than one strategy, making them difficult to categorize. Approximately 8,000 macrofungi (visible to the naked eye) are saprophytic, around 2,000 to 3,000 are mycorrhizal, and the remaining are either endophytic or parasitic, although more species are constantly being discovered and categorized. The balance of populations can vary drastically with environmental change, however: deforestation causes a rise in saprophytes and a decline in mycorrhizal mushrooms, for example. Now let's take a short tour through the 4 major categories of mushrooms.

Saprophytic Mushrooms: The Decomposers

Saprophytic mushrooms, the decomposers, steer the course for proliferating biological communities, shaping and forming the first menus in the food web from dead plants, insects, and other animals. Most gourmet and medicinal mushrooms are wood decomposers, the premier recyclers on the planet; building soils is the primary outcome of the activities of these saprophytic fungi, whose filamentous mycelial networks weave through and between the cell walls of plants. When organic matter falls from the canopy of trees and plants overhead onto the forest floor, the decomposers residing in the soil process this newly available food. (Competition is intense: on the forest floor, a single "habitat" can actually be matrices of fungal networks sharing one space.) These fungi secrete enzymes and acids that degrade large molecules of dead plants into simpler molecules, which the fungi can reassemble into building blocks, such as polysaccharides, for cell walls. From dead plants, fungi recycle carbon, hydrogen, nitrogen, phosporus, and minerals into nutrients for living plants, insects, and other organisms sharing that habitat.

▲ FIGURE 28

Turkey tail *(Trametes versicolor)* fruiting on a conifer log deep in old-growth forest in Olympic National Park.

➤ **FIGURE 29**

These towering old-growth trees near Mount Rainier, grow out of thin soil but gather nutrients from afar from their mycelium-supported roots. In fact, most plants are supported by vast and complex colonies of fungi working in concert. Here I point to *Bridgeoporus nobilissimus* (for a closer view see figure 50), a mushroom exclusive to old-growth habitat and the first fungus to be listed as an endangered species.

As decomposers, saprophytic mushrooms can be separated into 3 key groups: primary, secondary, and tertiary, although some mushroom species can cross over from one category to another, depending upon circumstances. Primary, secondary, and tertiary decomposers can all coexist in one location. Primary and secondary decomposers such as oyster and meadow mushrooms are the easiest to cultivate.

Primary Decomposers

These saprophytes are typically the first to grow on a twig, a blade of grass, a chip of wood, a log, a stump, or a dead insect or other animal. Primary decomposers are typically fast growing, sending out rapidly extending strands of mycelium that quickly attach to and decompose plant tissue. These woodland species include oyster mushrooms (*Pleurotus* species), shiitake (*Lentinula edodes*), and maitake (*Grifola frondosa*). However, species employ different sets of enzymes to break down plant matter into varying stages of decomposition.

Secondary Decomposers

Secondary decomposers rely on the activity of primary fungi that initially, although partially, break down plant and animal tissues. Secondary decomposers all work in concert with actinomycetes, other bacteria, and fungi, including yeasts, in soil in the forest floor or in compost piles. Heat, water, carbon dioxide, ammonia, and other gases are emitted as by-products of the composting process. Once the microorganisms (especially actinomycetes) in the compost piles complete their life cycles, the temperature drops, encouraging a new wave of secondary decomposers.

Cultivators exploit this sequence to grow the white button mushroom (*Agaricus bisporus*), the most widely cultivated mushroom in the world. Other secondary

▼ **FIGURE 30**

David Arora, author of *Mushrooms Demystified* and *All That the Rain Promises and More* is positioned to take a photograph of a family of ambiguous Stropharias, *Stropharia ambigua,* near my home.

My daughter, La Dena Stamets, sits beside the garden giant *(Stropharia rugoso annulata)*, which is deep burgundy in color when young and fades as it matures, sometimes achieving a majestic stature. This mushroom can be both a primary and a secondary saprophyte but is dependent upon soil microbes for fruiting.

▲ **FIGURE 32**

Commercial button mushroom *(Agaricus bisporus)* cultivation in Holland. This mushroom is a classic secondary saprophyte, growing on compost.

saprophytes that compete with compost-grown mushrooms are inky caps (belonging to the family Coprinaceae, which includes the choice, edible shaggy mane *[Coprinus comatus]* and others including the hallucinogenic *Panaeolus subbalteatus* and *Panaeolus cyanescens*); and, in outdoor wood chip beds, the ambiguous Stropharia *(Stropharia ambigua)*. Industrial growers try to thwart these undesired invaders by heat steaming their composts to temperatures inhospitable to their spores.

Secondary decomposers, as a group, seem more versatile than primary decomposers for dealing with complex assortments of microorganisms, since they have evolved in direct contact with microbially rich soils. Secondary decomposers typically grow from composted material. The best culinary *Stropharia* species, the garden giant, or king Stropharia, *(Stropharia rugoso annulata)* is an example of an intermediary between primary and secondary decomposers since this species first digests fresh debris and then continues to thrive as complex communities of microbes join with it to create soil.

▲ **FIGURE 33**

The honey mushroom *(Armillaria ostoyae)* fruiting from a stump.

Tertiary Decomposers

This difficult-to-categorize group includes fungi found toward the end of the decomposition process. They thrive in habitats created by primary and secondary decomposers over a period of years, often popping up from soils holding little decomposable material. Tertiary decomposers include species of *Conocybe, Agrocybe, Mycena, Pluteus,* and *Agaricus.* Tertiary decomposers rely upon highly complex microbial environments. The division between secondary and tertiary decomposers is often obscure; mycologists simply call tertiary decomposers "soil dwellers," for lack of a better description. Some mushrooms initially act as parasites, and once they have killed their hosts, they act like saprophytes, growing on their dead remains. Honey mushrooms belonging to the genus *Armillaria* are good examples of species that grow both parasitically and saprophytically.

Parasitic Mushrooms: Blights of the Forest or Agents for Habitat Restoration?

Parasites are predators that endanger the host's health. In the past, foresters saw all parasitic fungi as hostile to the long-term health of forests. Although they do parasitize trees, they nourish other organisms. Parasitic fungi such as the honey mushroom, which can destroy thousands of acres of forest, are stigmatized as blights. However, more foresters are realizing that a rotting tree in the midst of a canopied forest is, in fact, more supportive of biodiversity than a living tree. Parasitic mushrooms may be nature's way of selecting the strongest plants and repairing damaged habitats. Ultimately, parasitic mushrooms set the stage for the revival of weakened habitats that are too stressed to thrive.

Of all the parasitic blight mushrooms that are edible by humans, the assorted honey mushrooms such as *Armillaria mellea* and *Armillaria ostoyae* are the best known. One mycelial mat from a honey mushroom (*Armillaria bulbosa*) made national headlines when a specimen was found in a Michigan forest that covered 37 acres, weighed at least 50 tons, and was estimated to be 1,500 years old. In Oregon, a far larger honey mushroom (*Armillaria ostoyae*) mycelial mat found on a mountaintop covers more than 2,400 acres and is possibly more than 2,200 years old (see figure 60). Each time this fungus blight sweeps through, nurse logs are created, soil depth increases, and centimeters of soil accumulate to create ever-richer habitats where once only barren rock stood. (For further discussion of *Armillaria* blights, see page 47.) What makes mushroom mycelia different from the mycelia from mold fungi is that some mushroom species can grow into massive membranes, thousands of acres in size, hundreds of tons in mass, and thousands of years old.

Many saprophytic fungi can be weakly parasitic, especially if a host tree is dying from other causes, such as environmental stress or parasite infestation. Saprophytes that can take advantage of a dying tree are termed *facultative* parasites. For example, oyster mushrooms (*Pleurotus ostreatus*) are classic saprophytes, although they are frequently found on dying cottonwood, oak, poplar, birch, maple, and alder trees. And although reishi (*Ganoderma lucidum*) is considered a true saprophyte by most mycologists, the Australian Quarantine Inspection Service has classified this medicinal species as a parasite and has banned its importation. Authorities on other islands including New Zealand and Hawaii also consider this mushroom a threat to their native trees. Some parasitic fungi behave like saprophytes, such as honey mushrooms (*Armillaria mellea* and *Armillaria ostoyae*), which may be found thriving on the corpse of their tree host.

Most parasitic fungi, however, are microfungi, barely visible to the naked eye, but en masse they inflict cankers and lesions on the shoots and leaves of trees. Often their prominence in a middle-aged forest is symptomatic of other imbalances in the ecosystem, such as acid rain, groundwater pollution, and insect damage. After a tree dies, parasitic fungi may inhabit the tree, competing with saprophytes for dominance. Since the hosts for some parasites can be short-lived, natural selection sometimes favors fast growers. Foresters have observed this with *Phytophthora ramorum,* the cause of

Matsutake, which are mycorrhizal mushrooms known to mycologists as *Tricholoma magnivelare,* growing deep in the old-growth forest of Washington State.

sudden oak disease; this downy mildew pathogen can kill an ancient oak tree in days and an ancestral forest in a few weeks, and remain viable on the dead carcasses of its victims, allowing a new staging platform for infection further into the forest.

Mycorrhizal Mushrooms: Fungus and Plant Partnerships

Mycorrhizal mushrooms (*myco* means "mushroom"; *rhizal* means "related to roots"), such as matsutake, boletus, and chanterelles, form mutually beneficial relationships with pines and other plants. In fact, most plants from grasses to Douglas firs have mycorrhizal partners. The mycelia of fungal species that form exterior sheaths around the roots of partner plants are termed *ecto*mycorrhizal. The mycorrhizal fungi that invade the interior root cells of host plants are labeled *endo*mycorrhizal, although currently the preferred term for these fungi is *vesicular arbuscular mycorrhizae* (VAM). Both plant and mycorrhizae benefit from this association. Because ectomycorrhizal mycelium grows beyond the plant's roots, it brings distant nutrients and moisture to the host plant, extending the absorption zone well beyond the root structure. The mycelium dramatically increases the plant's ingestion of nutrients, nitrogenous compounds, and essential elements (phosphorus, copper, and zinc) as it decomposes surrounding debris. David Perry (1994) postulates that the surface area—hence its absorption capability—of mycorrhizal fungi may be 10 to 100 times greater than the surface area of leaves in a forest. As a result, the growth of plant partners is accelerated. Plants with mycorrhizal fungal partners can also resist diseases far better than those without. Fungi benefit from the relationship because it gives them access to plant-secreted sugars, mostly hexoses that the fungi convert to mannitols, arabitols, and erythritols.

One of the most exciting discoveries in the field of mycology is that the mycorrhizae can transport nutrients to trees of different species. One mushroom species can connect many acres of a forest in a continuous network of cells. In one experiment, researchers compared the flow of nutrients via the mycelium between 3 trees: a Douglas fir (*Pseudotsuga menziesii*), a paper birch (*Betula papyrifera*), and a western red cedar (*Thuja plicata*). The Douglas fir and paper birch shared the same ectomycorrhiza, while the cedar had an endomycorrhiza (VAM). The researchers covered the Douglas fir to simulate deep shade, thus lowering the tree's ability to photosynthesize sugars. In response, the mycorrhizae channeled sugars, tracked by radioactive carbon, from the root zone of the birch to the root zone of the fir. More than 9 percent of the net carbon

▲ FIGURE 35

Dusty Yao happily holds her harvest of wild porcinis, the mycorrhizal *Boletus edulis,* collected in the mountains above Telluride, Colorado.

▲ FIGURE 36

Jim Gouin is pleased to find these delicious matsutakes *(Tricholoma magnivelare),* a mycorrhizal mushroom, in the mountains somewhere within 200 miles of Seattle, Washington.

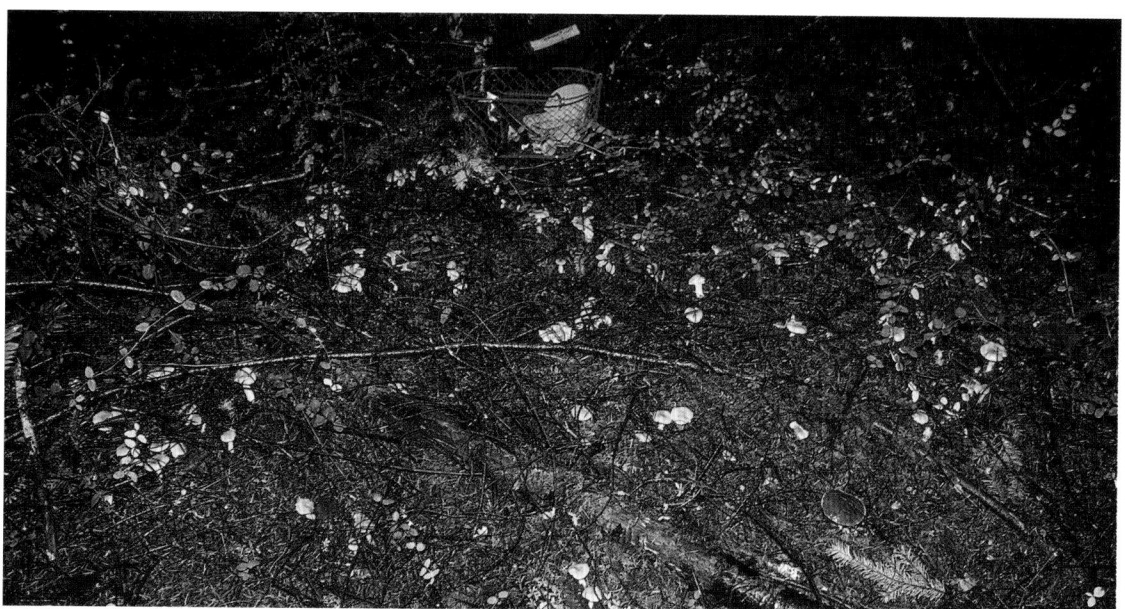

▲ FIGURE 37

Eureka! My basket awaits a bountiful collection of these apricot-smelling chanterelles, probably *Cantharellus formosus,* a mycorrhizal mushroom species growing in a 40-year-old Douglas fir forest near Olympia, Washington. My practice is to pick no more than 25 percent of the mushrooms of a wild patch, leaving young ones, and when encountering pairs of mushrooms, only pick one of them. Chanterelles tend to form as twins, so cutting one mushroom near to the ground saves the other twin, allowing it to mature, sporulate, and spread.

compounds transferred to the fir originated from the birch's roots, while the cedar received only a small fraction. The amount of sugar transferred was directly proportional to the amount of shading (Simard et al. 1997). An earlier study by Kristina Arnebrant and others (1993) showed a similar bidirectional transfer of nitrogen-based nutrients from alder *(Alnus glutinosa)* to pine *(Pinus contorta)* through a shared ectomycorrhizal mycelium.

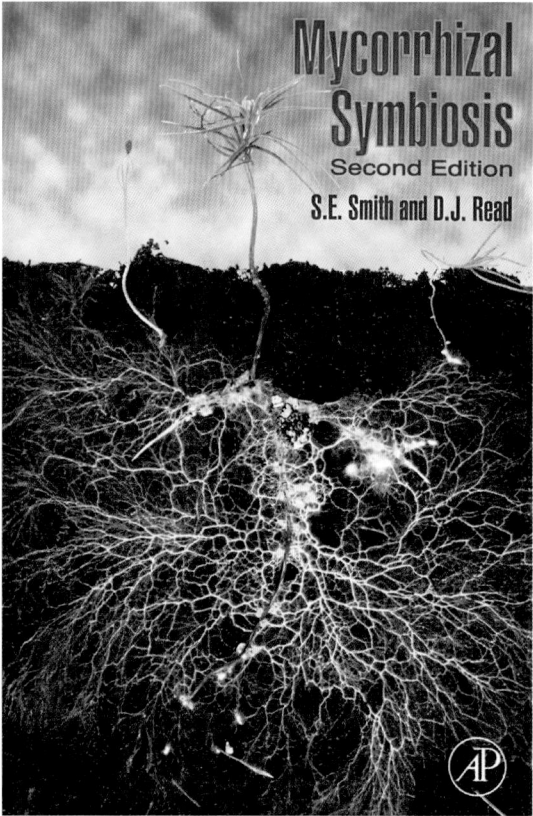

▲ FIGURE 38

On the cover of this excellent book, the roots of a young pine tree *(Pinus sylvestris)* are enveloped with the mycelium of the mushroom *Suillus bovinus*. The mycelium extends the tree's range for absorbing nutrients and water while conferring a fungal defense against invasive diseases. This symbiotic pairing is the norm in nature, not the exception.

The Simard experiment showed that a common mycelial net could unite 3 species of trees and underscored a remarkable ability of mycorrhizal fungi: mycorrhizae can keep diverse species of trees in forests fed, particularly younger trees struggling for sunlight. Now we have a better understanding of how saplings survive in the shadows of elder trees that tower overhead and block out essential light. The fact that a single mycorrhizal mushroom nutritionally supported 2 different trees—one a conifer and the other deciduous—shows that the mycelium guards the forest's overall health, budgeting and multidirectionally allocating nutrients.

Another example of a fungus and plant partnership is the matsutake, which has a unique relationship with the non-chlorophyll-producing candystick plant *(Allotropa virgata)*. The candystick gains virtually all its sugars from the matsutake mycelium and the western hemlock and/or Sitka spruce with which it associates (Hosford et al. 1997; Trudell et al. 2003). One mycologist I know speculates that the spot fruitings of matsutake *(Tricholoma magnivelare)* on a slope of Oregon's Mount Hood may, in fact, be from a vast interconnected mycelial colony extending over thousands of acres. A further example is the bigleaf maple *(Acer macrophyllum)*, which projects vinelike aerial roots that ascend to the canopy of Pacific Northwest rain forests and are teamed with mycorrhizae.

Our understanding of the role of fungi in the forest is far more advanced than the simplistic views held just a few decades ago. Most ecologists now recognize that a forest's vitality is directly related to the presence, abundance, and variety of mycelial associates. A large portion, one-tenth to one-fifth, of the total biomass in the topsoil of a healthy Douglas fir forest in the Pacific Northwest may be made up of mycelium, and even more if we include the endomycorrhizae and ectomycorrhizae that thrive in the canopy. I doubt a forest can be defined without its fungi.

Growing mycorrhizal mushrooms has proved to be a greater challenge than first anticipated due to the complex interdependencies in which fungi play a critical role. Once the hurdle of establishing mycorrhizal

Bigleaf maples *(Acer macrophyllum)* grow in the rain forest of the Olympic Mountains, in Washington State. Research by Cobb et al. (2001) showed that this maple extends roots on its outer trunk that climb into the canopy, essentially creating a biosphere high above the forest floor. The biomass of these aerial roots is similar to the biomass of the subterranean roots. Upon these aerial roots, a complex habitat has evolved, including mosses (nonvascular epiphytes) and licorice ferns (*Polypodium glycyrrhiza;* a vascular epiphyte), once thought to be parasitic to the tree but now known to be part of the tree's healthy ecosystem.

▲ **FIGURE 40**

The Perigord truffle *(Tuber melanosporum),* is one of the most sought-after and highly regarded gourmet mushrooms in the world. This mushroom is mycorrhizal, growing in association with filberts and oak trees.

mycelium has been overcome, decades may pass before a single mushroom forms. Nuances of climate, soil chemistry, and predominant microflora limit our success in cultivating mycorrhizal mushrooms in natural settings. The challenge we face is to tilt the balance so that a species of our choice can take up residence in such a complex natural setting—to design habitats in which it can grow. Species native to a region are more likely than imported species to adapt readily to these designed habitats.

Many American growers hope for huge profits when they try to grow European truffles, mycorrhizal mushrooms that sell at very high prices. In an attempt to duplicate the well-established truffle orchards in France, Spain, and Italy, where the renowned Perigord black truffle *(Tuber melanosporum)* fetches up to $500 per pound, dozens of growers have tried to cultivate nonnative European truffles around the American oaks or filberts on their land. Capitalizing on this desire, sev-

eral companies now market truffle-inoculated trees for commercial use, and calcareous (high in calcium) soils in Texas, Washington, and Oregon have been suggested as ideal sites for these. One company (www.truffle-tree.com) that seems on top of its game confirms that the tree, inoculated with truffle mycorrhizae, is absent of competitor fungi before shipment (although it makes no promises about yield). However, I know of only a few successes—one from North Carolina and one from Northern California—that have produced European truffles, and only after more than a decade of effort. In the past 30 years tissue culture techniques have increasingly replaced the tradition of transplanting truffle-supporting trees. Despite this development, most plantings or inoculations of European truffles beyond their native habitat still fail to produce mushrooms. Showing that growing native species is far more successful than growing nonnative ones, a trufflateur in Washington recently produced the Oregon white truffle (*Tuber*

▲ **FIGURE 41**

Truffle "brule" surrounds this filbert tree. As the mycelium of *Tuber melanosporum* consolidates its domain, the surrounding vegetation dies, creating a noticeable zone in the calcareous soils, a telltale sign that truffle mycelium has taken root.

gibbosum), after patiently waiting for 20 years until the first truffles could be harvested. Nevertheless, commercialization of mycorrhizal gourmet mushrooms has seen little success outside of the European truffle orchards, particularly those in France and Italy.

The reality is, though, that our native species of mycorrhizae quickly outcompete the foreign European truffles. Since European truffles like basic (high pH) soil, the addition of calcium diminishes competition from native mushrooms, but this alone will not assure success. In New Zealand, where the repertoire of competing mycorrhizae is limited to just a few species, inoculated trees are likely to do better than in regions of North America that are resplendent with hundreds of competing mycorrhizal varieties.

One method of inoculating mycorrhizae calls for planting young seedlings near the root zones of proven truffle trees. The new seedlings acquire mycorrhizae from a neighboring tree, and a second generation of trees carrying the mycorrhizal fungus is produced. After a few years, the new trees are dug up and replanted in new locations. This method has had the longest history of success in European sites where the soils, trees, and fungi are compatible.

Another approach, simple and elegant but not guaranteed, is to dip the exposed roots of seedlings into water enriched with the spore mass of a mycorrhizal candidate. First, mushrooms are gathered from the wild, and the spore-bearing surfaces are removed from the fruiting bodies, crushed, and immersed in water. Thousands of spores are washed off, resulting in an enriched broth of inoculum. A spore-mass slurry from a single mushroom, diluted in a 5-gallon bucket of water, can inoculate a hundred or more seedlings. Mycorrhized seedlings are healthier and grow faster than nonmycorrhized ones (see figure 42). Even if you are not successful in growing truffle mushrooms, the trees benefit from this pairing with the introduced mycelium.

Tossing spores using water as a carrier on the ground above the root zones of likely tree candidates is another method that takes little time and effort. Habitats should be selected on the basis of their parallels in the wild. For instance, chanterelles can be found in oak forests in the Midwest and in Douglas fir forests in the Northwest. Casting a spore mass of chanterelles into a forest similar to one where chanterelles naturally proliferate is obviously the best choice. However, the success rate is not high: even tree roots confirmed to be mycorrhized with gourmet mycelia will not necessarily yield harvestable mushrooms. Fungi and their host trees may have beneficial associations for long periods of time with no edible fruiting bodies appearing. Inoculations of mycorrhizae by one generation of mycologists may not see fruition until the next generation.

▲ **FIGURE 42**

Comparison of big leaf maples *(Acer macrophyllum)* without (smaller) and with (larger) mycorrhizae.

Chanterelles are one of the most popular collected mushrooms. In the Pacific Northwest, harvesting chanterelles is a controversial, multimillion-dollar business. Unfortunately, the gourmet mycorrhizal mushroom species are not readily cultured. Chanterelles demonstrate an unusual interdependence on soil yeasts, making tissue culture difficult. At least 4 organisms must be cultured simultaneously: the host tree, the mushroom, pseudomonas bacteria, and soil yeasts (red soil yeast, *Rhodotorula glutinis*, is needed for stimulating spore germination and healthy mycelial development). Not only do other microorganisms play essential roles, but the timing of their introduction is also critical to success in the fungal theater. Many experts believe that decades will pass before the plantations growing mycorrhizal species like chanterelles mature to a productive state.

No one has yet grown chanterelles to the fruiting body stage under sterile laboratory conditions, although greenhouse-grown pines have produced chanterelles after inoculation. In 1997 Eric Danell (accompanied by F. Camacho) was the first to successfully cultivate a chanterelle, fruiting mushrooms with a potted 16-month-old pine seedling in a greenhouse. Soon thereafter, Danell patented this particularly vigorous strain, which showed commercial potential. Field tests in 24 locations revealed chanterelle mycelium in the seedlings' root zones 2 years after inoculation. Unfortunately, he could not stop grazing animals, such as deer, squirrels, and beetles, from foraging and disturbing his crops. More recently, Danell started a Swedish company called Cantharellus AB to commercialize this breakthrough mycotechnology in the creation of chanterelle orchards. His group has planted thousands of trees with the chanterelle mycelium in an attempt to create mushroom plantations that produce mushrooms within a decade of planting. For the time being, only the patient might want to invest in mycorrhizal plantations.

Given the long time involved in honing laboratory techniques, I favor the low-tech approach and traditional method of planting seedlings adjacent to known producers of chanterelles, matsutake, truffles, and boletus and then replanting the seedlings several years later. In this way, we can value the forest not for its quantity of harvestable lumber but for its potential to harbor mushroom colonies.

Mutualistic Species: Fungal Partnerships

Mutualism occurs when 2 or more organisms work directly together for their mutual benefit, usually to prevent infestation by parasites and gather nutrients. Many organisms, from plants to ants, seek fungi as protectors, and vice versa. This rapidly expanding field of study is one of the most exciting in mycology.

Insects and Mushrooms

Many insects use mushrooms as platforms for incubating and feeding their larvae. One of the first cases of mutualism to be noticed was the interrelationship between mushrooms and termites. The mushroom genus *Termitomyces* includes several species of mushrooms associated with terrestrial termite colonies, especially in the tropics. Termites construct their nests with organic matter and cultivate mycelium. When abandoned, these mycelium-rich nests sprout delicious mushrooms.

Since certain fungi function as natural bactericides and fungicides, some insects engage them as allies in an effort to counter infections from hostile bacteria and other fungi. Studies at Oregon State University (Currie et al. 2003) show that attine ants, which include leaf-cutters, grow *Lepiota* mycelium as a host for a benign bacterium that produces an antibiotic against destruc-

▲ **FIGURE 44**

This giant mushroom, *Termitomyces* sp., is highly favored by the people of central Africa as a delicious edible. The primordium begin several feet underground, deep in a termite nest, extending upward as a long "taproot," and then forming a mushroom on the surface, especially when the nests are abandoned. How termites cultivate this mushroom has befuddled the best mycological minds in the world.

▲ **FIGURE 43**

Termitomyces robustus, a delicious choice edible mushroom, sprouts from an aged, abandoned termite colony.

tive microfungal parasites (*Escovopsis* sp.), and they also feed *Lepiota* mycelium to their larvae. This complex partnering has allowed lineages of attine ants to survive for more than 50 million years, and to establish massive colonies numbering in the many millions of inhabitants. (See chapter 14 for a discussion of *Lepiota procera* and *Lepiota rachodes* growing from anthills inoculated with spawn; also see chapter 8 regarding mycopesticides.)

The mutualism between ants, mushrooms, and bacteria is a useful model for how we humans can live in closer harmony with our environment. Both ants and people benefit from the guardianship of mycelium—by partnering with fungi, many organisms, including humans, can resist disease.

Snails as Fungus Farmers

Snails and slugs love mushrooms—an unfortunate situation for many of us mushroom lovers. Some snails enlist fungi to help them digest plants. Silliman and Newell (2003) found that a seaside snail, the marsh periwinkle *(Littoraria irrorata)*, damages and then defecates on certain grass *(Spartina alternifolia)*, where a particular fungus soon grows. Days later, the snails return to the grass, now overgrowing with fungus, and consume both fungus and plant. Grasses without the snail-enabled fungus grew 50 percent faster but were less appealing to feeding snails, whereas the plants covered with fungus were more palatable and nutritious for the snails. As you can see, the snail and fungus relationship affects other species in the marsh environment, such as grasses.

Endophytes: Mutualistic Symbionts

Endophytes are primarily benevolent, nonmycorrhizal fungi that partner with many plants, from grasses to trees. Their mycelia thread between cell walls but don't enter them, enhancing a plant's growth and ability to absorb nutrients, while staving off parasites, infections, and predation from insects, other fungi, and herbivores. Generally, endophytes are not true saprophytes or parasites but are in a class of their own. In contrast to mycorrhizal fungi, many endophytes grow well under laboratory conditions, so we can make spawn by using methods like those used for saprophytic mushrooms (Stamets 2000a).

The vast majority of endophytes are undescribed, and some appear to have lost the ability to produce spores, living vegetatively in a continuous mycelial state. Most endophytes described thus far are ascomycetes. One example is *Pezicula aurantiaca*, a small cuplike mushroom that lives on healthy alder trees. Like many endophytes, this fungus is dimorphic,

⋀ FIGURE 45

Trevon Stamets is excited to have his picture taken with the parasol mushroom *(Lepiota procera)* which is cultivated by ants to help them stave off infections. Many ants and termites farm fungi.

expressing itself in two forms, with one being an asexual mold.

Endophytic fungi are especially skilled at producing specialized mycotoxins (often alkaloids), a class of compounds that includes toxic cyclopeptides and serotonin-like tryptamines such as psilocybin. Endophytes hosted by grasses are similar to the ergot fungi whose alkaloids prevent their hosts from insect attack. Endophytes in large crabgrass, for example, appear to produce toxins that kill fire ants, and those in grasses such as the darnel weed (*Lolium temulentum*) cause sleepiness in cattle and horses, a fact long known to ranchers in Central America. However, Stanley Faeth (2002) suggests that varying levels of alkaloids in plants may not yet afford consistent protection against herbivores. Because some grasses produce more mycotoxins than others in the same habitat, cattle may sometimes get a chemical cocktail but other times not, making it more difficult for them to learn which grasses to avoid.

Nevertheless, endophytes, which were once thought to be pathogens, are increasingly viewed as engaging the plant in a mutually beneficial relationship. In a 2003 experiment in Panama, researchers found that when endophyte-free leaves from the chocolate-producing cocoa tree (*Theobroma cacao*) were inoculated with endophytes, leaf necrosis and mortality declined threefold, suggesting a biodefensive effect is possible against other pathogens such as *Phytophthora*, the genus responsible for sudden oak death—a disease devastating California's native oak population. (Arnold et al. 2003).

Spores from endophytes compete with many other free-flying fungal spores. According to one estimate, more than 10,000 spores of fungi land on each leaf per day. Amidst such competition, friendly fungi taking up residence is actually an asset to plants otherwise subject to pathogenic assault. Increasingly, mycologists believe that endophytic fungi may have coevolved with hospitable plants (Arnold and Herre 2003).

Wheat farmers benefit from the endophyte *Piriformospora indica*. The basidiomycete of this species has

yet to be identified, so it's referred to as *imperfect* (in the mycological world, this means that the fungus has no sexual phase or the sexual phase has not yet been discovered). This species is a root-based endophyte that promotes the growth of wheat shoots and roots and is capable of increasing leaf and seed production by more than 30 percent while shielding roots from infection by pathogenic microbes. Furthermore, seedlings paired with this mutualist successfully germinated 95 percent of the time, compared to only 57 percent for seedlings without this species. Root and shoot mass also doubled (Varma et al. 1999). This species has also demonstrated growth-enhancing properties when paired with maize (*Zea mays*), tobacco (*Nicotiana tobaccum*), and parsley (*Petroselinum crispum*). This fungus is easy to cultivate in the laboratory and widely coexists with many grasses. Clearly, pairing this and other endophytes with agricultural crops can increase yield, decrease disease, and reduce the need for fertilizers and insecticides.

Endophytic fungi may have other practical applications in agriculture. Joan Henson and other researchers (2004) filed a patent application using a *Curvularia* species isolated from grasses in the geothermal zones of Yellowstone and Lassen Volcanic national parks. This fungus qualifies as an *extremophile*—a thermally tolerant species that grows at the far fringe of temperatures where life can be found—and confers some tolerance to drought and heat to the host plant. Henson's research showed that grasses inoculated with this endophyte survived temporary exposure to extraordinarily high temperatures—158°F or 70°C—while those without shriveled and died. Although wheat did not survive in their experiment, it did demonstrate increased drought resistance. When watermelon seedlings and mustard seedlings were dusted with *Curvularia* spores, the spores germinated and inhabited the young plants. After the endophytic fungi became established on their hosts, researchers exposed the seedlings to extremely high temperatures (122°F or 50°C). The seedlings with the endophytic spores survived prolonged exposures, but the same types of seedlings without endophytic spores died. The discovery of the *Curvularia* spores'

⋀ **FIGURE 46**

The tinder or amadou mushroom *(Fomes fomentarius)*, a species found predominantly on birch, is distributed throughout the boreal forests of the world.

effects on plants may expand the biological tool set for mycorestoration, possibly even drastically expanding oasis environments and countering desertification.

Some wood conks once seen as parasites on trees may in fact be symbiotic endophytes. Baum and others (2003) report that the basidiomycetous polypore *Fomes fomentarius*, the tinder polypore or ice man mushroom, can operate as a nonsaprophytic endophyte in beech trees *(Fagus sylvatica)*. This well-known polypore (see figure 46 and page 225) is commonly found on beech, birch *(Betula)*, poplar *(Populus)*, alder *(Alnus)*, maple *(Acer)*, cherry *(Prunus)*, and hickory *(Carya)*—trees that thrive in the boreal regions of the world. In the Baum experiment, healthy beech wood was cut into sections and incubated in culture dishes for 8 weeks, whereupon fungal cultures of this conk emerged. A number of isolates of *Fomes fomentarius* were identified from the wood sections, and some proved to be genetically different strains.

The fact that one species can perform separate but complementary functions in the forest suggests that the species may play a larger role in the forest than is presently understood. How many other perennial wood conks do the same? I once received a call from a manager of a chestnut orchard in Quebec, who told me

⋀ **FIGURE 47**

Mycologist Jim Gouin in Quebec, Canada, with chaga, the aerial sclerotium of *Inonotus obliquus*.

that trees sporting chaga (see figure 47), the aerial sclerotium of *Inonotus obliquus*, were resistant to chestnut blight. (*Sclerotium* is a compact mass of hardened mycelium, with stored food, that can sometimes become a detached entity.) When he made a poultice of ground chaga and packed it into the lesions of infected chestnut trees, the wounds healed over and the trees recovered free of the blight. This leads me to wonder whether *Inonotus obliquus* can operate as an endophyte, as does *Fomes fomentarius*, and whether these species, or others like them, could defend host trees against invading parasites.

Many other mushrooms may be endophytes, including gilled species. For decades mycologists have been mystified how *Psilocybe cyanescens*, a psilocybin mushroom, can suddenly appear when trees are chipped into landscape mulch. A mycologist friend of mine had a truckload of fresh, mostly alder chips delivered to his house in the spring. Soon thereafter, his mostly conservative friends took some home for mulching. That fall most of their mulched beds were fruiting with hundreds of potent *Psilocybes*. Where could they have come from, my mycologist friend wondered? The only plausible explanation is that the mycelia were already in the wood, aboveground, while the trees were alive. Perhaps *Psilocybe* mycelium can be endophytic but delays fruiting with mushrooms until the trees make ground contact.

When engaging fungal allies—the mycorrhizal, saprophytic, and endophytic mushrooms—plants benefit in 3 ways. These complementary mycological systems help plants survive starvation, dehydration, and parasitization. The richer the fungus-plant partnerships, the more organisms the habitat can support. Masanobu Fukuoka, the farmer-scientist, ecological visionary, and author of *The One-Straw Revolution* (1978), understood that scientific reductionism failed to reflect biological synergisms—processes that are

still far beyond the most sagacious scientists today. He invented "seedballs," clay-soil pellets rich in microbes and seeds, designed to jump-start weakened habitats. Although Fukuoka did not select the populations of fungi in his seedballs, he appreciated their contribution. One of our tasks is to better Fukuoka's pioneering method and customize it for mycorestoration projects. Beginning with one straw, the saprophytes lead the charge, followed by mycorrhizae and endophytes.

As caretakers for future generations, mushroom communities surrounding trees govern habitat progression. I believe fungi have evolved to support habitats over the long term, protecting generations hundreds of years into the future. Saprophytic mushrooms gobble up debris fallen from the trees and prevent invasion by parasites. The mycorrhizae channel nutrients, expand root zones, and guard against parasites. Similarly, endophytic fungi, less well understood, chemically repel bacteria, insects, and other fungi. After hundreds of millions of years of evolution, fungal alliances have become part of nature's body politic. It is time for our species to partake in this ancient mycological wisdom.

◄ **FIGURE F**

Psathyrella aquatica nom. prov. Until recently, conventional wisdom held that gilled mushrooms did not exist underwater. In 2005, this Psathyrella, a new species, was discovered in the clear, flowing, pristine waters of the Rogue River near Crater Lake, Oregon. (See Coffan, Southworth, and Frank, 2008, in press.) This discovery opens up a new branch of aquatic mycology, and raises many questions. How many other mushroom species grow underwater?

CHAPTER 4

The Medicinal Mushroom Forest

Forest dwellers long ago discovered the value of medicinal mushrooms for the healing of both the body and the forest. Sadly, most of our ancestors' empirical knowledge is lost, but what little survives hints at a rich, albeit vulnerable, resource. The science of soils—mapping the matrix of plant, animal, and microbial communities in a habitat—remains in its infancy. Researchers have shown, however, that the forest is thoroughly interlaced with fungal nets of mycorrhizal, saprophytic, parasitic, and endophytic species. Mushrooms are forest guardians. A forest ecosystem cannot be defined without its fungi because they govern the

transition between life and death and the building of soils, all the while fueling numerous life cycles. Primary saprophytes initiate the decomposition process, and what the saprophytes don't break down, the mycorrhizal fungi do. I suspect that the overlying saprophytic fungi on the forest floor also influence the diversity of mycorrhizal fungi through their selection of trees to associate with, and that they stream nutrients to the root zones. Other groups of fungi (including endophytes and parasites) also work in concert. With a complex interplay of partnerships, mutualism, and parasitism, fungi build the soils beneath our feet.

As loggers cut down the old-growth forests, many fungi lose their foothold in the ecosystem. Whether these fungi remain as mycelia, not resurfacing in fruiting for decades or centuries, is a

◀ **FIGURE 48**

The health of a forest ecosystem's foundation is an interplay of mycelial networks from saprophytic, mycorrhizal, endophytic, and parasitic fungi.

matter of debate. When the forest returns to its previous majestic state, do the same mushroom strains also return, having lain latent in the landscape?

Fungi as Allies of People and the Planet

Let's look at the environment I know best: the rainforests of Washington State where I once worked as a logger. The dominant trees in the Olympics and the Cascades of the Pacific Northwest are Douglas firs, western and mountain hemlocks, red cedars, maples, alders, and various true firs—western, Pacific silver, noble, grand, white, red, and subalpine. More than 2,000 species of mushrooms live symbiotically with Douglas firs. Randy Molina and others (1997) estimate that 250 species of mycorrhizal fungi associate with hemlocks. Of the more than 527 mushroom species growing in old-growth forests, at least 109 of them are native to the Pacific Northwest (USDA 1993).

One of the rarest old-growth-forest mushroom species is *Bridgeoporus nobilissimus* (formerly known as *Oxyporus nobilissimus*) the noble polypore (Stamets 2002a; Redberg et al. 2003). This mushroom once held the record for the largest in the world but was bumped by a more massive individual of the species,

Rigidoporus ulmarius, estimated to weigh more than 660 pounds (about 300 kg). Other rare species are likely to thrive in old-growth forests, but they may go undiscovered for decades to come.

Most of the mushrooms collected in the forest are gathered for food. And most of these varieties are mycorrhizal, dependent on trees. The most recent data I

Some Commonly Collected Wild Edible Mushrooms from Northwestern North America*

Mushroom	Common Name
*Boletus edulis***	King bolete
*Cantharellus cibarius***	Chanterelle
*Cantharellus formosus***	Chanterelle
*Cantharellus subalbidus***	White chanterelle
*Coprinus comatus**	Shaggy mane
*Cortinarius caperatus***	Gypsy mushroom
*Craterellus cornucopiodes***	Horn of plenty
*Hydnum repandum***	Hedgehog
*Hypomyces lactifluorum****	Lobster
*Leccinum insigne***	Aspen bolete
Leccinum seabrum	Birch bolete
*Morchella elata**	Black morel
*Morchella esculenta**	Yellow or white morel
*Pleurotus ostreatus**	Oyster
*Polyozellus multiplex***	Blue chanterelle
*Sparassis crispa****	Cauliflower
*Tricholoma matsutake***	Pine mushroom
*Tuber gibbosum***	Oregon white truffle

* Species are saprophytes unless otherwise indicated.

** Mycorrhizal species, difficult to cultivate.

*** Nonmycorrhizal species, parasitic.

▲ **FIGURE 49**

Dusty Yao hunts medicinal mushrooms in the Olympics.

▲ **FIGURE 50**

The author squats beside a massive noble polypore *(Bridgeoporus nobilissimus)* growing on a stump in the Oregon Cascades. This mushroom, 53 inches in diameter and estimated to weigh more than 300 pounds, is perhaps the largest of its kind in North America.

▲ **FIGURE 51**

An unusual mushroom, the noble polypore *(Bridgeoporus nobilissimus)* hosts other plants and fungi. This young specimen, weighing several hundred pounds, is covered with a luxuriant coat of moss.

➤ **FIGURE 52**

Dusty Yao with gargantuan *Phaeolepiota aurea* mushrooms deep in an old-growth forest in Washington State. These mushrooms stay erect and firm and resist rot for many weeks, suggesting to me that they could possess some powerful antibiotics.

have seen on the harvesting of wild mushrooms comes from a species survey in which the British Columbia Ministry of Forests tabulated 40 mushroom species of commercial interest (Berch and Cocksedge 2003). The most commonly collected mushrooms are chanterelles, matsutake, and hedgehogs.

We face escalating challenges to our health from pollution and disease. Even in this era of high-tech genomics, natural compounds still provide a baseline for synthesizing drugs. Estimates are that two-thirds of our pharmaceuticals still originate from nature. For example, natural medicines such as taxol, discovered in the bark of Pacific yew trees *(Taxus brevifolia)*, give chemists clues to manufacturing similar potent compounds for treating deadly diseases, including ovarian and other cancers. Andrea Stierle, Gary Strobel, and Donald Stierle (1995) discovered that an endophytic fungus, *Taxomyces andreanae*, inhabiting the yew tree synthesized taxol, and for this discovery they were awarded several patents. Synthesizing taxol requires numerous steps, and despite advances, it's still more economical to derive taxols from natural sources, such as the English yew *(Taxus baccata)* or North American ground hemlock *(Taxus canadensis)*.

Another example of the potential medicinal value of old-growth-forest fungi is my discovery that an extract of the mycelium from the agarikon polypore mushroom *Fomitopsis officinalis* (see figures 53–55) protects human blood cells from infection by orthopox viruses, the family of viruses that includes smallpox (Stamets 2005b). Strains of agarikon varied in their potency.

A moldy cantaloupe, sent to an army research lab in 1941 by a housewife from Peoria, Illinois, gave rise to the strains of *Penicillium chrysogenum* that allowed for the commercial production of penicillin. This discovery saved millions of lives and billions of dollars, and helped us win WWII, since the Germans and Japanese did not have effective antibiotics. In contrast, agarikon *(Fomitopsis officinalis)* does not enjoy the luxury of this *Penicillium* mold's widespread cosmopolitan habits. Agarikon is restricted to an endangered habitat in rapid decline. Less than 5 percent of our northwestern old-growth forests survive today, in the

▲ **FIGURE 53**

Stainless steel tree cork borers can be used for removing a thin cylinder of tissue from a mushroom, in this case, a small agarikon mushroom *(Fomitopsis officinalis)* leaving the mushroom in the woods. The culture is taken from the layer just above the spore-producing polypored hymenial layer. For conks high in the air, lightweight tethered arrows whose tips are equipped with hollow metal shafts can be shot into the underside of the conk, to retrieve tissue for cloning with minimum impact.

▲ **FIGURE 54**

Agarikon *(Fomitopsis officinalis)* a mushroom found in the old-growth forests of the Olympic Peninsula in the Pacific Northwest. Extracts from the culture I generated from cloning this conk produced compounds very active against two pox viruses when screened through the Biodefense BioShield program administered jointly through the U.S. National Institutes of Health (NIH) and the U.S. Army Medical Research Institute of Infectious Diseases (USAMRIID), a coordinated effort to combat potentially weaponized viruses. The genome of this species may give rise to novel antivirals and hence should be protected. Although the mushrooms were not active when boiled in water, specially prepared extracts from living mycelium showed potent activity against vaccinia pox and cowpox viruses.

aftermath of 150 years of logging. No doubt strains of agarikon living in these ancestral forests will prove to be more potent against pox viruses than what I have recently discovered. The old-growth mycoforests suddenly become more valuable not as a timber source but as a remedy against natural or weaponized diseases.

What other mycomedical remedies await discovery in our ancient forests? I have little doubt that many other mushrooms will provide us with antiviral or anticancer drugs—provided our forests survive the effects of short-sighted political and corporate agendas. With the increasing threat of bioterrorism—especially from viruses like smallpox and bacteria like anthrax— protecting our fungal genetic diversity, especially in old-growth forests, is a matter of national defense. Most importantly, the survival of future generations may be at stake. (For a further discussion, see Mushrooms That Prevent and Heal Viral Disease on page 42.)

Wild Medicinal Mushrooms of North America

Preliminary studies on mushrooms have revealed novel antibiotics, anticancer chemotheuropeutic agents, immunomodulators, and a slew of active constituents. The following charts list a few of them. For more information on the medicinal properties of mushrooms, please consult *MycoMedicinals: An Informational Treatise on Medicinal Mushrooms*, by Paul Stamets and C. Dusty Yao (2002), and *Medicinal Mushrooms*, by Christopher Hobbs (2003).

▲ **FIGURE 55**

Another agarikon *(Fomitopsis officinalis)* collected in the central Cascades of Washington State. A specially prepared extract from this strain was very active against pox viruses. Strains appear to differ in their antipox activities.

▲ **FIGURE 56**

Dusty Yao with necklace of soma *(Amanita muscaria).*

Cross-Index of Mushrooms and Targeted Therapeutic Effects

Each mushroom species has a unique chemistry and molecular architecture. Many species are now known to have medicinal properties useful for improving human health. Here is a short list of some of those properties.

	Antibacterial	Anti-Candida	Anti-inflammatory	Antioxidant	Antitumor	Antiviral	Blood Pressure	Blood Sugar Moderator	Cardiovascular	Cholesterol Reducer	Immune Enhancer	Kidney Tonic	Liver Tonic	Lungs/Respiratory	Nerve Tonic	Sexual Potentiator	Stress Reducer
Agaricus brasiliensis (Himematsutake)				X	X				X	X	X						
Cordyceps sinensis (Cordyceps)	X			X	X	X	X	X	X	X	X	X	X	X	X	X	X
Flammulina velutipes (Enokitake)				X							X						
Fomes fomentarius (Ice Man Polypore)	X					X											
Ganoderma applanatum (Artist Conk)	X		X	X											X		
Ganoderma lucidum (Reishi/Ling Chi)	X	X	X	X	X	X	X	X	X	X	X	X	X	X	X		X
Ganoderma oregonense (Oregon Polypore)	X			X					X		X		X	X			
Grifola frondosa (Maitake/Hen of the Woods)	X	X			X	X	X	X			X			X	X		X
Hericium erinaceus (Yamabushitake/Lion's Mane)	X	X	X	X											X		
Inonotus obliquus (Chaga)	X		X	X	X		X				X		X				
Lentinula edodes (Shiitake)	X	X			X	X	X			X	X	X	X			X	X
Phellinus linteus (Mesima)	X		X		X												
Piptoporus betulinus (Birch Polypore)	X		X		X						X						
Pleurotus ostreatus (Hiratake/Pearl Oyster)	X				X	X			X	X					X		
Polyporus sulphureus (Chicken of the Woods)	X																
Polyporus umbellatus (Zhu Ling)	X		X		X	X					X		X	X			
Schizophyllum commune (Suehirotake/Split-Gill)		X			X	X											
Trametes versicolor (Yun Zhi/Turkey Tail)	X			X	X	X					X	X	X				

Mushrooms with Activity Against Specific Cancers

For the past 30 years, researchers have studied mushrooms' effectiveness against cancer. Some of their findings are summarized below.

	Breast	Cervical / Uterine	Colorectal	Gastric / Stomach	Leukemia	Liver	Lung	Lymphoma	Melanoma	Ovarian	Pancreatic	Prostate	Sarcoma
Agaricus brasiliensis		X	X										X
Clitocybe illudens* (Omphalotus olearius)	X						X			X	X		X
Cordyceps sinensis					X		X	X					
Flammulina velutipes								X				X	
Ganoderma lucidum					X	X	X					X	X
Grifola frondosa	X		X		X	X	X					X	
Hericium erinaceus				X		X							
Inonotus obliquus		X											
Lentinula edodes	X					X			X			X	
Phellinus linteus		X	X	X		X			X				
Piptoporus betulinus									X				
Pleurotus ostreatus													X
Polyporus umbellatus					X	X	X						
Schizophyllum commune		X		X									
Trametes versicolor	X	X		X	X	X	X					X	

* Poisonous species, not edible.

Antimicrobial Properties of Mushrooms

Despite recent medical advances, microbes, especially viruses, continue to kill millions of people, stimulating the search for new antimicrobial agents that are safe for human use. Mushrooms, which naturally produce a surprising array of antibiotics, may provide the answer. Mushrooms share a deeper evolutionary history with animals than with any other kingdom, so humans and mushrooms share risks of infection from some of the same microbes, for instance the bacteria *Staphylococcus aureus* and *Pseudomonas fluorescens*. Although mycelium has just a single cell wall protecting it from hundreds of millions of hostile microbes in every gram of soil, it manages to form networks extending, in some documented cases, thousands of acres and weighing thousands of tons. Nutrient-rich mushrooms, before sporulation, resist infection and rot. After sporulating, mushrooms rot, and I believe each mushroom species predetermines which bacterial colonies can live upon it.

⋏ **FIGURE 57**

Cortinarius caperatus (formerly known as *Rozites caperata*) the gypsy mushroom, is a choice edible that contains powerful antiviral compounds (Piraino and Brandt 1999). Widely distributed in temperate conifer forests, this mycorrhizal mushroom cannot be readily cultivated but often forms great colonies.

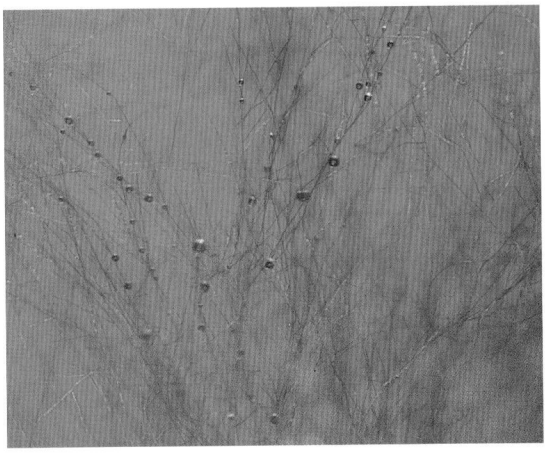

⋏ **FIGURE 58**

Mushroom mycelia exude droplets containing enzymes and antibiotics and profuse water. The enzymes digest lignin and cellulose, petroleum products, and many molecules held together by hydrogen-carbon bonds. The antibiotics stop microbial parasites. Mushrooms resist bacterial and fungal rot until they release spores, age, and die.

How do mushrooms and mycelium do this? The cell surface of mycelium "sweats" out antibiotics that are known in the field as *exudates* or *secondary metabolites* (see figure 58). A Merck team in Spain published an extensive survey—*Screening of Basidiomycetes for Antimicrobial Activities* (Suay et al. 2000)—assessing the antibacterial activities of 204 mushroom species. The results showed that many mushrooms target specific species of bacteria.

Useful antibiotics isolated in mushrooms include calvacin from giant puffballs (*Calvatia gigantea*), armillaric acid from honey mushrooms (*Armillaria mellea*), campestrin from meadow mushrooms (*Agaricus campestris*), coprinol from inky caps (*Coprinus* species), corolin from turkey tail mushrooms (*Trametes versicolor = Coriolus versicolor*), cortinellin from shiitake (*Lentinula edodes*), ganomycin from reishi (*Ganoderma lucidum*), agaricin from agarikon (*Fomitopsis officinalis*) and sparassol from cauliflower mushrooms (*Sparassis crispa*). With a diversity estimated at over 140,000 species, mushrooms are a promising resource for new antibiotics. That mushrooms inhibit some bacteria but not others shows that mycelium influences the makeup of microbial populations in its immediate ecosystem. For more information on the mushroom-microbe nexus, consult *Novel Antimicrobials from Mushrooms* (Stamets 2002b) and *New Anti-viral Compounds from Mushrooms* (Stamets 2001c).

Since most of the mushrooms described in this book are primary saprophytes—the first to consume fresh debris—their antibiotics dramatically influence bacterial populations. I hypothesize that mushrooms select those bacteria that ultimately favor their fungal lineage. I sense that these interrelationships are critical to an ecosystem's health and fungal evolution.

Mushrooms That Prevent and Heal Viral Disease

That medicinal mushrooms have been ingested for hundreds and, in some cases, thousands of years, strongly suggests most are not toxic, and research supports them

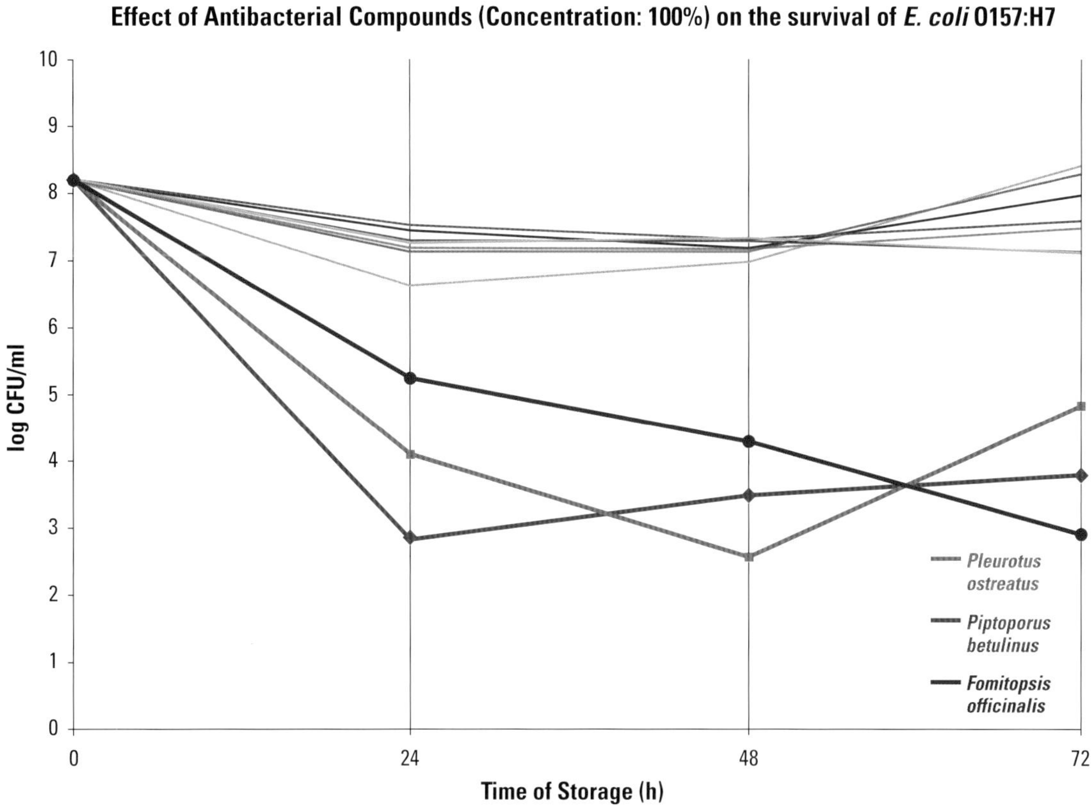

Effect of Antibacterial Compounds (Concentration: 100%) on the survival of *E. coli* O157:H7

Legend:
- *Pleurotus ostreatus*
- *Piptoporus betulinus*
- **Fomitopsis officinalis**

▲ **FIGURE G**

This chart and the one on the following page show the antimicrobial activity of cold-water extracts created from washing exudates secreted from living mycelia from ten mushroom species. Vertical scale is log 10. With both *Escherichia coli* and *Staphylococcus aureus*, the number of colonies forming units (CFUs) per gram of water plummeted from more than 100,000,000 to the 1,000–10,000 CFU range in 48–72 hours, equivalent to more than 99.99% inhibition. The most antibacterially active species were an oyster mushroom (*Pleurotus ostreatus*), the birch polypore (*Piptoporus betulinus*), and agarikon (*Fomitopsis officinalis*).

as likely candidates in our search for natural antiviral agents. Suzuki and others (1990) discovered an antiviral water-soluble lignin in an extract of the mycelium of shiitake mushrooms *(Lentinula edodes)* isolated from cultures grown on rice bran and sugarcane bagasse. Another mushroom recognized for its antiviral activity is *Fomes fomentarius*, a hoof-shaped wood conk growing on trees, which inhibited the tobacco mosaic virus in a study (M. Aoki et al. 1993). Collins and Ng (1997) identified a polysaccharopeptide from turkey tail *(Trametes versicolor)* mushrooms inhibiting HIV type 1 infection, while Sarkar and others (1993) identified an antiviral substance extracted from shiitake *(Lentinula edodes)* mushrooms.

More recently, derivatives of the gypsy mushroom *(Cortinarius caperatus)* were discovered by Piraino and Brandt (1999) to inhibit the replication and spread of varicella zoster (the shingles virus), influenza A, and the respiratory syncytial virus (RSV) that causes colds. Eo and others (1999, 2000) found

Effect of Antibacterial Compounds (Concentration: 100%) on the survival of *Staphylococcus aureus*

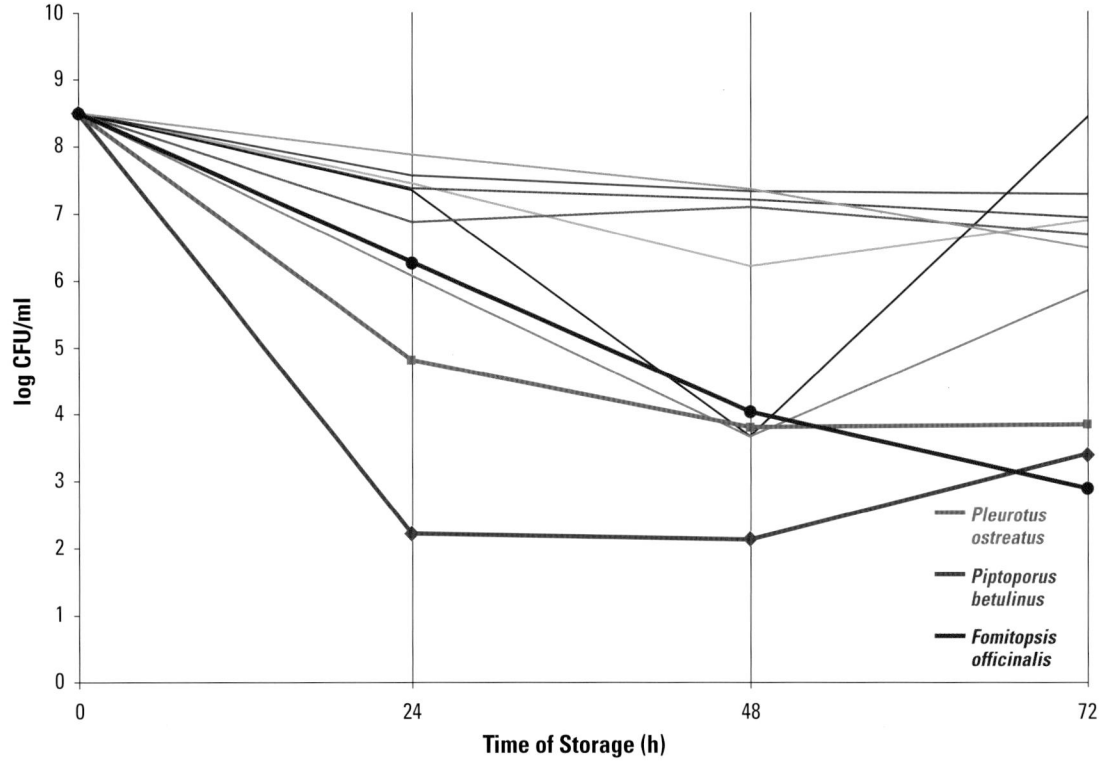

▲ **FIGURE H**

See caption, page 43.

antiviral activity in the methanol-soluble fractions of reishi mushrooms *(Ganoderma lucidum)* that selectively inhibited herpes simplex 1 and 2, and the vesicular stomatitus virus (VSV). Wang and Ng (2000) isolated a novel ubiquitin-like glycoprotein from oyster mushrooms *(Pleurotus ostreatus)* that inhibited HIV. Mushroom derivatives also activate natural immune response in mammalian cells, in effect boosting an organism's resistance to microbial infection (Stamets 2003b). Summaries on the antiviral properties of mushrooms were published by Brandt and Piraino (2000) and Stamets (2001c), and reports on antimicrobial properties were published by Suay and others (2000) and Stamets (2002b).

People whose immune systems are compromised by a respiratory virus can become infected by bacteria such as *Streptococcus pneumonia*. Mushrooms having both antiviral and antibacterial properties may prevent such opportunistic infections. Mushrooms also influence populations of bacteriophages—viruses that use bacteria as incubators and vectors for further infection. I hypothesize that studying the interrelationships between mushrooms and their related bacteria, viruses, and bacteriophages will reveal medically significant antibiotics in the near future.

Virologists are concerned about the threat of viral infection from animals. For example, the 2003 sudden acute respiratory syndrome (SARS) epidemic may have originated from human contact with captive civet cats in rural China. Viruses and bacteria can also spread when

Mushrooms with Direct Antiviral Activity

Mushrooms are being actively explored by virologists for new sources of antiviral medicines.

	Hepatitis B	Herpes simplex I	Herpes simplex II	HIV	Influenza	Pox	Respiratory syncytial virus	Rous sarcoma virus	Tobacco mosaic virus	Varicella zoster	Vesicular stomatitis
Agrocybe aegerita									X		
Cordyceps sinensis	X										
Cortinarius caperatus (=Rozites caperata)		X	X				X	X		X	
Fomes fomentarius									X		
Fomitopsis officinalis				X	X						
Ganoderma lucidum		X	X	X	X						X
Grifola frondosa				X							
Inonotus obliquus				X	X						
Lentinula edodes		X		X	X						X
Piptoporus betulinus				X		X					
Pleurotus ostreatus				X							
Polyporus umbellatus	X										
Trametes versicolor				X							

birds, dogs, prairie dogs, bats, vermin, and other animals, including primates and humans, concentrate their populations. Of particular concern to me are animal "factory farms," wherein thousands of chickens, hogs, cows, or other animals are aggregated, providing a prime breeding environment for microbes. Feedlots and factory farms could possibly be used by bioterrorists as launching platforms for pandemics. Hence, these sources pose a significant microbial threat to human health.

With airline passengers from remote regions of the world concentrating in airports and traveling to far-flung destinations, contagious passengers are likely to infect others. Similarly, the Norwalk virus, which has been appearing on cruise ships, should be a wake-up call reminding us that transportation vessels are effective vectors for the spread of disease. Virtually anywhere humans concentrate provides opportunities for contagions to spread, whether by air or by physical contact. With the threat of bioterrorism from weaponized viruses, a readily available, inexpensive, broad-spectrum antiviral antidote would serve the public's health. Mushrooms, especially combinations of mushrooms, offer protection from infectious diseases in at least three ways: first, directly as antimicrobial agents (antibiotics); second, by increasing your immune system's natural defenses—what physicians call the *host-mediated response* (Stamets 2003b); and third, the custom construction of mycelial mats for mycofiltration can reduce the risk of infection from environmental sources such as sewage from feedlots and slaughterhouses. The key is to match the mushroom with the pathogen.

Mushrooms are a hot topic right now with medical researchers. They are the subject of clinical studies that examine their usefulness in adjunct therapies used as a complement to conventional medicine. Researchers have found that mushrooms contain polysaccharides, glycoproteins, proteoglycans, ergosterols, triterpenoids, enzymes, acids, and antibiotics that when used individually and in concert can stop infection. Scientists have also found that each species of mushroom has a signature architecture and defense against microbes. The medicinal properties of mushrooms are covered in further detail in my book *MycoMedicinals* (2002) and in Christopher Hobbs's *Medicinal Mushrooms* (2003).

Mobilizing Mushrooms against AIDS

Several mushroom species—oyster, shiitake, maitake, turkey tails, and other polypores—have shown anti-HIV activity under certain circumstances. The National Institutes of Health (NIH) funded a small clinical study using oyster mushrooms in conjunction with protease inhibitors. The application, entitled "Anti-Hyperlipidemic Effects of Oyster Mushrooms in the CAM Therapy in the Treatment of HIV/AIDS," was approved in late 2003 with trials scheduled to start in mid-2005 and to be supervised by the School of Medicine at the University of California, San Francisco (Abrams 2004). Our mushroom farm, Fungi Perfecti, grew a strain of oyster mushrooms for these trials that were frozen, freeze-dried, and powdered for consumption. This may be the first clinical trial funded by the NIH to study the medicinal effects of ingesting mushrooms. Updates will be posted at www.fungi.com.

Protease inhibitors, commonly prescribed to combat HIV, interfere with lipid metabolism in the liver, causing an accumulation of "bad" cholesterol, LDL. From the hyperaccumulation of LDL, many patients suffer from arteriolosclerosis, endocrine disruption, and heart disease. Oyster mushrooms contain a natural isomer of lovastatin (an inhibitor of 3-hydroxy-3-methylglutaryl coenzyme A reductase), an FDA-approved cholesterol-lowering drug (Gunde-Cimerman 1999; Gunde-Cimerman and Cimerman 1995). At an international mycological conference, Nina Gunde-Cimerman reported that a small pool of people who had ingested "15 grams of oyster mushrooms per day for 30 days reduced LDL cholesterol by up to 30 percent" (Gunde-Cimerman and Plemenitas 2002). Our recently funded, small, clinical study is designed to confirm or disprove these results. The cholesterol-reducing properties of oyster mushrooms, combined with their anti-HIV glycoproteins (Wang and Ng 2000), suggest that this mushroom may be one that can dually mitigate the side effects of protease inhibitor therapies while fighting AIDs.

Imperiled immune systems increase susceptibility to bacterial and viral infections like HIV. In many regions of the world, such as Africa and India, poor nutrition, bad sanitation, and impure water are among the factors exacerbating the effects of HIV. As mushroom cultivation enterprises spread to developing countries in order to combat hunger, they are also well positioned to help fight HIV. Urgent needs in these areas—the need to fight famine, recycle waste, build soil, protect crops, create jobs, and combat HIV—are well served by mushroom cultivators. Mushroom farms could reinvent themselves as healing arts centers.

Rarely in the natural world are there organisms whose use can be pivotal in addressing the many causes of disease. Mushrooms stand out. Not only are they essential for bolstering the food web by increasing sustainability of soils and helping to integrate communities, but their mycelia and fruitbodies produce a gamut of highly potent products, medically beneficial to the environment and the organisms living within. Our mandate is to engage these fungi as allies. Mushrooms can rescue us from our current spiral toward ecological collapse and massive extinction.

Medicinal Mushrooms for Healing Forests and Fighting Fungal Blights

A blight is a species-specific parasitic invasion by a fungus that kills many members of the target species in a community. Fungal blights can fell a forest of firs and oaks in a matter of months. Nonblighting fungi, which also have medicinal or nutritional uses for humans, may be the best defense against blighting fungi. The introduction of select saprophytic or endophytic species can forestall the spread of parasitic species that cause blights. Since live trees contain much dead tissue, saprophytic and endophytic communities thrive upon them and guard against invading parasitic fungi.

Disease blights can inflict massive economic damage on the timber value of forests (Ferguson et al. 1998), but they may actually be beneficial, especially when viewed over the long term. Honey mushrooms (*Armillaria ostoyae*, *Armillaria gallica*, and *Armillaria mellea*) will attack a tree, causing devastating root rot and hollow brown core rot. As the diseased trees in the forest die, the wood dries and may catch fire if struck by lightning, especially if located on ridgetops. The forest fire often cauterizes the soil, killing the *Armillaria* that originally killed the forest. The result may be high mountain meadows inhabited by grass until a new forest regenerates. Fires help create meadows which, due to their low wood content, provide firebreaks and forest disease–free zones. This cycle of forest to meadow to forest may be healthier for the ecosystem in the long run because with each succession the soil biosphere is enriched as soils thicken.

In Western Australia, inoculating karri (*Eucalyptus diversicolor*) stumps, Pearce and Malajczuk (1990) found that *Hypholoma* proved to be competitive against *Armillaria*, in some cases excluding it entirely. They have been among the first researchers to test the idea of fighting a mushroom blight with another mushroom. In British Columbia, Chapman and colleagues (2004) buried *Hypholoma fasiculare* in the form of sawdust spawn in the root zones of stumps in clear-cuts. They found that after 5 years *Armillaria* root rot disease in seedlings of lodgepole pine (*Pinus contorta*) and Douglas firs (*Pseudotsuga menziesii*) was significantly reduced—up to 67 percent. I prefer a sister species *Hypholoma capnoides* which is edible but otherwise very similar. I more often find *H. fasciculare* than *H. capnoides* but recently, while on a hike up the Sol Duc

River on the Olympic Peninsula, I was surprised to see the valley was dominated by *H. capnoides* almost to the exclusion of *H. fasciculare*, but growing in close proximity to Armillaria honey mushrooms. I suspect these two species, *H. capnoides* and *Armillaria ostoyae*, battle each other in mycelial combat for dominance in the same ecosystem. As mycoforesters, we benefit from understanding how mushroom species compete and cooperate, giving us new tools for ecological management.

Other mushrooms—including gourmet edible species—might prove useful in defending forests from *Armillaria* blights. Cauliflower mushrooms (*Sparassis crispa*) secrete the antifungal antibiotic sparassol, or orsellinic acid, which is also produced by *Armillaria* fungi. Mushrooms are naturally immune to their own antifungal secretions and are unaffected by the same secretions from competitors. However, cauliflower mushrooms also secrete other antifungal agents that allow them to parasitize *Armillaria* mushrooms. This suggests an antifungal strategy for foresters to consider. In principle, mushrooms like cauliflowers could defend forests against blights by *Armillaria*; inoculating stumps at the perimeter of an *Armillaria* blight could limit further spread of this destructive forest disease.

Some mycologists believe that the cauliflower mushroom is a root parasite that like *Armillaria*, can

▲ **FIGURE 59**

A forest parasite, probably *Phaeolus schweinitzii*, fruiting from a stump, causes core rot. Note the band of white mycelium.

grow saprophytically after the host tree dies. However, I have yet to see a forest blighted by cauliflower mushrooms, whereas I have seen thousands of acres of forests in Washington, Oregon, and Colorado that were killed by *Armillaria* fungi. I have found cauli-flowers only at the base of very large trees or stumps, so if it is a root parasite, the cauliflower doesn't seem to become destructive until their aged hosts are already in decline. For this reason, I'd be willing to bet that most people would prefer to see cauliflower

---

Mushrooms and Their Known Medicinal Agents

Mushroom	Common Name	Derived Medicine or Medicinal Properties
Agaricus campestris	Meadow mushroom	Campestrin, antibiotic
Amanita muscaria	Soma	Muscimol, ibotenic acid
*Amanita pantherina**	Panther cap	Muscimol, GABA-mimicking compound for neurological research
Calvatia gigantea or *Calvatia booniana*	Giant puffball	Calvacin, antibiotic
Clitocybe illudens	Jack-o'-lantern	Illuden-S, irofulven, anticancer, antibacterial
Coprinus comatus	Shaggy mane	Coprinol, antibiotic
Cordyceps subsessilis	Beetle Cordyceps	Cyclosporin, immunosuppressant
*Cortinarius caperatus**	Gypsy mushroom	Antiviral
Fomes fomentarius	Amadou	Antibacterial, fire-starting tinder
*Fomitopsis officinalis***	Agarikon	Antibacterial, antiviral, anti-inflammatory
Ganoderma applanatum	Artist conk	Antibacterial, anticancer
Ganoderma lucidum	Reishi	Triterpenoids, beta-glucans, anticancer, anti-inflammatory, ganomycin, antibiotic
Ganoderma tsugae	Hemlock reishi	Triterpenoids, beta-glucans, anticancer, anti-inflammatory, ganomycin
Grifola frondosa	Maitake	Anticancer (grifolan), antidiabetic
Piptoporus betulinus	Birch polypore	Betulinic acid, anticancer, antiviral
Pleurotus ostreatus	Oyster mushroom	Antiviral, anticancer (pleuran), cholesterol-lowering properties
Sparassis crispa	Cauliflower mushroom	Antibiotic (sparassol)
Trametes versicolor	Turkey tail	Anticancer, polysaccharide-K (PSK), polysaccharide-P (PSP), anticancer, corolin, antibiotic

* Mycorrhizal mushrooms that are difficult to cultivate.

** This species is on the Red List of nearly extinct species in Europe. Although still surviving in the Pacific Northwest of North America, its habitat is shrinking; as a result, widespread harvesting of this species is discouraged. Cultures can be obtained using cork borers without harvesting this rare species (see figure 53). I do, however, encourage the inoculation of trees with this species, especially wind-topped old-growth or mature second-growth Douglas firs.

mushrooms rather than honey mushrooms in their forests.

Laboratory tests show the promise of this blight-blocking scenario. When a strain of parasitic honey mushrooms is cultured in the same petri dish with a strain of cauliflower mushroom, the cauliflower outcompetes and appears to parasitize the honey mushroom (see figure 62). By prefilling the susceptible forest niche with a chosen species, a landowner can forestall or prevent invasion by blight fungi such as *Armillaria*. For forestland managers, I suggest erecting mycelial perimeters of beneficial mushrooms as species barriers against devastating parasites (For techniques for inoculating stumps with cauliflower mycelium, see pages 183 to 186.)

Other mushrooms that outcompete *Armillaria* root rot are the medicinally active turkey tails *(Trametes versicolor)*; smoky gilled woodlovers *(Hypholoma capnoides)*, a culinary mushroom; and its close sibling *Hypholoma fasciculare*, a beautiful although poisonous

mushroom. *Hypholoma* fungi produce white rhizomorphic (ropelike) mycelia that overwhelm *Armillaria* rhizomorphs, which are often black. These features help you zero in on which species may be digesting wood even when mushrooms have not yet formed. Turkey tail mycelium grows far faster than that of *Armillaria* and *Hypholoma*. When smoky gilled woodlovers and turkey tails confront honey mushrooms in a culture, they overgrow this parasite.

I think these insights may be one of the most significant to date in the use of nonmycorrhizal fungi in mycoforestry practices. By occupying the niche with selective species in advance, invasive fungi cannot take root. Although turkey tail mushrooms *(Trametes versicolor)* had a similar protective effect against *Armillaria* blight, this species did not compete as well as the *Hypholoma* species belowground in the root zones. I find this particularly interesting, since *Hypholoma* mushrooms often form at the interface where the stump emerges from the ground, whereas turkey tails tend to stay above the ground. Perhaps *Hypholoma* species are better subterranean competitors in complex

▲ **FIGURE 60**

Is this the largest organism in the world? This 2,400-acre site in eastern Oregon had a contiguous growth of mycelium before logging roads cut through it. Estimated at 1,665 football fields in size and 2,200 years old, this one fungus has killed the forest above it several times over, and in so doing it has built deeper soil layers that allow the growth of ever-larger stands of trees. Mushroom-forming forest fungi are unique in that their mycelial mats can achieve such massive proportions.

▲ **FIGURE 61**

Along a trail up the Sol Duc River valley in the Olympic Peninsula rain forest, I came across this log fruiting honey mushrooms, probably in the *Armillaria ostoyae* group. Many logs and stumps sported this mushroom and smoky gilled woodlovers *(Hypholoma capnoides)*.

soil environs, producing cordlike mycelial growth similar to that of the garden giant, a relative from the same family, the Strophariaceae. We may be able to promote constellations of species, with communities of turkey tail, cauliflower, and *Hypholoma* fungi that can work in concert as mycobarriers to corral *Armillaria* infestation by inoculating stumps with sawdust, plug, or rope spawn peripheral to the advancing blight. By encircling the encroaching mycelial mat of blight, not only might we stop the advance of blight fungi, but these techniques may be refined over time to customize mushroom species populations to include the best of gourmet and medicinal mushrooms.

If you're concerned about spreading a parasitic species, then using a nonparasitic native woodlover *(Hypholoma)* or turkey tail *(Trametes versicolor)*, enokis *(Flammulina velutipes)*, oysters *(Pleurotus* species), or psilocybes may be more satisfactory. A variation of these strategies would be to inoculate multiple species on a stump, as is often seen in nature: *Hypholoma, Trametes, Armillaria,* and cauliflowers naturally occur together in the same habitat. The array of species could be custom selected by you for your habitat.

The species listed below are the fungi that I believe can be used to benefit stressed woodland ecosystems. Only experimentation will find the best matches since

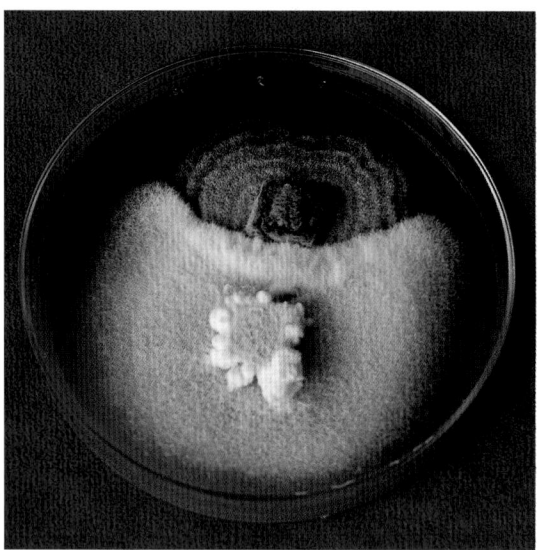

▲ **FIGURE 62**

The mycelium of the cauliflower mushroom *(Sparassis crispa),* overpowers the honey mushroom *(Armillaria ostoyae)* in culture. Similar dominance over *Armillaria* occurs with woodlovers like *Hypholoma capnoides,* and most notably turkey tails *(Trametes versicolor).* Although such observations from a petri dish test do not necessarily translate into the natural world, it is nevertheless a good indicator of competitiveness between species living in the same ecological niche. See also figure 288.

➤ **FIGURE 63**

Azureus Stamets holding a cauliflower mushroom *(Sparassis crispa),* collected near Bagby Hot Springs, Oregon. This mushroom is often associated with very large old-growth trees and their stumps. In culture, it fights *Armillaria* root blight. See figure 62.

▲ **FIGURE 64**

The smoky gilled woodlover *(Hypholoma capnoides)* is an edible mushroom, although not yet popular. This species, I believe, has many beneficial properties helping mycorestoration efforts.

▲ **FIGURE 65**

A cauliflower mushroom *(Sparassis crispa)* fruiting from a coil of hemp rope. The mycoforester can lasso stumps with this myceliated rope, inoculating them, and help prevent the spread of devastating parasitic fungi like the honey mushroom *(Armillaria ostoyae).*

some strains perform better outdoors than in the lab. I feel that these species pose minimal risk to healthy forests but are aggressive enough to dominate many fungal parasites. See chapter 14 for growth parameters and recommended courses of inoculation for each of these species:

- clustered woodlover (*Hypholoma capnoides*)
- oyster (*Pleurotus ostreatus* and *Pleurotus pulmonarius*)
- garden giant (*Stropharia rugoso annulata*)
- psilocybes (*Psilocybe cyanescens* and allies)
- turkey tail (*Trametes versicolor*)

Turkey tails, woodlovers, oysters, garden giants, and psilocybes are perhaps the best saprophytic sentinels in our mycological armamentarium for helping an injured forest ecosystem recover. These aggressive mushrooms love bacteria, and they grow with so much vigor that they suppress parasitic invaders such as honey mushrooms, protecting and benefiting forest growth.

Since we have changed the environment so radically in such a short time, nature needs our help in order to mend. Under ordinary circumstances, nature

▲ **FIGURE 66**

A delicious cauliflower mushroom *(Sparassis crispa),* a mild parasite, fruits at the base of an old-growth Douglas fir. This tree has given rise to annual fruitings of cauliflower mushrooms for two decades, perhaps longer.

self-prescribes fungi for its own healing. But since we have accelerated the forests' natural destruction and renewal cycles, thereby creating massive debris fields for instance, through clear-cutting, we ought to help the forests accelerate the decomposition cycles by introducing mycelium in key areas—in essence by running mycelium. Otherwise our ecosystems will lose their equilibrium, destabilize, and crash, possibly becoming overrun by disease. By encouraging selected saprophytes in this stressed terrain, we can increase

Chicken of the woods *(Laetiporus conifericola)*, is an edible polypore. When slices of this specimen were grilled on a barbecue, the flavor was just like white chicken meat. A non-parasitic but aggressive species, this group of mushrooms has species with interesting antimicrobial properties. I often find them in river valleys in the summertime. Being a brown rotter, this species plays a unique role in ecosystems infused with white rot fungi.

the carrying capacity with greater nutrient flows, improve moisture absorption, bolster disease resistance, reduce erosion, and provide friendly niches for fauna and flora. Once the vanguard saprophytes enter the landscape, subsequent generations of other organisms will be able to thrive in the soil created by fungus. Ultimately, trees will grow and bequeath debris streams for more fungal cycles.

Mycorestoration strategies can also help landscapes whose immune systems have been harmed by pollution. Fortunately, mushrooms like turkey tail are multibeneficial—preventing blights, fighting bacteria, and breaking down toxic chemicals like PCBs and dioxins. Later in the book we will see how we can use this mushroom to fortify our own immune systems and break down toxins. I believe habitats—our body's and our environment's—have immune systems that can benefit from using certain mushroom species.

You Can't Do That!

If you are planning to introduce saprophytic mushrooms to a forest for healing, you should brace yourself for criticisms from foresters and mycophobes who warn against unleashing a dangerous fungal plague that could wipe out entire forests, saying that my strategies are untested and risky. Their fears are primarily based on the proven devastation of California oaks by *Phytophthora* species and Oregon firs by *Armillaria* species. I would counter by saying that promoting these saprophytic mushrooms conforms to the precautionary

principle, adopted by many cities and organizations across America in the past few years (Raffensperger and Tickner 1999). This principle advocates thinking of the future as much as the present—a blending of long-sighted intention and environmentally rational strategy. Like the Hippocratic oath taken by physicians to first do no harm as healers, the precautionary principle suggests that doing nothing is often better than doing something if there are substantial unknown risks inherent in an action. However, the precautionary principle advocates action in the face of impending disaster, and this is where I think mycorestoration strategies fit well. One could argue that introducing the cauliflower mushroom *(Sparassis crispa)* a weakly parasitic root rot fungus, might violate the precautionary principle. However, I know of no cases of *Sparassis crispa* blight, nor have I found any reports by mycologists or foresters. My in vitro tests (see figure 62) show that *Sparassis crispa* outcompetes *Armillaria mellea*, suggesting that these mycelia may actually limit this known virulent root pathogen, thereby preventing a devastating blight by introducing a negligible one. (Similarly, forest fires are often stopped by starting smaller backfires.) As you can see, this strategy is not risky and therefore does not violate the precautionary principle.

A potential downside of the precautionary principle, which is used to protect against negative unforeseen consequences, is that it puts the brakes on progressive thinking and may limit the search for existing nature-based remedies, tested in the theater of evolution. I believe that it is better to search our planet's existing genetic diversity for naturally resistant crops instead of birthing GMOs (genetically modified organisms), the Frankensteinian creatures of our era. This book encourages exploration of fungal biodiversity for indigenous remedies to many of the problems we have created for ourselves.

This issue was exemplified when a friend tried to introduce the woodlover (*Hypholoma capnoides*) to wood chips in a decommissioned road reclamation project in northern Arizona (see chapter 6). Although he had found reports that the woodlover was a native of that area and was following my principle of amplifying native strains rather than importing nonnative ones, a forester threatened to fight his plan because she incorrectly claimed that this woodlover was not native and had parasitic potential. The saprophytic genome is no longer geographically restricted because spores easily travel by attaching themselves to hikers, birds, automobiles, pieces of mail, agricultural shipments, and winds. In this case however, her mycophobia clouded rationality as this woodlover mushroom is a natural resident of the forests she helps manage. Even the best of new ideas are often met with passionate resistance.

Many plant pathologists' niche of expertise is fungal parasites. Saprophytes do not "infect"; they degrade already dead tissue. Just walking through a forest, a forester or anyone else transports a pathogenic and saprophytic payload of spores hitchhiking on his or her body that far outstrips the environmental impact of the spore-free spawn from an implanted saprophyte. We need to weigh the balances of potential costs versus benefits to the environment. Enlightened forest pathologists can aid mycoforestry research by testing some of the principles set forth in this book. Together, we can create strategies to use fungi to improve forest health while minimizing risk. We must continually reevaluate the benefits of mycorestoration strategies in contrast to their risks. By the same token, we must also continuously reevaluate the costs and risks of doing nothing.

Part II

MYCORESTORATION

Habitats, like people, have immune systems, which become weakened due to stress, disease, or exhaustion. Mycorestoration is the use of fungi to repair or restore the weakened immune systems of environments. Whether habitats have been damaged by human activity or natural disaster, saprophytic, endophytic, mycorrhizal, and in some cases parasitic fungi can aid recovery. As generations of mycelia cycle through a habitat, soil depth and moisture increase, enhancing the carrying capacity of the environment and the diversity of its members.

On land, all life springs from soil. Soil is ecological currency. If we overspend it or deplete it, the environment goes bankrupt. In either preventing or rebuilding after an environmental catastrophe, mycologists can become environmental artists by designing landscapes for both human and natural benefit. The early introduction of primary saprophytes, which are among the first organisms to rejuvenate the food chain after a catastrophe, can determine the course of biological communities through thoughtfully matching mycelia with compatible plants, insects, and others. The future widespread practice of customizing mycological landscapes might one day affect microclimates by increasing moisture and precipitation. We might be able to use mycelial footprints to create oasis environments that continue to expand as the mycelium creates soils, steering the course of ecological development.

Mycorestoration practices can be implemented in the following ways:

- mycofiltration
- mycoforestry
- mycoremediation
- mycopesticides

Mycorestoration involves using fungi to filter water (mycofiltration), to enact ecoforestry policy (mycoforestry) or co-cultivation with food crops (mycogardening, see part III), to denature toxic wastes (mycoremediation), and to control insect pests (mycopesticides). Mycorestoration recognizes the primary role fungi play in determining the balance of biological populations.

We are in constant molecular communication with fungi, but our interactions are at such a subtle level that most people fail to notice fungi's talents. Each mushroom species has a mycelium that degrades organic matter by secreting unique mixes of extracellular enzymes and acids. Since unique suites of enzymes are generated by each species, using a plurality of species can have a synergistic effect for the more complete degeneration of toxins than could be achieved with one species alone. The art of this emerging science is in the selection of species and, of equal importance, their timely introduction. If we use fungi as keystone species, we can enable subsequent populations and create unique habitats and species mixes. Although these courses are not yet charted, observations of the recovery of disturbed habitats by such mycological sages as Roy Watling (1998) can serve as natural guidelines for restoration. Nature teaches by example. We must learn from her teachings to envision the mycological paths best leading to habitat restoration strategies.

Using mushroom mycelia as tools for ecological restoration is a new concept borrowed from the age-old methods of nature. After forest fires, when burned habitats begin to recover, the species that appear amid the ash and cinders are mushrooms, particularly morels *(Morchella)*, and cup fungi *(Auricularia)*, which can appear in a matter of weeks. These fast-growing and quick-to-decompose mushrooms emerge where seemingly no life could survive. As these succulent mushrooms mature and release spores, they also release fragrances that attract insects and mammals, including mushroom hunters. A biological oasis emerges as new species gather around the postfire fungus. Flies deposit larvae in morels, and as the larvae mature they attract birds and other maggot lovers. Birds and mammals coming to eat morels defecate seeds of plants eaten far from the fire zone. All these critters scour the burnt wasteland searching for mushrooms. Each mushroom-seeking organism imports hitchhiking species from afar with every visit, essentially carrying its own universe of organisms, an ecological footprint of flora and fauna. Then, with every mushroom encounter, each animal is dusted with spores, leaving an invisible trail of them as they wander on. As animals crisscross the barren terrain, the layering of ecological footprints creates interlacing biological pathways. Morel mushrooms, for instance, are pioneers for biodiversity, first steering animate vessels of genomic complexity into an otherwise near-lifeless landscape.

Similarly, mushrooms can help restore environments damaged by pollution or other toxins. Targeting toxic environments is the first step in mycoremediation. In profiling these complex environments, mycoremediators can determine which combinations of species would be most beneficial to a particular environment. Integration of companion planting strategies then sets the stage for an emerging oasis in a lifeless landscape. Each succeeding mycelial mat provides different components as the ecosystem is steered toward recovery.

Surveying the Habitat

In order to determine what method of mycorestoration should be used, a damaged habitat should first be surveyed for its species mix. The resident species are nature's recommendations for habitat restoration. If the habitat is mycologically neutral—not showing any evidence of mushroom growth—then we have a cast of characters that might be suitable but must be imported in the form of spawn. We must match the habitat with the species with the carrier debris field. Whether or not the extant debris field is sufficient for mycoremediation will affect the mycorestoration strategy. For

instance, let's consider a microbial pathogen. It matters whether the microbe appears seasonally or is continually active. For example, rain patterns have a major influence in the outflow rates of contaminated surface water. Around cattle and pig feedlots, the flow of coliforms, nitrates, and phosphorus affects the downstream watershed, especially during the rainy season. If wheat or cornfields are nearby, cornstalks and wheat straw can furnish substrate for carrying mycelium. Otherwise, you'll need to import this type of agricultural roughage or wood chips for use as a medium for infusing mycelia. The essential idea is to grow mats of mycelium matched to the cubic size of the contamination source. By creating a sheet mulch, a shallow compost bed 6 inches to 2 feet thick, the mycofiltration properties of the mycelium and surface areas of the substrate particles will capture the microbial outflow.

Oftentimes, a native fungus can correct the biological imbalance. A field survey of the affected site may turn up mushrooms that could counter the pathogen's effects. Placing myceliated burlap bags, or "bunker spawn," stacked like a rim of sandbags is one simple method of introducing those beneficial mushrooms (see page 151 for a discussion of how to make burlap bag spawn). You can expand a helpful mushroom's domain by using methods such as stem butt transfers or spore slurries (see chapter 9). More intensive techniques of cloning and propagating pure culture spawn are described in my books *The Mushroom Cultivator*, coauthored with Jeff Chilton (1983), and *Growing Gourmet and Medicinal Mushrooms* (2000a). The goal of these processes is to create enlarged mycelial masses of the desired species.

Since mushrooms seasonally grow fruiting bodies from their mycelia, visiting the contaminated habitat during the mushroom-forming season of a particular species is the best time to survey the site for naturally occurring mycoflora. Someone knowledgeable in the areas of taxonomy and tissue culture will greatly aid your search for a mycological solution. Mushrooms proliferating there already tolerate the toxin. These species are naturally selected and predominate, to the disadvantage of species that are not as well equipped. In the laboratory, cultures of mycorestorative strains found in the damaged habitat are easily tested using standard agar well assays for antimicrobial activity or toxin sensitivity.

Additionally, if a toxin contaminates a habitat, mushrooms often appear that not only tolerate the toxin but also metabolize it as a nutrient or cause it to decompose. Some mushrooms even live in heavy-metal-contaminated sites intolerable to others. You can pinpoint the tolerance of a fungal decomposer by titrating dilutions of the toxin into media-filled culture dishes growing a strain of mycelium. Working with Battelle Marine Science Laboratories in Sequim, Washington, a team of scientists and I identified a fungus that broke down dimethyl methylphosphonate (DMMP), a key ingredient in the deadly neurotoxin VX (and sarin). Over the course of a few weeks, the mushroom thrived in a petri dish eating nothing but the DMMP. We essentially trained the strain to focus on DMMP as its sole nutrient source. Subsequent analysis of the culture media showed that the majority of the VX surrogate had been metabolized by the mycelium into unstable subderivatives that soon became nontoxic. This particular strain demonstrated tolerance to the VX at levels that would be toxic to other mushroom strains, showing that species vary substantially in their ability to adapt to specific toxic loads.

If native mushrooms suitable for mycoremediating the toxin are not found at the contaminated site, then an appropriate species can be imported. Using the following charts (pages 62, 96, and 106), see if any of the mushroom species listed matches the pathogen or toxin of concern. If you see a match, then propagating that mushroom species might neutralize the threat. Allow for some strain variability and site-specific nuances that may alter outcomes. Over time, you will learn to identify and deploy the most appropriate mycelia.

For 20 years, I have been visiting a rhododendron garden lovingly cared for by a now elderly couple for more than 4 decades. Each year, they would distribute wood chips around the plants, building pathways and for general landscaping. The past 2 years, they've been no longer physically able to replenish the soil with top-dressings of wood chips as they'd done previously. As a result, there has been a sudden transition in the mycoflora; *Hypholomas*, *Psilocybes*, and other species that were once prominent are now scarce. Mycological landscapes must be replenished with carrier materials and sometimes recharged with spawn to preserve the saprophytic mushroom communities. In woodlands, the constant falling of overhead debris feeds the saprophytic mushroom laying upon the forest floor. Throughout this process, soils deepen underneath.

Mycorestoration is an infant science to humans, but a highly refined method used by nature for millions of years. As we open our eyes to the fungal opportunities—literally underfoot—we soon see many mushrooms in their roles as environmental healers. In my mind, mushrooms are shamanic souls, spiritually tuned into their homelands. We, as cocreators, will benefit from listening to their voices.

CHAPTER 5

Mycofiltration

Mycofiltration is the use of mycelium as a membrane for filtering out microorganisms, pollutants, and silt. Habitats infused with mycelium reduce downstream particulate flow, mitigate erosion, filter out bacteria and protozoa, and modulate water flow through the soil. More than a mile of threadlike mycelial cells can infuse a gram of soil. These fine filaments function as a cellular net that catches particles and, in some cases, digests them. As the substrate debris is digested, microcavities form and fill with air or water, providing buoyant, aerobic infrastructures with vast surface areas. Water runoff, rich in organic debris, percolates through the cellular mesh and is cleansed. When water is not flowing, the mycelium channels moisture from afar through its advancing fingerlike cells.

Mycofiltration has many applications. Mycofiltration membranes can filter the following:

- pathogens including protozoa, bacteria, and viruses
- silt
- chemical toxins

They can be installed around the following types of sites:

- farms and suburban and urban areas
- watersheds
- factories

- roads
- stressed, harmed, or malnourished habitats

Installation of mycofiltration membranes can utilize debris from the following sources, which is then inoculated with toxin-specific mushrooms (some of which are listed on pages 96 and 106):

- forests (brush, tree trimmings, wood chips)
- pulp, paper mills (cellulose, fiber, cardboard, or paper waste)
- city and rural recycling centers (yard waste)
- farms (straw, corncobs, cotton, coffee wastes, and so on)
- breweries (washed grains) and other industries

Mycofiltration: Germination of the Idea

In 1970s, while studying at the Evergreen State College near Olympia, Washington, I peered at fungi through a scanning electron microscope and found the patterns of mycelia fascinating. I imagined that this fabric of fine cells could act as a biological filter. At that time, mycologists typically described mycelium as growing *on* habitats. I saw them as growing *through* the environment, filtering water after rains. I made filters by peeling mycelia from petri dishes and comparing their filtration properties to that of cotton. Mycelia's absorbency of tobacco smoke, ink, and water was astonishing. So was the tenacity that held the mat tightly together. But my idea of mycelium as a filtration system fell dormant until I bought a small waterfront farm on Kamilche Point in Skookum Inlet, Washington.

After years of collecting mushrooms on other people's property, I now had property of my own on which to do field studies. On my land, 2 swales gradually narrowed over the course of about 800 feet, dropping 120 feet from my uplands to the saltwater bay downstream. A ravine led to a small waterfall directly above a bay where my neighbor grows clams and oysters for commercial purposes. My property came with a small herd of 6 Black Angus cows; chickens and pigs soon followed. I installed outdoor wood chip beds of garden giants (*Stropharia rugoso annulata*) and other mushrooms at the top of one of these parallel sloping

basins. First, I dumped several truckloads of wood chips into the depression. The utility company trimming tree branches away from the power lines along my county road had provided the wood chips. On top of each dump load, I spread several bags of *Stropharia rugoso annulata* spawn and then raked out the pile into a foot-deep layer. Springwater saturated the wood chips—a perfect environment for running mycelium. Several months later, I had a garden giant mycelial bed about 50 feet wide and 200 feet long. The next summer, enormous mushrooms grew, providing delicious fare for many warm-weather barbecues (see figure 31).

At that time, most of the septic systems in the area were primitive cedar barrels, pits, or poorly constructed

▲ FIGURE 68

In this scanning electron micrograph of mushroom mycelium, these cells are about .5 to 2 microns thick. Chains of these cells can extend in length from a few microns to a few miles, forming an integrated, netted fabric of interconnected cells.

▲ FIGURE 69

Skookum Inlet, where mycofiltration in the form of a patch of garden giant mushrooms, cleaned up flow of fecal coliform contaminants from my land.

Rhizomorphic mycelium of the garden giant *(Stropharia rugoso annulata)* tenaciously holds wood chips together. This mushroom thrives when it comes into to contact with bacteria, compared to its slow growth behavior under sterile, bacteria-free conditions in the laboratory. Until it makes contact with microbes, this species does not produce rhizomorphs.

▲ **FIGURE 71**

Rhizomorph of a caerulescent *Psilocybe* grasping a cluster of dowels enveloped in a sheath of silky white rhizomorphs. The length of the pictured thread of mycelium weighed .002 grams and held dowels weighing 6.079 grams, meaning that this rhizomorph supported 3,029 times its mass. When 90 percent of this rhizomorph was cut away, it still supported the wooden dowels, meaning that it can hold more than 30,000 times its mass. This places into perspective how tenacious mycelial mats can be when they infuse habitats with their cellular networks. They grip a habitat and hold it tightly, stabilizing and protecting it from erosion.

drain fields installed by the original settlers. Fecal coliform pollution directly threatened the shellfish industry on the inlet. The livestock on my property was just one source of bacteria jeopardizing these family businesses. A few months after I moved in, the sheriff visited me and all the upland property owners along Skookum Inlet, serving us with court orders to install new septic systems within 2 years or vacate the land. But just 1 year after I had installed my beds of mycelium, before I had even repaired my septic system, analysis of my outflowing water showed dramatic improvement: a hundredfold drop in coliform levels despite the fact I had more than doubled my population of farm animals.

The anomalous decline in fecal coliforms surprised the water quality inspectors monitoring the inlet. We walked over to the headlands of my basin, where a tiny spring trickled water, and I showed them the mycelial bed I had installed. As we walked across this thick layer of wood chips, it felt like spongy duff. Our feet sank softly into the wood chips, which bounced back with each step. I explained to the inspectors that the contaminated water seeped from our livestock pasture, entered this mycofilter, and fed the myceliated wood matrix with nutrients and bacteria. As the fungus grew, the wood chips became infused with white, silky mycelium. The water that exited our wood chip bed was largely cleansed of bacteria, which had been consumed by the mycelium of the garden giant.

My neighbors, who worked for the Washington Department of Fish and Wildlife, gave me a score of silver salmon fingerlings. I put them in a 20-foot-long outdoor stainless steel tank filled with water. How should I feed them? I didn't have to look far. I discovered that when I threw mature garden giant mushrooms into the tank, they floated. Struggling for air, the squirming fly larvae living in the garden giants soon emerged from the soggy mushrooms. The fish learned that bumping the floating mushrooms would dislodge larvae into the water for easy feeding. Later, whenever I tossed a mushroom into the tank, a feeding frenzy ensued. Salmon rammed the garden giants, preying on the helpless, succulent larvae. So the mycelium not only acts as a biofilter but also provides food for me, my salmon, and other organisms.

That my backyard mycofiltration experiment reduced populations of coliforms drew the attention of researchers at the Battelle Marine Science Laboratories in Sequim, Washington. More-formal studies ensued, demonstrating that other mushrooms—oysters and wood conk mushrooms—also functioned as microbiological filters. Eventually, this collaborative research showed that mycofiltration could have wide applications in the destruction of biological as well as chemical toxins.

Mycofiltration of Microbial Pathogens

Mushroom mycelium has an unquenchable appetite for organic debris. Taking advantage of this appetite, the mycological landscaper can select mushroom species that target and consume the bacteria and protozoa in a habitat. For example, Lovy and others (1999) found that my strain of zhu ling (*Polyporus umbellatus*), a polypore mushroom I had cultivated, was 100 percent effective in vitro in inhibiting the malarial parasite *Plasmodium falciparum*. Mycomulching—infusing a layer of wood chips with mycelium—with this species around a malaria-infected swamp could reduce the background population of malaria, since the fungus likely consumes the parasite and secretes antibiotics into the habitat. Although this temperate species is not known to inhabit the same ecosystems as malaria, other polypores native to malarial habitats could possibly be effective. Combining these polypore species with pesticidal mycelia that could attract and attack plague-carrying mosquitoes could amount to a double-pronged response to malaria. Mycofilters and, indeed, entire landscapes, can also be customized with mosaics of mycelial mats to prevent infection from coliform or staph bacteria and protozoa. The mats can even trap disease-carrying insects, thwarting disease vectors and protecting our health.

Pathogens passing through the cellular nets of mycelia are digested by the fungi. If enzymes and antibiotics secreted by the mycelium don't kill all the pathogens, the bacteria, such as *Bacillus subtilis*, are blocked from reproducing and are suspended in a state of dormancy. These types of bacteria may later revive when the mycelial mat dies and a hospitable habitat reemerges. (Mycologists call this *dieback*, a process that is analogous to the decay of a fishing net that loses its cohesiveness and unthreads.)

In another field trial, I built a pond to catch the surface water flowing from the high part of our

▲ **FIGURE 72**

This mycofiltration test chamber, sized to fit a bale of straw, myceliated or not, can be sectioned off, and samples of water can be taken both upstream and downstream.

Mushrooms versus Microbes

This chart describes mushroom species found to have specific antimicrobial effects on the corresponding microbes. Most microbes listed here are pathogens to both animals (humans) and mushrooms. (For a list of pathogenic fungi and their classification, see www.pfdb.net/myphp/database_eng.php. For a list of pathogenic bacteria and their classification, see www.bmb.leeds.ac.uk/mbiology/ug/ugteach/icu8/classification/head.html.)

	Aspergillus niger	Bacillus spp.	Candida albicans	Escherichia coli	Listeria monocytogenes	Mycobacterium tuberculosis	Plasmodium falciparum	Pseudomonas aeruginosa	Pseudomonas fluorescens	Staphylococcus aureus	Streptococcus pneumoniae	Streptococcus pyogenes
Agaricus brasiliensis				X								
Armillaria mellea		X								X		
Chlorophyllum rachodes										X		
Coprinus comatus	X	X	X	X				X		X		
Flammulina velutipes										X		
Fomes fomentarius				X				X				
Fomitopsis officinalis				X		X		X	X	X		
Ganoderma applanatum		X		X						X		
Ganoderma lucidum	X	X	X	X								
Grifola frondosa			X									
Hericium erinaceus	X	X	X									
Hypsizygus ulmarius										X		
Laetiporus sulphureus		X		X						X		
Lentinula edodes			X		X	X				X	X	X
Lepista nuda			X							X		
Macrolepiota procera										X		
Merulius incarnatus										X		
Piptoporus betulinus		X		X				X	X	X		
Pleurotus ostreatus	X	X		X			X	X	X	X		
Polyporus umbellatus				X			X			X		
Psilocybe semilanceata										X		
Schizophyllum commune			X	X						X		
Sparassis crispa		X										
Stropharia rugoso annulata				X								
Trametes versicolor	X		X	X						X	X	

property. Measuring about 150 feet by 50 feet and around 10 feet deep, our pond contained mostly rainwater, supplemented with groundwater pumped from a well. Testing the water that flowed from it, I found that the predominant bacterium was *Pseudomonas fluorescens*, the probable cause of the rash I got after taking a brisk swim there on a hot summer day. When we placed straw bales inoculated with oyster mushroom mycelia into the slough, channeling the runoff water through the straw, lab analysis found that the benign and omnipresent *Bacillus subtilis* reigned supreme while the upstream dominant *Pseudomonas fluorescens* failed to register in the top 5 bacteria downstream. In just this way, matching the mushroom species to the problematic bacteria, ecological engineers can customize mycofilters to prevent upstream pathogens from passing into downstream environments. Several factors affect the efficiency of mycofiltration: slope, flow rate, turbidity, straw shaft diameter, mushroom species, degree of mycelial colonization, and microbial populations. Given the numbers of mushroom species that have specific antibacterial properties, we already have the ability to grow hundreds of mushroom species in mycomulches to buffer or eliminate threats posed by upstream microbes.

Mycofiltration around Farms

News from the heartland is not good. Manure ponds—commonly amassing around cattle, hog, and chicken factory farms—are bursting with fecal-rich effluent leaching into the watersheds. River ecosystems are imperiled. *Pfiesteria, Listeria, Streptococcus, Escherichia coli*, amoebic parasites, and viruses are posing increasing human health risks as corporate farming policies aim for profit at the expense of environmental health. Factory farms, which crowd livestock into tight quarters for efficient feeding and slaughtering, are causing an overly focused and growing outflow of waste products that threaten the health of all. This outflow may have exceeded the amount that our habitats can absorb. And

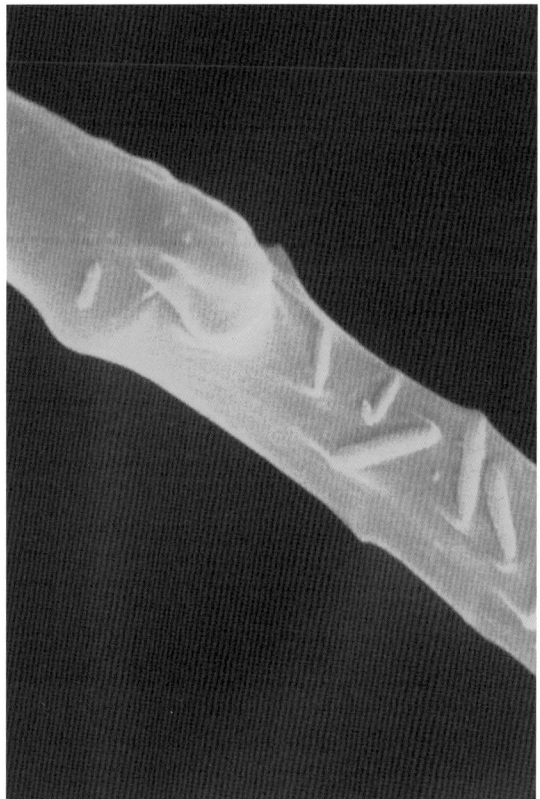

▲ **FIGURE 73**

A scanning electron micrograph of mushroom mycelium carrying rods of *Bacillus subtilis,* a bacterium generally thought to be friendly to humans and mammals but antagonistic to many fungi. Also depicted is a clamp connection bridging 2 cells of mycelium. (See also page 128.)

so our waste streams run into the waterways of our nation, wreaking havoc. In addition, new housing projects encroach upon former farmlands, where the threat of biological disaster from untreated livestock wastes represents a virulent threat.

Hog farms are particularly worrisome to environmental scientists. When hurricane Floyd hit North Carolina in 1999, the monsoonlike rains caused dikes to burst and manure ponds to overflow, flooding thousands of acres with animal feces and causing incalculable health problems both on and off the farms. Residents in Charlotte were rudely awakened to the enormity of the problem by the fouling on their doorsteps. Filth filled the streets and flooded basements. The collateral damage included contaminated wells, fisheries, and crops. Many diseases spread, including ones pathologists are still at a loss to identify. This ecological mess eroded the public trust in farmers as good neighbors.

Toxic levels of zinc and copper accumulate in livestock feedlots, a by-product of manure production. A particular study in North Carolina showed that the mycelium of a mold fungus, *Aspergillus niger*, removed 91 percent of the copper and 70 percent of the zinc from treated swine effluent (Price et al. 2001). Other mushroom-forming fungal species may be used to remediate sites contaminated with heavy metals and other pollutants (see chapter 7).

Corporate giants responsible for this dangerous situation play political football by demonizing opponents, especially supporters of government regulations, and by pitting farmers against environmentalists. The theater of conflict features a growing cultural divide between corporate and green philosophies. Practical proposals for fixing these problems, let alone proven remedies, have been few and far between. And so we exist perilously close to the edge. As physician Andrew Weil ruefully warned me, now we must be more careful, since "nature bats last and the bases are loaded."

A variety of forms of mycelial mats can prevent downstream pollution. I am keen on using bunker spawn—mycelium in burlap sacks—to build mycelial buffers to capture microbes and nutrients. This subject is discussed in greater detail on pages 151 to 154. Please also see figure 140.

Mushrooms Munch Rocks

Researchers looking at the weathering of monuments discovered that mycelium has the surprising ability to break through rocks, including granite and marble (Burford et al. 2003). Mycelium consumes granite and loosens soil creating microcavities that can retain water and, when drained, fill with air. From the tips of emerging mycelium, polysaccharides and glycoproteins are secreted, along with powerful chelating enzymes and acids, opening paths for its flow into solid rock.

The fact that mycelium is able to penetrate granite is clear evidence of the enormous pressure exerted at its cell tips—forces equivalent to tens of atmospheres that also allow the hyphae to penetrate through plants and insect exoskeletons. Mycelia's invasive physical strength is coupled with its ability to solubilize inorganic matter using metabolic acids, metal-chelating anions, protons, and enzymes (Money 2004; Gadd 2001). This ability to mineralize substrates— to make minerals available by removing them from a tightly bound matrix—helps mycelia encroach into barren habitats, disintegrating rocks and setting the stage for lichens (a partnership between algae and fungi) and succeeding populations of diverse organisms. As mycelia advance onto arid habitats, the water carrying capacity of these myceliated environments steadily increases and the forces of erosion are kept at bay—to a degree. The soil can retain moisture and yet breathe through the membranous lungs of mycelium. An ecosystem's ability to withstand massive loss of life-sustaining soils is greatly influenced by the infusion of mycelium into topsoils. When impacted by sudden changes in weather conditions, the mycelium can be taxed beyond its abilities, losing its grip, so to speak, on its homestead. As the mycelium dies back, its cellular architecture breaks apart, and soil cohesion also declines.

Mushroom Mycelium, Soil Preservation, and No-Till Farming

Because water propels fungal life cycles, droughts effectively shut down the nutrient-return pathways. The balance of major and minor species shifts. When a habitat loses its mycelium due to drought or fire, soil debris crumbles and blows away. For example, thousands of years of wind lifted rich topsoils from eastern

Oregon and deposited them as loess in eastern Washington during Holocene time, after the last ice age. Settlers, including my great-grandfather Charles Davis, discovered that these rolling hills near St. John, Washington, made superb land for growing grain. The contrast between these environments, even today, is remarkable: eastern Oregon is a sagebrush desert, while the Palouse country of eastern Washington is covered with low, undulating, verdant hills planted with waving fields of wheat, barley, and oats.

Wheat and barley still grow on our ancestral Davis farm, although a century of farming has depleted the soil bank to a mere fraction of its original 100-foot depth. While farmers have increasingly relied upon fertilizers to sustain crop yield, in the late 1980s my cousin Jim Davis, in St. John, adopted the "no-till" method of farming, drastically reducing the need for externally introduced fertilizers, despite skepticism from his neighbors. When I recently visited in October after the harvest, he showed me his wheat fields adjacent to his neighbor's. Chopped stubble, left for nature to recycle, covered his fields, while his neighbor's fields were marked by deep grooves from erosion.

The no-till method succeeds largely due to an unseen ally—beneficial fungal mycelium. The downturned stubble of my cousin's farm harbored native fungi, which had both stopped erosion and replenished the soil. Since water doesn't run off as quickly, the resident soil moisture seen in no-till fields is naturally higher due to the spongelike effects of the mycelium gobbling up the crop stubble and swelling with water. The coarse soil structure embedded with stalks from crops is perfect for mycelium to run upon. Tilling breaks the stubble into finer fragments, compacts the soil, and encourages growth of anaerobic organisms to the detriment of the oxygen-starved mycelium. Then, the carbon cycle stalls; natural nutrients are not rereleased; and importation of fertilizers is required to continue profitable farming.

A 21-year study in Germany found that no-till organic farming methods were superior to conventional methods in energy use and effects on wildlife.

Organic farming practices used one-half to two-thirds of the energy consumed in conventional methods. In addition, they cut pesticide use by 97 percent, resulting in healthier soils with better diversities and numbers of beneficial organisms such as fungi, earthworms, beetles, and wild plants. Although initial yields may be 10 to 20 percent less than those from conventional methods, a subsequent increase of 15 percent was seen as the soils adapted to the no-till nutrient cycles (Mader et al. 2002). In addition, the researchers reported that less fertilizer is needed with the no-till method. Soil is built, water infiltration improves, and less runoff and erosion occurs. Researchers believe that these organic practices are more sustainable over the long term, paying a net ecological dividend and making organic farming more profitable for future generations.

Plowing the stubble into the soil releases 41 percent more carbon dioxide into the atmosphere than no-till practices do, impairing the soil's carbon return cycle. In contrast, the no-till method keeps the carbon dioxide with the soil's biosphere. Government subsidies have been put in place to encourage farmers in the United States to practice the no-till method. Currently in the United States, according to Rattan Lal and others (2004), 37 percent of the cropland is worked using the no-till method, compared to 5 percent elsewhere in the world. Lal estimates that if all farmers in the United States would adopt the no-till method, 300 million tons of carbon could be kept in the soil.

Not only does the mycelium unlock natural nutrients, it holds soils together while providing aeration. Without being exposed to nutrients released by mycelium, the roots of farm crops are undernourished and underdeveloped. Researchers at Montana State University led by TheCan Caeser-TonThat (2000) discovered that the resident mycoflora, particularly the higher fungi (sexual fungi that produce mushrooms), aggressively decompose the stubble in no-till farms and increase water-stabilizing aggregates (glomalins) in soil structure. These saprophytic allies extend their fine, nearly invisible cellular filaments in between plant fibers. Fungal enzymes break down plant cells into basic nutrients and also synthesize polysaccharides that sponge moisture.

Caeser-TonThat (2002) found that polysaccharides manufactured by the mycelium act as mucilaginous soil-binding agents. (Coincidentally, these same polysaccharides boost the human immune system; see Stamets and Yao 2002.) I was awestruck to learn that nearly a mile of mycelium can entangle a gram of pasture soil (Ritz and Young 2004) and that a cubic inch of soil can weigh more than 13 grams, I realized that a cubic inch can be intermeshed with a staggering 8 miles of mycelium. Once I understood that the fabric of these mycelial cells makes up the architectonics of soil's food web, I knew that the influence of mycelium and its binding agents on soil aggregation was beyond anything I had previously imagined. When the mycelium infuses soil, the internal space is framed in architecture of dense interconnecting hyphal networks. Microstructural cavities hold water and provide life to diverse microbial populations. Growth of the mycelium is focused on the tips of the emerging, forking hyphae, where polysaccharides, glycoproteins (glomalins), enzymes, antibiotics, and messenger molecules are secreted. Mycelium gives soils porosity, aeration, water retention, and ultimately a platform for diversifying life-forms. It is truly a networking organism, adding cohesion to vast biological communities.

Many temperate mushroom species produce antifreeze glycoproteins that protect the mycelium from the harmful effect of water crystallizing into ice. These antifreezing agents also help prevent the soils from freezing, conferring protection to plants during extreme cold. In 2003, Hoshino and others filed a patent on the antifreezing polypeptides from several mushrooms. Additionally, soils infused with actively growing mycelium benefit from thermogenesis—the natural escalation of temperature—as the mycelium decomposes organic matter and releases heat, water, and carbon dioxide.

Fungi flourish with seasonal feedings of downturned stubble. Furthermore, agricultural crops, especially grasses like wheat, benefit from resident endophytic fungi that enhance growth and promote disease resistance. A couple of saprophytic mushroom species I have found to promote the growth of several crops are the garden giant (*Stropharia rugoso annulata*) and the elm oyster (*Hypsizygus ulmarius*). See chapter 12 for information on incorporating these mushrooms using companion planting strategies.

Gardeners and farmers using the no-till method can select crop-enhancing fungi that have antinematodal, pesticidal, and antiblight properties. In effect, you can customize the mycosphere for your land. As the mycelium decomposes compost or crop stubble, it projects a fine network of cells, a food web that draws in nutrients from great distances. Not only is the mycelial network exquisitely efficient at recycling plant debris, it is also just as good at gobbling up bacteria, catching nitrogen-based nutrients, and modulating the flow of water (and effluents). Proven in the laboratory of life, this method is adaptable to agriculture. A natural offshoot of leaving stubble for no-till is the use of agricultural debris as a biological filter in and around farms.

Farmers can build soils while creating mycofiltration membranes for trapping pollutants by using thick sheet mulch inoculated with mycelium. Farms are generally well equipped to adapt fungal-filter solutions to pollution, especially where wood chips and straw are abundant.

For example, corncobs are perfectly structured for hosting mushroom mycelia. Their fiber and small cavities allow for aeration and provide food for rapid colonization. Chopped corn stalks combined with fragmented, husked corn makes a suitable medium both for growing mushrooms and then for mycofiltration. Corn farmers can first profitably grow oyster mushrooms on corncobs and then use the spent substrate, after mushroom production, as inoculum into sheet mulch. Such combinations of methods are one of the amazing aspects of mushrooms: they can be used with versatility in a number of applications.

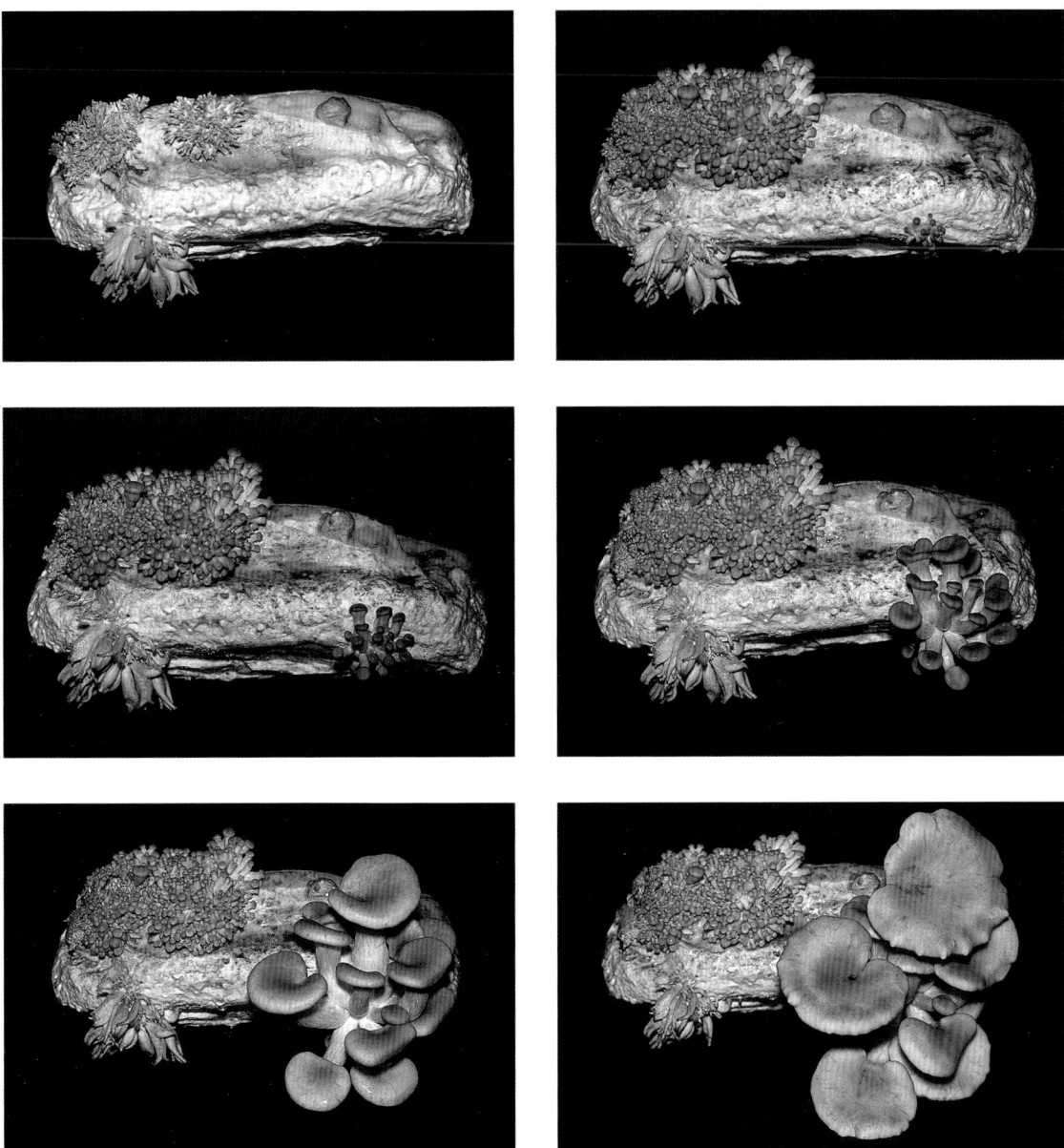

▲ **FIGURE 74**

Daily time-lapse photos of oyster mushrooms fruiting from corncobs.

Installing a Mycofilter

A gently sloped area below a feeding lot or manure pond, where effluent from the lot or pond continually seeps through, is an ideal site to install a mycofilter, essentially a myceliated organic drain field. For the bottom layer, scatter sawdust or wood chips to a depth of 3 to 4 inches. For the first of 2 layers of spawn, on top of the sawdust or wood chips, spread inoculated sawdust by hand or by silage spreader; use $1/4$ pound inoculated sawdust per square foot of the site. Next, add a layer of corncobs 4 inches deep, and follow with a second layer of spawn. Because high winds and harsh sun can dry out mycofiltration beds, cover the site with waste cardboard before adding the last layer of straw. The finish layer of straw should be 4 to 6 inches deep to provide shade, aeration, and moisture to layers below. (If natural rains do not provide sufficient moisture, sprinklers can be set up for the first few weeks until the site becomes charged with mycelia.) The surface area of the mycofilter should be at least several times larger than the surface area of the manure pond or feeding lot, depending upon depths, slope, and flows.

Mycofilters are best built in the early spring. Once established, the mycofilter will mature in a few months and remain viable for years, provided that fresh organic debris is periodically added to the top layer and covered with more straw. For some, the best time for this may be after the fall harvests, when agricultural debris is plentiful. After some time, red worms will arrive and transform the mycelium, cardboard, and debris into rich soil. Every 2 to 3 years, the newly emerging material can be scooped up using a front loader tractor and used elsewhere as soil; the timing of this cycle will vary. (Incidentally, gourmet mushrooms may form after rains depending upon temperatures. These mushrooms "reseed" the beds, provided there is enough food.) For more information on cultivating mycelium in outdoor settings, please refer to pages 187 to 200 and my book *Growing Gourmet and Medicinal Mushrooms* (2000a).

I encourage farmers to try this method. The amount of time to install a mycofilter is minimal, only a couple hours. Spawn will probably be your biggest expense, but once established and cared for, the mycelium can regenerate itself until the debris base has been reduced to soil. As these areas mature, they usually become covered with native grasses, which also play remediative roles. A universe of compatible organisms shares this habitat, with mushroom mycelia reigning as the pioneering organisms.

Mycofiltration is a natural upland fit to John Todd's "living machine"—the use of estuary ecosystems to break down toxic wastes (Todd and Todd 1993). Mycelial systems also bridge nicely with Bill Mollison's pioneering permaculture practices, which strengthen sustainability using natural systems (1990). All these systems use synergism of biological communities and gain strength from biodiversity. As we advance toward a better understanding of sustainability, I see these three systems—mycorestoration, permaculture, and living machines—as being essential components in a new model of habitat restoration. In essence, future perfection will be founded upon synergizing these three systems.

CHAPTER 6

Mycoforestry

Without fungi, there are no forests. Mycoforestry is the use of fungi to sustain forest communities. Mycoforestry can be used to help accomplish the following goals:

- preservation of native forests
- recovery and recycling of woodland debris
- enhancement of replanted trees
- strengthening sustainability of ecosystems
- economic diversity

We have a simplistic view of the interrelationships between mushrooms, the forest, and its inhabitants. For instance, in the 1940s through 1960s, timber companies commissioned the wide-scale slaughter of bears in a misguided attempt to protect the lumber industry. My neighbor was hired by a timber company to kill more than 400 black bears in Mason County, Washington. Bears love mushrooms and actively spread their spores. The conventional thinking was that when bears scratched trees in search of grubs, they created wounds that soon became infected by polypore mushrooms. Now we now know that bears and other animals actually help lowland old-growth forest ecosystems by fishing salmon and trout from streams, replenishing the stream banks with essential sea salts and nitrogen-rich nutrients. Furthermore, the spawning fish feed upon the grubs growing in fly-infested mushrooms that are washed into streams by heavy rains. The fish carcasses pulled from the streams by

bears transport trace phosphorus and nitrogen, nutrients essential for tree growth. Migrating fish whose carcasses are further spread by animals are one of the few ways sea minerals and nitrogenous nutrients are carried into upland forests. Scavenging animals like bears, raccoons, birds, and insects eat the carcasses, allow the minerals to move through their digestive systems, and deposit them in locations far from the streams. Today, the significance of this mineral pathway is demonstrated by the policy of returning uneaten fishery-run salmon carcasses to spawning grounds to help nutrify the ecosystem, benefiting salmon fry, the riparian habitat, its trees, and other nitrogen- and mineral-dependent species.

Mushrooms contribute phosphorus and confer other ecological benefits to the riparian and forest ecosystems. Mushrooms become launching platforms for explosive growth of bacterial populations, many of which are critical for plant health. Mushrooms have a preselecting influence on the bacteria sharing their habitat (Tornberg et. al 2003). Bacteria beneficial to trees regulate inputs and outputs of nitrogen and are phosphorus limited (Sundareshwar et al. 2003). Mycelium absorbs phosphorus from its surroundings, moving these mineral salts over distances and later releasing them when mushrooms rot or the mycelium dies. Fungal-decomposing bacteria then absorb the phosphorus. As the mushrooms rot, the ecosystem benefits from this cycling in which the bacteria allow

▲ **FIGURE 75**

Several miles deep in the old-growth forest of the Hoh River valley of the Olympic National Forest, we encountered this tree scratched by a bear. Such wounds give birth to polypore mushrooms. Five years later, we returned to this tree, which had died, and had evidence of cubical brown rot. Presumably, the bear scratch was an entrance wound for infection from a brown rot fungus.

phosphorus, zinc, potassium, and other essential minerals to be redeposited back into the nutritional bank.

Like salmon carcasses, mushroom carcasses fertilize the ecosystem. Other organisms quickly consume the dying and rotting mushrooms. As plants grow, their falling leaves, branches, and flowers enter into the fungal cycle of decomposition. This response—a highly energized state of regrowth—is nature's safe-

guard for rapid, adaptive habitat renewal. After catastrophes strike, the saprophytes lead the way toward renewal, supporting the construction of complex life-supporting soils. Unfortunately, humans often disrupt these cycles, largely because of ignorance or greed.

The Undervaluing of Biodiversity

Reforestation efforts are greatly enhanced when mycorrhizae are introduced to sprouting seeds or to the roots of young trees before or at the time of planting (see figure 81). The value of fungi is being increasingly recognized when economists assess forests. Many researchers sincerely believe that secondary products from woodland ecosystems, in addition to the other benefits they offer, provide strong economic incentives to leave the forests intact. And beyond short-term economic incentives, forests affect climate and prevent desertification. Placing an economic value on an aesthetic like unspoiled landscapes, an incalculable asset like biodiversity, or yet-undiscovered mycomedicines confronts the simplistic conclusions of conventional economic models that compare timber and mushrooms as two mutually exclusive commodities. From an economic perspective, many biologists rightly believe the biosphere is underappreciated.

Of course, the window of opportunity for harvesting timber is relatively short, and with each harvest the soil is depleted, impairing the value of future timber harvests successively over time. I know of no logging companies in the Pacific Northwest that espouse the belief that the fourth replanting will provide yields comparable to those of the first 2 replantings. As a result, many logging companies are selling their lands because their return on investment has declined to an unattractive ratio. When taking into consideration high-value mushrooms such as matsutake, the economic benefits of preserving the woodlands increasingly outweighs the short-term gains of logging, assuming the economic ratios remain on par. A study by Alexander and others (2002) using a soil evaluation equivalent (SEV) formula found that, in south-central

Oregon, harvesting timber and matsutake mushroom each yielded close to the same economic benefit from the land over the same time period. This evaluation, however, does not factor in many other positives in not cutting the timber—thickening soils, reducing erosion, helping the health of streams, increasing biodiversity, improving air quality, effecting regional cooling (thereby offsetting global warming), and some particularly incalculable aesthetics: the experiential pleasure a person or family has in the pursuit of wild mushrooms while walking through verdant natural landscapes. In fact, wild mushrooming, according to a United Nations report by Eric Boa (2004), has worldwide socioeconomic significance.

Selective harvesting of developing second- and third-growth forests, however, when done with the intention of preserving other secondary forest products such as mushrooms, may prove to be the best practice for sustaining profits. These principles are the cornerstone of an emergent new management strategy called *ecoforestry*.

The manner in which mushrooms are harvested, too, has a dramatic impact on subsequent crops. For instance, a study in southern Oregon showed that when matsutake patches were raked and the divots from the harvested mushrooms were not re-covered with duff, matsutake crops in subsequent years plummeted 75 to 90 percent; if harvesters re-covered the divots after shallow raking, yields were not adversely affected (Eberhart et al. 1999). Similarly, no adverse effects were noticed over a decade of harvesting chanterelles in a forest west of Portland, Oregon (Pilz et al. 2003), where the preferred practice was to cut the mushrooms while picking, leaving the stem butts undisturbed. We now know that chanterelles often come up in pairs, and if harvesters cut only one partner, then the other mushroom, often hidden from view as a resting primordium, can grow to maturity. Perfecting harvesting practices to protect future harvests dramatically improves the profitability of the mushroom-harvesting industry. To be viable, the industry's management practices must reconcile long-term ecological costs and benefits.

Destroying the last vestiges of the ancient old-growth forests for lumber and paper is shortsighted and may put our country at risk The natural capital that potentially flows from these forests exceeds short-term economic gains. We can't yet accurately assess the value of our old-growth forests, but the valuations continually increase directly in proportion to our knowledge. If a mushroom species exclusive to the old-growth forest prevents a viral epidemic that could kill millions and cost billions, how do we value it? Mushroom species producing enzymes to break down VX and antibiotics to protect cells from pox and HIV viruses dramatically increase the value of old-growth forests. Losing these mycoremedial species in order to improve the quarterly reports of lumber companies may cost our civilization far more in economic terms than it gains for a single industry: the future of our species may literally be at stake.

Forests Are *Not* America's Renewable Resource

I killed old-growth trees and clear-cut virgin forests. In the 1970s, I worked in the woods in Washington's northern Cascades, in the lumber and shingle mills, and on steep slopes where helicopters hauled away cedar salvaged from old clear-cuts. After a time I graduated to logger, a job that had me cabling logs in the woods, bundling them in slings, and hauling the logs via a carriage to a yarder anchored at a cliff's edge.

Since I was the only long-haired hippie in my crew, my bosses tried to break me by assigning me the most dangerous tasks. However, my resolve and my body only grew stronger. I followed the advice of a logger friend who told me to say little; remember that trees always fall downhill; and always run, staying uphill of the mayhem. I worked underneath the "skyline"—a mile-long cable suspended from a 70-foot-tall tower ("yarder") erected on a cliff's edge, anchored downslope by a massive stump and further secured to an equally redoubtable tree at the forest's edge. A winch on the yarder would slack or tighten this main line. When the cable slacked, a wheeled carriage dangling with smaller cables was lowered into position over fresh-cut trees. Our crew would lasso individual logs with one of the 4 available individual choker cables on the underside of the carriage. A winch operator in the yarder would tighten the cable, lifting the 50-ton cargo hundreds of feet into the air and then winch the log-heavy carriage along the skyline toward the tower, where the logs would be off-loaded onto trucks. Doing this harrowing work, I soon developed the spatial intuition of a combatant: as long as nothing trespassed on the immediate 2 cubic feet occupied by my body I could not be harmed. (In truth, I *was* working in a war zone, in a unilateral war against nature.)

One August day, after cabling our last load and sending it up valley, my 3 crewmates and I walked downhill. As we passed below the mighty 4-foot-diameter stump that anchored the skyline, we heard an ominous crack behind us, and then a loud "crack-crack-crack." This was a very bad sound. Succumbing to the tension of the skyline, the overstressed stump imploded, splintering into old-growth shrapnel. We ran like hell downhill, violating one of the cardinal rules of logger safety. In fear of losing our lives, we hurdled down the steep slope, flying with every step.

Hunks of stump, some weighing hundreds of pounds, sailed over our heads. These missiles vibrated with so much kinetic energy that the slightest contact could kill you. Meanwhile, uphill the loaded skyline convulsed, yanking out the standing anchor tree, which came crashing downhill right on the same path on which we had unwisely fled. This towering hemlock, which unbeknownst to us had a rotted core, quickly gave way, shearing the tops of other trees as it fell, showering us with a chaotic storm of debris. The 4 of us, cowering in fear, huddled behind a huge old-growth Douglas fir on the far edge of the clear-cut. This old giant saved our lives, for which I am eternally grateful. That day, I questioned my future as a logger and decided to go back to college to study botany.

My experiences as a former logger and now as a protector of the forest ecosystem give me an unusual perspective. Faced with the corporate logging propaganda promoting the idea that "trees are America's renewable resource," I wondered: with hundreds of tons of trees harvested per acre in one week from soils that have been built in the last 10,000 years, how could this loss of biomass be called renewable in our lifetime? This philosophy defies common sense. Carbon cycles that fuel the food chain and build the forest soils move at a much slower pace than logging companies. With each generation of trees we cut, soils increasingly shallow and we further jeopardize the health of forests. The richness and depth of soil is our legacy from centuries of mycelial activity. And with each harvesting and replanting, the soil loses nutrients and gradually becomes overtaxed, no longer able to support the growth of healthy trees. Trees prematurely climax, falling over as the root wads can't hold the trees upright. Current "sustainable" logging practices strive to balance the impact of overharvesting with ecological restoration, potentially irreconcilable objectives. The bottom line is that we need to focus on carbon cycles and raise the nutritional plateau in timberlands by accelerating decomposition of wood debris and restarting plant cycles.

Recycled Wood Debris Helps the Forest Recover

As a mushroom grower, I have seen habitats whose decomposition cycles influence subsequent successions of organisms. Nowhere is this more apparent than in clear-cut forestlands. Once the trees are killed, mycorrhizal fungal communities die back. After loggers haul trees away, vast debris fields remain behind: stumps, brush, and downed small-diameter or otherwise unmarketable trees. Until this wood debris decomposes, its biomass is locked away from the food web and is therefore unavailable to bacteria, protozoa, insects, plants, animals, and fungi, some of which would dismantle the cellular structure of the wood, freeing nutrients. In order to stimulate decomposition and trigger habitat recovery, we can selectively introduce key-

◀ **FIGURE 76**

Piles of branches sit undecomposed after more than 20 years on a farm on the border of Mason and Thurston counties in Washington State. Had this wood been chipped, placed into contact with the ground, and/or shaded, saprophytic fungi would have flourished and the debris would have reentered the food chain.

stone mushroom species such as saprophytic fungi, the first species to feed on dead wood.

Making wood debris fields more fungus friendly speeds up decomposition and helps the decomposition cycles become more balanced. To help nature recalibrate after logging, fungi must be brought into close contact with the dead wood so that the forest floor can act as a springboard for saprophytic and other fungi, which are instruments of the forest's immune system, ready to heal its wounds. For several years after a forest has been cut, the mycosphere survives underground, with an increasing loss of diversity over time unless plant communities and debris fields are renewed.

In forestlands, mycelium follows trails of fallen wood. Sticks and branches making ground contact are soon consumed by mycelium from existing fungal communities. Mycelia literally reach up from the ground into the newly available wood.

Whether wood is whole or fragmented affects the rate at which nutrients return to the soil: wood chips are quickly consumed by fungal mycelium, whereas logs decompose much slower. I recommend creating a matrix by chipping wood into variably sized fragments in order to let mycelium quickly grab and invade the wood. Wood fragments with greater surface areas are more likely to have contact with spores or mycelium; this is especially true in the cultivation of mushrooms, where spawn growth is integral to success. The fungal recycling of wood chips lessens reliance on fertilizers, herbicides, and pesticides. So leaving the chips in the woods helps recovering forest soils just like leaving stubble on farmed land helps agricultural soil.

However, if the wood is reduced to too fine a dust and piled too deeply, it suffocates aerobic fungi, including beneficial saprophytes, and anaerobic organisms flourish. From my experiences, I have learned that chips should be no smaller than 1/8 inch and piled no more than a foot deep.

Thinning replanted forests is a tenet of ecoforestry, reducing crowding and fire danger. Several companies have developed mobile machines that can grind whole trees into chips, clearing up to an acre per hour. A particular device called the Hydro-Ax represents a new generation in mobile logging equipment that can greatly aid mycoforestry and reduce fuel load by thinning and chipping on-site.

The Guiding Principles of Mycoforestry

Mycoforestry is a newly emerging science, an offshoot of ecoforestry practices with an emphasis on the role of beneficial fungi. As with any new scientific path, guidelines help steer the course of research and the development of new implementation strategies. These are the guiding principles I foresee:

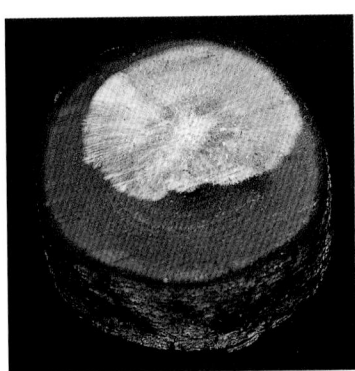

▲ **FIGURE 77**

When spored oil from a chain saw made contact with this slice of alder *(Alnus rubra)*, a mycelial colony of oyster mushrooms *(Pleurotus ostreatus)* soon developed. By using spored oils in chainsaws and chipping equipment, the decomposition process and therefore habitat recovery can be jump-started immediately upon cutting. Furthermore, by choosing an aggressive saprophytic mushroom species such as oysters, turkey tails, or woodlovers, parasitic fungi are confronted in a form of mycelial combat, thus lessening their resurgence.

▲ **FIGURE 78**

Heart rot from a brown rot fungus, probably the velvet polypore *(Phaeolus schweinitzii)*. Such wood has low timber value and is usually left on-site, a motivation stated by some foresters to justify burning.

- Use native species of fungi in the habitats needing restoration.
- Amplify saprophytic fungi based on available wood substrates.
- Select species known to help plant communities.
- Select mushroom species that attract insects whose larvae are food for fish and birds.
- Select fungal species according to their interactions with bacteria and plants.
- Choose species that compete with disease rot fungi (such as *Armillaria* species and *Heterobasidion annosum*) by using mycorestorative saprophytes like *Hypholoma, Psilocybe, Trametes, Ganoderma, Sparassis,* and allies.
- Choose species of known medicinal or culinary value if economically valuable mushrooms help tilt the balance in favor of preservation.
- Promote ground contact with fallen trees so they can reenter the soil food chain.
- Leave snags to sustain bird and insect populations.

- Use spored oils in chain saws, chippers, and cutting tools so that wood debris is immediately put into contact with fungal spores, speeding up decomposition.
- Retain wood debris on-site, and place debris around newly planted trees, along roads, or wherever erosion control is needed.
- Only burn wood debris as a last-ditch measure for disease control.
- Use mycorrhizal spore inoculum when replanting forestlands. (Seedlings cultivated in pasteurized or constructed soils on tree nurseries typically lack mycorrhizae.)

Matching Mycorrhizal Mushroom Species with Trees

Thousands of mushroom species form mycorrhizal relationships with trees, and most vascular plants, especially trees, can host diverse populations of mycorrhizae. It's been estimated that in the lifetime of an individual Douglas fir *(Pseudotsuga menziesii)*, more than 200 species of mycorrhizal fungi live symbiotically in and on its roots. Those mushroom species that mycorrhize with the largest number of host tree species have the greatest mycoforestry potential. However, only a few species have been commercialized as mycorrhizal spore inoculum and made available from a number of sources (see the Resources section). The species list is limited primarily by a trio of factors. First, the method of collecting spore mass limits availability. Mushrooms have to be collected and the spores separated from the source mushrooms before rot begins, and separation of the spores from the background tissue is easier with some species than with others. Second, since the economy for this industry is still small, only a few varieties are offered. Third, with the exception of some truffles, many of the best mycorrhizal varieties for mycoforestry, such as *Glomus intraradices*, have virtually no market value as an edible or medicinal mushroom. Far better to have a gourmet or medicinal mycorrhizal mushroom like a chanterelle, matsutake, or boletus. Unfortunately, I know of no significant successes of harvesting these three mushrooms from inoculated tree seedlings even though a decade

has passed since outdoor plantings. There have been some reports of *Leccinums* naturalizing in pine forests from the spreading of spores and fruitbodies, but the sites lacked control noninoculated, comparative test plots. Good minds are pursuing methods paralleling the successes with other mycorrhizal varieties and I expect entrepreneurial mycologists will eventually succeed. Just as we have tree plantations today, we hope to have mycoforestry parks in the future. Balancing biodiversity while encouraging the predominance of select species will be a challenge but not one that should be an obstacle to experimentation. We will succeed!

By far the preferred mycorrhizae for trees are *Rhizopogons*, *Pisolithus*, and *Glomus* species. These inedible mushrooms, which resemble little puffballs, mycorrhize with about 80 percent of all trees in temperate climates. In fact, these native puffball-like mushrooms are so ubiquitous that they compete with the truffles, chanterelles, or matsutake inoculated by wishful growers. Many believe the best way to colonize the root zone of a tree is to grow its roots first in pasteurized soil, which is then inoculated with mycorrhizal mushroom spores. Another school of thought is

Some Notable Mycorrhizal Mushrooms and the Trees They Love

Nongourmet Mycorrhizal Mushrooms	Endomycorrhizae/ Ectomycorrhizae	Preferred Trees
Glomus intraradices	Endomycorrhizae	Cedars, redwoods
Pisolithus tinctorius	Ectomycorrhizae	Pines
Rhizopogon parksii	Endomycorrhizae	Deciduous
Gourmet Mycorrhizal Mushrooms	**Endomycorrhizae/ Ectomycorrhizae**	**Preferred Trees**
Boletus edulis and allies	Ectomycorrhizae	Pines
Cantharellus cibarius and allies	Ectomycorrhizae	Oaks, firs
Hydnum repandum	Ectomycorrhizae	Firs
Leccinum aurantiacum	Ectomycorrhizae	Pines
Tricholoma matsutake and allies	Ectomycorrhizae	Pines

that the seeds should be in contact with mycorrhizal spores immediately upon germination. (Spores are dusted over the seeds prior to being soaked for germination.) Either method is likely to have greater success than simply planting unadulterated seedlings into native soils outdoors.

Mushrooms and trees have love affairs. The table on page 75 lists mushrooms and the trees they like to partner with. This list is by no means exhaustive but is a glimpse into the intimate relationship shared by fungus and plant. Knowing pairings can help you in mycoforestry efforts. For instance, if you were planting trees grown in pasteurized nursery soils, you could collect wild mushrooms during the rainy season and use their spores as natural inoculum in the root zones of the trees you plant.

The Cortes Island Mycoforestry Research Project

In 2003, I bought land on Cortes Island, British Columbia, half of which had been clear-cut by the previous owner. What remains is a mixture of old-growth and second-growth Douglas fir, hemlock, and western red cedar forests. We saw this land as an ideal place for demonstrating mycoforestry strategies that offer an alternative to traditional forestry methods, which often involve burning the debris on clear-cut land. Traditional foresters' motives for burning such debris are to remove the obstructive brush and reduce fuel for future forest fires. Generally, I oppose slash and burn as a management policy for reforestation and propose a new, combined strategy. Our land will become the experimental theater for testing these hypotheses. This multi-lifetime-long experiment will compare the effects of introducing mycorrhizae and the effects of a topdressing of wood chips as

▲ **FIGURE 79**

On the Olympic Peninsula and elsewhere, the conventional practice after clear-cutting is to stack up the brush and burn it. First, the forest food chain must withstand the sudden loss of carbon and nutrients from cutting and removing the trees. Adding insult upon injury, the remaining brush is stacked and burned, further undermining the resident carbon return cycles. Is it any wonder our forest yields are in rapid decline, with thin soils and increased erosion after 3 generations of such practices?

a source of delayed-release nutrients and to help retain moisture. From this devastated landscape, we saw an opportunity for demonstrating novel mycoforestry strategies, with the end goal of reestablishing the old-growth forest for future generations.

For this long-term mycoforestry experiment, I divided the clear-cut portion of my property into 4 types of test plots. We picked 2 native tree species to reforest the land: Douglas fir and cedar, 35,000 seedlings in all. Half of all the tree seedlings were root-dipped in a mycorrhizal spore-mass slurry containing spores from 2 species of puffballs. The root-balls of half of the cedars were each exposed to approximately half a million spores of the endomycorrhizal species *Glomus intraradices*, while the root balls of half of the

⋀ **FIGURE 80**

An aerial view of our mycoforestry research site on Cortes Island, British Columbia. Having logging roads as ecological barriers bisecting the clear-cuts helped us set up comparative test plots with and without mycological enhancements. Approximately 50 percent of the brush and small-diameter trees left after logging were chipped and left on-site. Contrary to the site in figure 79, we didn't burn the brush, but chipped it so the debris could be recycled back into the land's food chain.

⋀ **FIGURE 81**

Two Douglas fir trees, one with and one without introduced mycorrhizae. Note the increase in root, shoot, and needle development.

Douglas firs were exposed to a similar quantity of the ectomycorrhizal species *Rhizopogon parksii*. Half of the cedars and Douglas firs did not receive any spore-mass treatment. Around the base of half of the trees in each treated and untreated group we used a topdressing of wood chips—wood chips made using a mobile wood chipper. I hired a crew who chipped about 50 percent of the brush left over from logging. From these piles, we carried buckets of wood chips to the trees spread over 60 acres, specifically a gallon per tree (approximately 4 pounds) around half of the spore-treated and non-treated trees. The potential benefits from the collars of wood chips include a regional cooling of the soil, enhanced moisture retention, and the slow streaming of nutrients to the root zones as saprophytic fungi decompose the wood chips. In my prior experiences, I've observed foot-deep beds of wood chips decompose into 1 to 2 inches of rich soil in 2 to 3 years when inoculated with mycelium, or in 4 to 5 years from natural mycoflora. Mushroom mycelium is the grand demolecularizer of plant fibers (lignin and cellulose), creating soil as an end consequence. My goal is to make use of fungi's appetite for wood chips to increase soil depth so that the soil has a greater carrying capacity for the tree successions that spring from it. I see wood chips as valuable ecological currency that should be reinvested into forest's ecobank to enhance sustainability. Since not all the wood chips made were used for topdressing, we used the remainder as natural spawn and spread them in depressions and near roads to reduce silt flow and erosion. (See chapter 5 and page 81.)

The breakdown of methods and types of trees looks like this:

- Trees replanted without mycorrhizae using standard planting procedures

  - Without topdressing of wood chips
  - With topdressing of wood chips

▲ **FIGURE 82**

Members of a volunteer bucket brigade place wood chips around half of the trees inoculated with mycorrhizae and half of the noninoculated trees.

▲ **FIGURE 83**

Young Doulgas fir seedling without wood chips.

- Trees replanted with roots dipped into mycorrhizae
  - Without topdressing of wood chips
  - With topdressing of wood chips

We tagged 100 trees in each test plot, assigned database numbers, and plan to profile these by measuring growth, needle development, and overall health for the next 100 years or more. By September of 2004, we had tagged and measured the height and girth of approximately 700 trees, with and without the mycorrhizal treatments. After putting the data into a spreadsheet, we saw a net 8 and 7 percent increase in overall height and girth respectively, significant given that the trees had been in the ground for only 10 months. The goal is to see the effects of the 4 treatment combinations. A labor of love involving many volunteers, this experiment is large enough to provide data critical for comparing mycoforestation strategies.

When brush is burned from the land—the second sudden massive withdrawal from the forest's carbon bank after first removing the trees—long-term ecological recovery is impaired. My sense is that continual

➤ **FIGURE 84**

Young Douglas firs seedling collared with wood chips. The addition of wood chips cools the ground, increases moisture retention, and provides delayed-release nutrients as they decompose. Our experiment serves to prove how significant this practice is.

40-year tree crop rotations thin the soil faster than it can be built, with ever-diminishing returns. Such practices accelerate premature decline—trees climax in their life cycles prematurely—as the root

▲ **FIGURE 85**

▲ **FIGURE 85**

Arborists employed by many cities, including Seattle and Olympia, Washington, have instituted programs for placing wood chips from trimmings around trees for slow nutrient release.

wads cannot support the trees above them. As a matter of common sense, I do not believe you can harvest 3 generations of trees from the same land within 100 years, burn the brush each time, and not thin the soils. Such practices are not sustainable. This is a false premise espoused by the forestry and logging interests.

Logging companies could chip wood debris, leave it in the cut forestland, and disperse these nutrient fragments over the ground to help refuel the carbon cycle while reducing the danger of fire. This added expense does not thrill most investment loggers, who wish to maximize returns and minimize costs. The incentive is for those who take a long-term view of the forest and its ecological health. In future editions of this book, I will report the progress of the Cortes Island field tests.

Mycoforestry: Using Mushrooms to Prevent Forest Fires

In October 2003, while driving up the I-5 corridor in the Pacific Northwest and listening to National Public Radio, I heard a couple of curious discussions back-to-back. First I heard a news report about U.S. Forest Service employees who had filed a lawsuit against their employer in order to demand public hearings on the lives lost in fighting forest fires. These employees were raising questions about the service's use of funds: fully half of that agency's budget is intended to go toward firefighting, but most of it was being diverted for "salvage" logging and administrative costs. They were also questioning the loss of firefighters' lives in efforts to preserve property, especially trophy homes built deep in the woods. Next I heard a sponsor's announcement that "mulching saves water" (www.savingwater.org): mulching reduces aerial wood debris in forests and increases moisture retention, lessening the chance of fire and protecting forests, firefighters, and houses. These two concepts are natural fits, especially when mycelium-inoculated wood chips are used, further increasing moisture retention and reducing the threat of fire.

A secondary disastrous effect of fires is that the burnt landscape is vulnerable to erosion. Another ecological payoff of wood-decomposing mycelium is that it creates erosion-resistant soil.

For thousands of years, firestorms have been part of the woodland cycle. Until humans arrived on the scene, fires were primarily started by lightning strikes, volcanic eruptions, spontaneous combustion of peat bogs, and the rare meteorite impact. We have made the environment more susceptible to fires as water is diverted for agricultural and urban use as well as with our continual encroachment onto forestlands, not to mention our practice of purposefully setting fires. Rotating tree crops every 40 years thins the soil, and replanting clear-cuts leads to dense forests of homogenous age packing an enormous fuel load from dead aerial side branches (see figure 86). On the other hand, old-growth forests tend to have few side branches near the forest floor, where moist nurse logs give rise to succeeding generations. The lower canopy fuel load is in stark contrast. Monoculturing forests, which produces unnaturally high levels of flammable debris, makes fires more frequent and severe. No longer can we afford to view unchecked wildfires as the natural players they once were in natural woodlands. It's time to revaluate how we manage forests—and their fires.

Preventing fires by leveling forests, an idiotic tactic espoused by some politicians, is not in our long-term interest, although thinning certain forests, especially crowded monocrops planted on former clear-cuts, is beneficial. The current practice of burning or hauling wood debris robs biomass from the carbon cycle. The argument that wood debris fuels forest fires does not hold true if debris is brought in contact with decomposing mycelium already residing in the forest floor. Moist wood does not easily burn. Chipping wood and leaving it in place may increase the risk of fire for the first few years, but thereafter fire potential is drastically reduced as the wood is decomposed.

One argument I've heard is that growing mycelium on wood chips increases global warming. Carbon dioxide does cause global warming, but the output by fungi is nothing like that generated by fires. Combustion of a tree emits vast quantities of hot carbon dioxide in an hour or so. Yet fungal decomposition of that same tree releases cool carbon dioxide gradually over years. Since carbon dioxide is heavier than air and permeates soils, plants benefit from its close proximity. Plant cells absorb carbon dioxide and use it as raw material for creating cellulose, lignin, carbohydrates, proteins, sterols, and outgas oxygen. Burning sends carbon dioxide high into the atmosphere out of the reach of plant communities while also destroying the understory plants that would otherwise recapture this essential gas. The bottom line: Decomposition by fungi buffers carbon dioxide emission and cycles much of the gas back into the flourishing ecosystem.

In planted second- and third-growth forests, high density results in trees with numerous side branches that eventually die, dry, and become brittle and susceptible to fire. The current method of dealing with dense replanted forests with excessive brush and dead branches is to conduct selective thinning of trees to reduce the population per acre, and thus fuel load. Removing side branches and brush reduces forests'

▲ **FIGURE 86**

A problem with monocropping. This planted third-growth forest is extremely dense with dead branches, hence fuel load, posing a significant fire danger.

susceptibility to fire and so is preferred by forest managers. However, if the debris is burned or exported, the environment loses a nutritional layer.

As a mycoforester, I recommend chipping excess wood in replanted forests and leaving the chips on the forest floor. If mycoforesters spread mulch around young trees, along trails, or in mycofiltration buffer zones near watersheds, they can fortify the forest with beneficial fungi. Where wood chips touch the ground, fungi easily grow into them and transport moisture with their threadlike mycelium. The myceliated wood chips then become like a sponge, retaining water (which is needed by neighboring plants) and lessening fire danger.

Leaving wood chips on the forest floor provides many benefits, including these:

- delayed release of nutrients—to build soils
- supporting mycofiltration membranes that reduce erosion and siltation
- providing cavity habitats for diverse populations of bacteria, fungi, plants, insects, and animals
- moisture retention
- protection against forest fires
- substrates for decommissioned logging roads (see below)

Timberland that has lost topsoils is slower to recover with each successive crop of trees. In the near future, as Washington's third-, fourth-, and soon fifth-growth forests are harvested, the impact of thinning soils will become more catastrophic. Unless the depletion of the nutritional topsoil bank is reversed, the future economic return on timber will be bleak. Not surprisingly, this subject is rarely mentioned by the logging companies who distract the public's attention by blaming environmentalists for the loss of jobs. In fact, their own forecasts show diminishing returns, giving these large landowners incentive to cut as much as they can before the public wakes up.

Mycofiltration of Silt: Transforming Abandoned Logging Roads

Having spent much time in the woods, I've witnessed firsthand the enormous silt flow and erosion emanating from logging roads after torrential rains (see figure 89).

For economic and political reasons, many logging roads in northern California, Oregon, Washington, Colorado, and elsewhere are being closed. Roads are the primary vector of watershed siltation and pollution. Thousands of miles of these logging roads channel runoff from uplands. Rainwater courses down these roads and erodes life-sustaining topsoils. Sediment and silt (a finer grain of soil) clog downstream watersheds and streams where fish spawn. Logging roads are a concern in arid lands as well, especially those bordering rivers (Bagley 1999). Ecological damage caused by roads is a problem worldwide, from Brazil to Bosnia.

For every mile of paved road in Washington, there are more than 7 miles of unpaved roads. In 2001 the state legislature budgeted a mere $165,000 for the decommissioning of obsolete roads. Decommissioning unpaved roads means returning them to a near-natural state—an oxymoron. The U.S. Forest Service had budgeted $25 million for doing the same on federal lands 2 years earlier. Estimates for decommissioning these narrow, unpaved roads range from $4,100 to $15,500 per mile (Garrity 1995) in the Northern Rockies, and from $21,000 to $105,600 per mile in the Olympics and Cascades. At the same time,

◀ **FIGURE 87**

Logging roads channel silt into valleys, such as occurs along this northern California hillside etched with zigzagging logging roads.

the cost of building a road in Washington is approximately $32,000 per mile. So the cost of decommissioning a road is roughly the same as building a road, using current methods. When mycofiltration is used, however, the cost of decommissioning a road is lower, approximately 50 percent of the cost per mile. Restoration experts ought to consider the less costly route of mycofiltration, which involves placing a layer of bark and crude wood chips on logging roads and then inoculating them with a mosaic of carefully chosen native fungi in the form of spores and mycelia. A mesh of fungal networks lacing the wood chips then develops and slows the passage of fine silt. As the fungus breaks down the wood chips, it nourishes native flora and fauna.

Building roads compacts soil and banishes mycorrhizal fungi (Amaranthus and Trappe 1993; Amaranthus et al. 1996). This hinders later recovery of the preroad habitat. Adding wood chips to old roads has been explored in British Columbia, where a decision-making key for evaluating sites has been drafted (Allison and Tait 2000). Straight (2001) recommends the use of buffers such as wood chips to lessen the impact of nitrates, pesticides, and hydrocarbons in order to control pollution vectors channeling from roads, ditches, and other sources. Wood chips applied to road surfaces also reduce sedimentation (Hickenbottom 2000; Madej et al. 2001; Prescott 2001).

The key component of mycofiltration is the purposeful introduction of fungi—saprophytic and mycorrhizal—into the wood chip buffers. Adding spores to the oil used to lubricate the teeth of chain saws or chipping equipment is a method I highly recommend. As a result, the wood intended to be made into chips for spreading on disused roads will be exposed to the fungi as soon as it's cut, so fungi get a head start on decomposing stumps and brush. (Broadcasting spores on chipped wood also accelerates decomposition throughout the process.) This method jump-starts the road's recovery cycle. Through selective inoculations, we can steer the course of species succession on stressed woodlands and their disused roads.

The following list details the advantages of installing mycofiltration mats on logging roads over conventional techniques using heavy equipment to build tank traps and ditches and scarifying the surface:

- Sediment flow—reduction of silt erosion into streams, spawning grounds, and fisheries.
- Moisture—enhancing the moisture bank; remoistening of arid landscapes.
- Habitat enhancement—reestablishment of native mycoflora (soil-building mycorrhizal and saprophytic types).
- Hydrocarbon contaminants—reduction of diesel, oil, herbicides, pesticides, and other pollutants.
- Pathogenic microorganisms—mycofiltration of coliform bacteria, *Escherichia coli*, *Pfiesteria*, other bacteria, and protozoa.
- Temperature—cooling of water flowing into streams, benefiting fisheries and marine systems.
- Collateral disturbance—low impact on existing and adjacent ecosystems.
- Subsurface soil hydrophysics—subsurface growth of mycelium allows for water and mineral transport as well as aeration, minimizing flow of silt.
- Aesthetics—roads transformed into nature trails.

◀ **FIGURE 88**

Depicted here is the conventional approach to reclaiming a road: digging deep ditches (called *tank traps* in military speak; I call them "terra interrupti") that block vehicles so that, ostensibly, the land can recover from human activity by disturbing it further.

- Educational impact—accessible environment for habitat restoration education.
- Insectaries—creation of breeding ground for grubs helping fish, bird, bat, and other animal food chains; mycopesticidal barriers for wood-boring beetles and attractant zones for natural predator insects.*
- Long-term investment protection—road subsurface can be reused in future at reduced cost compared to that of new construction.

Tahuya State Forest Reclamation Test Site

Tahuya State Forest, located on the southern Kitsap Peninsula, Washington State, is a 23,100-acre multi-use forest available to hikers, motorcyclists, bicyclists, and horse riders. The revenues from this working forest help fund the Mason County school system.

In 2003, my friend David Sumerlin and I wondered if the Department of Natural Resources (DNR) would consider using fungi to reclaim a road that was closed while DNR was conducting a salmon stream rehabilitation project along the road using a variety of techniques. On a test site bordering the salmon stream, they had applied grass seed directly on the bare soil. After 7 years the site had recovered slightly with a sparse covering of grass, but the habitat lacked healthy soil, a forest understory, and biological diversity.

In response to our request, DNR set aside a swath of upland road destined for habitat recovery beside the salmon habitat. Our project, a hundred yards from the grass-remediated site, is a 500-foot stretch of unpaved road on a hill that slopes down to the salmon stream. This section of roadway is still used for overland recreational vehicles, but we were authorized to reclaim one lane of the road, leaving the other side for recreational users.

We set up 3 zones for our mycoremediation experiment. The lower section of the road, with the greatest slope, was most at risk for erosion. The middle section was fairly wide, over 30 feet, with a gentler grade. The upper portion of the road, the third zone, had the gentlest slope. Our intent was to prevent bank erosion and filter the silt-saturated runoff from rain.

We arranged for the delivery of 3 loads of waste wood—a crude mixture of bark, wood chips, and fir needles. North Mason Fiber, a local supplier to the pulp paper industry, donated these loads, and 7 Fungi Perfecti employees donated their time to spread the wood chip matrix 6 inches deep over the length of the road. Then we tossed handfuls of spawn of the native oyster mushroom *Pleurotus ostreatus* on top. After the chips were distributed, we spread 6 bales of wheat straw over the top to help hold in moisture for the spawn's benefit.

---

* *My research shows such that fungus gnats can be attracted by certain mycelia. Mycelia also attract phorid flies, a fungus gnat species and known predator of fire ants (*Solenopsis invicta; *invasive biting insects that transmit diseases and defoliate orchards). Some entomopathogenic (naturally insecticidal) mycelia can emit both a fragrance that attracts phorid flies and another that kills fire or leaf-cutting ants. A matrix of 2 fungal species could be mycoengineered for a double-pronged treatment. This is but a single example of hundreds of tactical predator-prey combinations described in my mycopesticide patents, both approved and pending. For other examples, go to www.uspto.gov, select "Search," below "Patents," and search under "Stamets."*

▶ **FIGURE 89**

Runoff saturated with silt flows directly into the Tahuya River, a salmon spawning ground.

▲ **FIGURE 90**

Wood chips are dumped on the road surface.

▲ **FIGURE 91**

Spreading straw over the wood chips that have been inoculated with oyster mushrooms.

On top of the straw, we spread 20 pounds of Regreen, a nonseeding wheat approved for erosion control, using a broadcast seeder stocked with 1 pound of Mycogrow, a mycorrhizal inoculum. (I recommend using native grasses over commercial varieties, but the simple fact is that native grass seed is difficult and expensive to acquire.) We completed the work in mid-April, when rainfall was intermittently heavy. A week later we returned to the site and found the habitat in its first stages of restoration, with seeds sprouting.

The reason we selected *Pleurotus ostreatus* as our keystone species is that this primary saprophyte is indigenous, aggressive, and adaptable to growing under

▲ **FIGURE 92**

David Sumerlin points to location where wood chips should be placed—up against the bank of the scar face of the road cut, thus eliminating erosion from the silt-producing exposed surface.

a variety of conditions and temperatures. Future trials will use a matrix of white rot, brown rot, and other mycorrhizal species. No parasitic species are contemplated.

Once a mycofiltration habitat is constructed, ecological recovery unfolds and nature guides the course. Spores are released as mushrooms grow to maturity, giving rise to more mycelium, which further colonizes the substrate. As the mycelium infiltrates the wood chips, more moisture is retained. The new mushrooms also attract native insects and the rotting mushrooms become breeding grounds for fly larvae and grubs, subsequently attracting animals from lizards to birds.

As the wheat grass climaxes and dies and the wood chips decompose, a rich soil is created, further nurturing recovering native species. For every 12 inches of wood chips, we estimate that 1 to 2 inches of soil are created after 4 years of decomposition by oyster mushrooms. After several years, a mantle of mycelium forms where the wood chips and gravelly sandy soil meet (see figure 93). This sheath of mycelium overlays and holds

▲ FIGURE 93

After 3 years, the reclaimed road showed a mantle of nearly contiguous mycelia at the point where wood chips and gravel meet, binding together the otherwise loose subterrain. Note that the upturned rock had rested upon a mantle of mycelium underneath it. My hypothesis is that the mycelium became resident in this subterranean zone, feeding upon the soupy, nutrified water that flows from above.

the gravel together—gravel that ends up actually being beneficial in that it adds structural resilience to the road's subsurface and provides porosity and microcavities that flourish with microbial life.

Through this simple, direct, and practical approach, ugly roads, the source of numerous ecological problems, are transformed into green, foot-friendly pathways for hikers and other fauna. Mycorestoration practices also offer a complementary management strategy for long-term sustainability of our forestlands in general. Wood chips are the ecological currency that we should bank for preventing erosion.

CHAPTER 7

Mycoremediation

As a species, humans are adept at inventing toxins yet equally inept at eliminating them from our environment. Bill Moyers (2001) reported in *Trade Secrets*, a Public Broadcasting Service program, that analysis of his blood by Mt. Sinai Medical Hospital revealed 84 of 150 known industrial toxins, many of them carcinogenic (13 dioxins, 31 PCBs, several pesticides, and numerous heavy metals). All but one are a legacy of the chemical revolution. Had his blood been analyzed in the 1930s, he noted, only lead would have been detected. Most citizens of this planet, unfortunately, are likely to have similar blood profiles. With current trends, our exposure to dangerous chemicals increases with time as our environments become more polluted. We face having to live with these toxins in our bodies and in our backyards. The threat from pollution, once characterized as a worry of "alarmist environmentalists," is now supported by the most sanguine of medical toxicologists. Simply put, pollution is an environmental disease that kills.

Mycoremediation is the use of fungi to degrade or remove toxins from the environment. Fungi are adept as molecular disassemblers, breaking down many recalcitrant, long-chained toxins into simpler, less toxic chemicals. Mycoremediation also holds promise for removing heavy metals from the land by channeling them to the fruitbodies for removal. Mycoremediation practices involve mixing mycelium into contaminated soil, placing mycelial mats over toxic sites, or a combination of these techniques, in one-time or successive treatments.

The powerful enzymes secreted by certain fungi digest lignin and cellulose, the primary structural components of wood. These digestive enzymes can also break down a surprisingly wide range of toxins that have chemical bonds like those in wood. Such mushrooms can be classified into 2 subgroups: brown rotters and white rotters. Only about 7 percent of mushrooms are brown rot fungi; of those, about 70 percent are polypores (Gilbertson and Ryvarden 1986–87). Brown rot fungi's extracellular enzymes break down the white, pulpy cellulose, leaving behind the brownish lignin (hence the name). These fungi cause checkered cubical cracking and shrinking in wood, which is commonly seen on downed conifer trees. Examples of brown rot mushrooms are multicolored gilled polypore (*Lenzites betulina*), large lentinus (*Lentinus ponderosus*), sulphur tufts (*Laetiporus sulphureus* and *Laetiporus conifericola*), velvet polypore (*Phaeolus schweinitzii*), split-gill polypore (*Schizophyllum commune*), agarikon (*Fomitopsis officinalis*), unzoned rusty gilled polypore (*Gloeophyllum trabeum*), and dry rot house wreckers (such as *Serpula lacrymans* and *Serpula himantiodes*).

White rot fungi, more numerous than brown rotters, produce enzymes that break down the recalcitrant brown fiber in wood, leaving the cellulose largely intact, thus giving the wood a white appearance. The oyster mushroom (*Pleurotus ostreatus*), maitake (*Grifola*

◄ **FIGURE 95**

After 3 weeks, oyster mushrooms fruit from the tub. As the oil-soaked straw is metabolized by the mycelium, it lightens in color, a direct reflection of the reduction of petroleum hydrocarbons. The large size of the mushrooms shows that the mycelium is undeterred and is digesting the oil as a nutrient.

⩕ **FIGURE 94**

In this version of a mycelial mat, 10 ml of car gearbox oil was poured onto 2 pounds of oyster mushroom mycelium that had colonized straw.

⩕ **FIGURE 96**

Two white rot polypore mushrooms, turkey tails *(Trametes versicolor)* and artist conks *(Ganoderma applanatum),* co-inhabiting the same alder *(Alnus rubra)* stump.

⩕ **FIGURE 97**

The remnants of a tree speak to the history of its fungal occupants. Here a brown rot fungus, causing cubical cracking of the remaining lignin, exists beside a white rot fungus that leaves the pithy cellulose behind.

Mushrooms and Their Habitats

	Grassland	Manured Soils	Grass & Leaf Litter	Wood Chips	Logs, Stumps & Snags	Brown (B) or White (W) Rot?
Agaricus bernardii	X	X				W
Agaricus brasiliensis	X	X				W
Agrocybe aegerita				X	X	W
Chlorophyllum rachodes			X			W
Coprinus comatus	X	X	X			W
Flammulina velutipes				X	X	W
Fomes fomentarius					X	W
Fomitopsis pinicola					X	B
Ganoderma applanatum					X	W
Ganoderma lucidum					X	W
Grifola frondosa					X	W
Hericium abietis					X	W
Hericium erinaceus					X	W
Hypholoma capnoides				X	X	W
Hypholoma sublatertium				X	X	W
Hypsizygus ulmarius					X	B
Inonotus obliquus					X	W
Laetiporus sulphureus & allies					X	B
Lentinula edodes					X	W
Macrolepiota procera	X	X	X			W
Morchella angusticeps				X		B?
Pholiota nameko				X	X	W
Piptoporus betulinus					X	B
Pleurotus ostreatus				X	X	W
Psilocybe cubensis		X	X			W
Psilocybe cyanescens & allies			X	X	X	W
Sparassis crispa					X	B
Stropharia rugoso-annulata			X	X		W
Trametes versicolor				X	X	W

frondosa), turkey tail (*Trametes versicolor*), reishi (*Ganoderma lucidum*), artist conk (*Ganoderma applanatum*), and crust fungus (*Phanerochaete chrysosporium*) are among the more powerful white rot mushrooms. Some species of mushrooms produce both types of rot, leaving a mottled white and brown discoloration in the wood. This makes identifying rots tricky. Bear's head (*Hericium abietis*) causes a white rot in the core of a stump while the outer regions become brown. The clustered woodlover (*Hypholoma capnoides*) grows on conifers, cedars, and redwoods; although the woodlover belongs to the Strophariaceae, a family of white rotters, I usually see this mushroom emerging from dark-colored wood.

White rot mushrooms are mycoremediators of toxins held together by hydrogen-carbon bonds. Enzymes secreted by this group's mycelia include lignin peroxidases, manganese peroxidases, and laccases (Schliephake et al. 2003). Only white rot mushrooms seem to produce manganese-dependent peroxidase, an enzyme that mineralizes wood and is particularly efficient in breaking hydrogen-carbon bonds. Lignases and cellulosic enzymes are also produced by the mycelium.

These complex mixtures allow the mycelium to dismantle some of the most resistant materials made by humans or nature. Since many of the bonds that hold plant material together are similar to the bonds found in petroleum products, including diesel, oil, and many herbicides and pesticides, mycelial enzymes are well suited for decomposing a wide spectrum of durable toxic chemicals. Because the mycelium breaks the hydrogen-carbon bonds, the primary nonsolid byproducts are liberated in the form of water and carbon dioxide. More than 50 percent of the organic mass cleaves off as carbon dioxide and 10 to 20 percent as water; this is why compost piles dramatically shrink and ooze leachate as they mature.

The Multi-Kingdom Approach to Decontaminating Toxic Waste Sites

Life springs from mycelium. Fungi control the flow of nutrients, and as a consequence they are the primary governors of ecological equilibrium. As ecosystems change, fungi adapt to steer the course of nutrient cycles. The strength and health of any ecosystem is a direct measure of its diverse fungal populations and their interplay with plants, insects, bacteria, and other organisms. Using fungi first in bioremediation sets the course for other players in the biological community to participate in its rehabilitation. When working with fungi on toxic waste sites, it soon becomes clear that many other organisms are being affected. The introduction of a single fungus, for instance oyster mycelium, into a nearly lifeless landscape triggers a cascade of activity by other organisms. A synergy between at least 4 kingdoms—fungus, plant, bacterium, and animal—denatures toxins into derivatives useful to myriad species and fatal to few.

The natural order of organisms sequencing through damaged ecosystems fluctuates according to each habitat's personality. Here's one I have seen repeatedly:

- fungi
- vertebrates (animals) and invertebrates (insects)
- bacteria
- plants
- vertebrates (animals) and invertebrates (insects)
- fungi

This hierarchy of organisms is one cyclical sequence of many. Synergistic waves of organisms quickly enter a habitat once its toxic barriers are removed with specific saprophytic mushrooms, such as oysters, leading the charge. Since most insects are fungus loving and are excited by spores, they appear as mushrooms ripen and overmature. Vertebrates from squirrels to bears to people seek mushrooms as food. Bacteria use rotting mushrooms as a rich base for growth, further freeing nutrients and releasing a cascade of microbes that destroy the structure of mush-rooms as they melt into the soil. This bacterial influx predisposes habitats for the emergence of plant communities. Ultimately, nature fosters complex partnerships of interdependence, with fungi blazing the path to ecological recovery.

One multi-kingdom method for decontaminating land is to use a wood chipper or chip blower to disperse spawn while making a layer of sheet mulch. Higher inoculation rates usually result in faster colonization. A preferred method is to disperse sawdust spawn in the stream of flowing chips equivalent to about one-fourth of the total mass of wood chips. The method and quantity of spawn is influenced by the site's particular toxic profile. Another method is to rake spawn into or lay it upon the contaminated site. The mulch layer should be $1^{1}/_{2}$ feet deep, or shallower if using smaller particles. The goal is to provide a layer of mulch where aerobic mycelia are not suffocated and displaced by anaerobic competitors. Cover the bed with cardboard, and then add a loose layer of straw. (For specifics on this method, see chapter 12.) After the residual levels decline to tolerable limits—which may take several reapplications—trees and plants infused with mycorrhizae can be planted. To maintain the site, you may need to reintroduce follow-up populations of mediating mycelia. Once the mushroom mycelium begins to unlock the nutrition from the wood chips, other organisms enter the landscape. Once these predecessor organisms are engaged, nature will steer the habitat on the path toward self-healing.

After the mycelium passes its prime and declines in vigor, it hosts other organisms that may metabolize remaining toxins. In particular, oyster mycelium growing on straw is prone to host more and more microbes as it ages. In the process of defending itself from parasitization, it becomes fortified against microbial attack. Also, other groups of bacteria proliferate alongside the mycelium and produce their own toxin-digesting enzymes. This form of mycelium is far better equipped for handling toxic waste sites than mycelium not pre-exposed to wild microbes. For this reason, using pure culture spawn for mycoremediation may not be the best choice. This realization—that aged mycelium from a mushroom farm has better mycoremediation properties than pure culture spawn—marks a major advancement in the understanding of how to project mycelium in mycoremediation strategies, and it may also cut the cost significantly. I call this "acclimated spawn" to denote that it has already become familiarized with the microbial population. Natural spawn (see chapter 9) or pure culture spawn that has made contact with habitat microbes before insertion has the best chance for successful mycoremediation.

In the course of my work with Battelle Pacific Northwest Laboratories in the late 1990s, in Sequim, Washington, I was given an opportunity to demonstrate mycoremediation beyond our previously successful bench-scale study. In lab experiments, we used one of my strains of oyster mushroom (*Pleurotus ostreatus*) to test its skill in breaking down diesel-saturated soil. In a series of mesocosm (midscale) tests, we obtained the best results by mixing sawdust spawn with soil and unsterilized alder chips. When combined with bunker C oil, the same petrochemical spilled by the *Exxon Valdez* tanker in Alaska, 97 percent of the oil's polynuclear aromatic hydrocarbons (PAHs) degraded after 8 weeks. However, if we first sterilized the alder chips before mixing them into the spawn, only 65 percent of the PAHs were degraded. For a control, we omitted the spawn and used unsterilized chips, which caused the PAHs to decline only 38 percent (Thomas et al. 1999).

What was surprising was that the overall effect was enhanced when mycelium was introduced to a microbially competitive environment of raw wood chips and soil. We had feared that competing microbes would attack the oyster mushroom mycelium and use it as a food. Instead, native microbes that fortified the oyster mycelia seemed to have been activated. Based on the results of our experiment, we hypothesize that coming into contact with microbes enhances the oyster mycelia's digestion of PAHs.

Around the same time, others studying mycoremediation published an article titled "Spent oyster mushroom substrate performs better than many mushroom mycelia in removing the biocide pentachlorophenol," showing another case wherein oyster mushroom mycelium working in concert with microbes performed better than oyster mycelium without them. These researchers noted, "Pleurotus . . . harboring both bacteria and fungi functioned over a wide range of initial PCP concentrations and reached a higher degradative capacity . . . in only three days" (Chiu at al. 1998). Eggen and Sasek (2002) also showed that spent oyster mushroom "compost" effectively reduces toxins in polluted soil. When a second treatment of spent oyster mushroom substrate was added 15 weeks after the initial inoculation, anthracene levels dropped from 87 percent to 50 percent, and fluorene levels went from 99 percent to 87 percent. I theorize that this second inoculation would have been more effective had researchers added unsterilized wood chips and covered the mycelia with cardboard and provided shade. (See chapter 12.)

House Wrecker Fungi Also Break Down Toxic Wastes

A couple of resupinate (flattened, stemless) annual polypore mushrooms, *Meruliporia incrassata* and *Antrodia radiculosa*, are well known to cause brown cubical rot. One of them, *Meruliporia incrassata*, an example of house wrecker fungi, causes millions of dollars of damage to the understructures of houses across North America. Another house wrecker, *Serpula lacrymans*, transports water from many meters away to moisten wood in advance of contact, allowing for the silent softening and consumption of a house's understructure, until it suddenly collapses from catastrophic structural failure.

Noting the house wrecker fungi's strong ability to degrade wood, in 2004 Illman, Yang, and Ferge, researchers at the U.S. Department of Agriculture published a patent application (U.S. Patent Application No. 6,727,087) titled "Fungal degradation and bio-remediation system for pentachlorophenol-treated

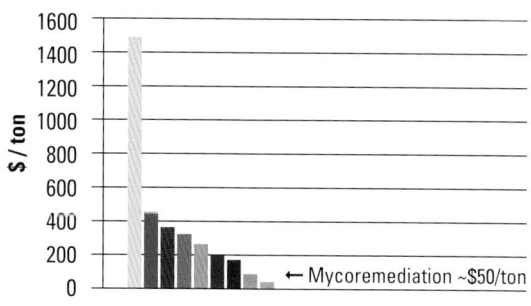

Soil Remediation Technologies Costs: Petroleum Hydrocarbons

← Mycoremediation ~$50/ton

Remediation Technology

- Incineration
- Solidification
- Soil Venting
- Bioremediation
- Mycoremediation
- Solvent Extraction
- Indirect Thermal
- Soil Washing
- Phytomediation

⋀ **FIGURE 98**

Cost comparison of remediation methods of polycyclic aromatic hydrocarbons.

⋀ **FIGURE 99**

Rhizomorphic mycelium running on cardboard channels water to moisten its habitat in advance of contact. Throughout nature, mycelia act as an active hydrological transport system. The mycelia increase the moisture retention ability of the habitats they colonize through secretions of water and sugars from the advancing, fingerlike hyphal tips.

wood." They created "choice tests" to see what cultures would grow toward and colonize wood treated with pentachlorophenol (PCP) and chromated copper arsenate (CCA). The vast majority of the cultures surveyed avoided the preservative-saturated wood, except for some aggressive strains of *Meruliporia incrassata* and *Antrodia radiculosa* that showed no reluctance and gobbled up the wood. This type of choice test could be used to screen fungi in order to find ones able to decompose other toxins. Ironically, the fungus that is the bane of a homeowner could be a boon to mycoremediators.

The Washington State Department of Transportation Diesel-Contaminated Maintenance Yard Experiment

During my research with Battelle, we learned of another opportunity to test mycoremediation. The Washington State Department of Transportation (WSDOT) in Bellingham, Washington, operated a maintenance yard for trucks for more than 30 years. Diesel and oil contaminated the soil at levels approaching 2 percent, or 20,000 parts per million (ppm) of total aromatic hydrocarbons, or TAHs. This is roughly the same concentration measured on the beaches of Prince William Sound in 1989 after the *Exxon Valdez* spilled its 11 million gallons of crude oil. In 1997 and 1998 the Washington State Department of Ecology (DOE) granted the Washington State Department of Transportation (WSDOT) a variance to permit our experiment. Our research group at Battelle Pacific Northwest Laboratories in Sequim, Washington (which included Susan Thomas, Meg Pinza, Pete Becker, Ann Drum, Jack Word, and others), decided to try mycoremediation techniques on this site.

In the spring of 1998, the WSDOT set aside 4 piles of diesel-contaminated soil, placing them onto 4 large sheets of 6 mm black plastic polyethylene tarps at the Bellingham site. Each pile measured about 3 to 4 feet in height, 20 feet in length, and 8 feet in width.

Into one of the piles (measuring about 10 cubic yards), we mixed about 3 cubic yards of pure culture sawdust spawn, an amount roughly equal to 30 percent of the pile. We placed the spawn in layers, sandwiching the soil in between, since my earlier studies had shown that mycelial colonization speeds up when the mycelium is concentrated in a layer rather than dispersed throughout a pile. (I have had great success using this method—*parallel sheet spawning*—for running mycelium outdoors. On each plane, the many constellation points of inoculation quickly grow to meet one another and fuse, and a strong mycelial mat, horizontally organized, predominates. Then the parallel planes of mycelium grow vertically, seeking each other out, and fuse.)

The other piles received no mycelia. Of these, 2 were given bacterial treatments and 1 was an untreated control. Our myceliated pile was covered with shade cloth, while black plastic tarps were pulled over the other piles to keep out rain. Approximately 4 weeks later, part of our team returned to the piles and pulled back the tarps on the non-mycelium-treated dirt. All 3 of the piles were black and lifeless and stank like diesel and oil. As the shade cloth of the myceliated pile was pulled back, onlookers gasped in astonishment. We were greeted by a huge flush of oyster mushrooms numbering in the hundreds, some more than 12 inches in diameter (see figures 100 and 101); such an abundant crop is seen only where the nutrition is especially rich. The pile, now light brown, no longer smelled of diesel and oil. By the ninth week, vascular plants had appeared and were flourishing. From weeks 4 to 9, the oyster mushrooms matured and sporulated and then died back.

When the oyster mushrooms were maturing and sporulating, they attracted insects that feasted upon the succulent fruiting bodies and laid their eggs in them. Soon squirming larvae attracted birds, which brought seeds and turned our pile into an oasis of life. (Seeds may have also blown in with the wind.) Other mushroom species—secondary decomposers—

appeared as the pile's biosphere diversified. When mushrooms rotted, bacteria and predator fungi rose to the occasion.

The WSDOT employee who operated the front-end loader, a collector of wild mushrooms, recognized the oyster mushrooms and eagerly wanted to eat them. We discouraged him, thinking that the mushrooms might contain harmful chemicals, but subsequent analysis of the mushrooms showed no detectable petroleum residues (heavy metal analysis was not conducted). Primary by-products from the mycelium were mushrooms, carbon dioxide, and water. The physical mass of the substrate had shrunk substantially compared to the other piles. We felt we had witnessed a mycomiracle: life was flowering upon a dead, toxic landscape.

Battelle researchers reported that total petroleum hydrocarbons (TPHs) had plummeted from 20,000 ppm to less than 200 ppm in 8 weeks, making the soil acceptable for freeway landscaping (Thomas 2000). Our mycelia had degraded the larger, more toxic PAHs into smaller, less toxic molecules.

Our group felt that this field trial had both succeeded in reducing toxins and supported the concept that oyster mushrooms could be introduced to toxic waste sites as a gateway to remediation. Known as primary decomposers to most, we saw oyster mushrooms are a vanguard species for habitat restoration. Increasingly, I see these primary decomposers as leading the way for habitat recovery. Our job is to set the stage. Nature will finish the act.

Warning: Many hazardous waste sites host a multitude of toxins. Although the enzymes from mushrooms break down many chemical contaminants, mushrooms can concentrate heavy metals. If a site also contains heavy metals, the mushrooms should not be eaten until they have been determined to be safe.

▲ **FIGURE 100**

A pile of diesel-contaminated soil under attack by oyster mushrooms.

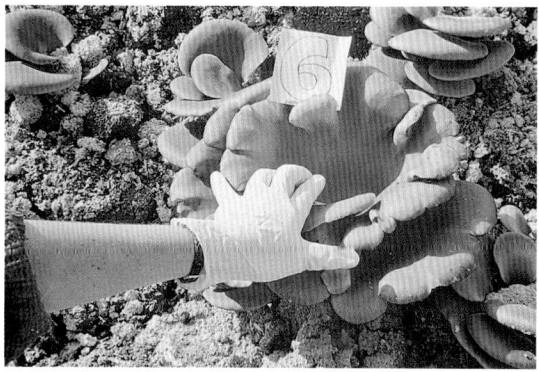

▲ **FIGURE 101**

Some of the mushrooms reached mammoth sizes, a testimonial to the nutrition they found in the petrochemicals.

◀ **FIGURE 102**

Near the end of the trial, as the mushrooms rotted away, plants appeared. Our pile regreened, becoming an oasis of life, while the other piles remained lifeless.

Mycoremediation of Chemical Contaminants: Mushrooms as Molecular Disassemblers

With mycoremediation, brownfields can be reborn as greenfields, turning valueless or even liability-laden wastelands into valuable real estate. Remediation with living organisms addresses several expensive issues. Foremost, bioremediation and mycoremediation eliminate the expense incurred in removing thousands of tons of tainted soil to a remote toxic waste storage site. Current policy prescribes burning, hauling, and/or burying toxic waste. These steps leave a lifeless environment that is ecologically crippled or inert.

Bioremediation has largely failed to be commercially practiced due to a complexity of factors, biological and legal. Technologically, I think the missing link has been *not* using mushroom mycelium as the starting species—starting the domino effect of biological decomposition. As is often the case with any new science, the promises first ballyhooed by optimistic scientists-turned-entrepreneurs failed to profitably be put into practice. Another problem with mycoremediation is a direct testimonial to its effectiveness. Oftentimes, while trying to break down diesel contamination for instance, PCBs (polychlorinated biphenyls) will be destroyed. In doing so, one mycoremediation method has succeeded in treating several targets, whether

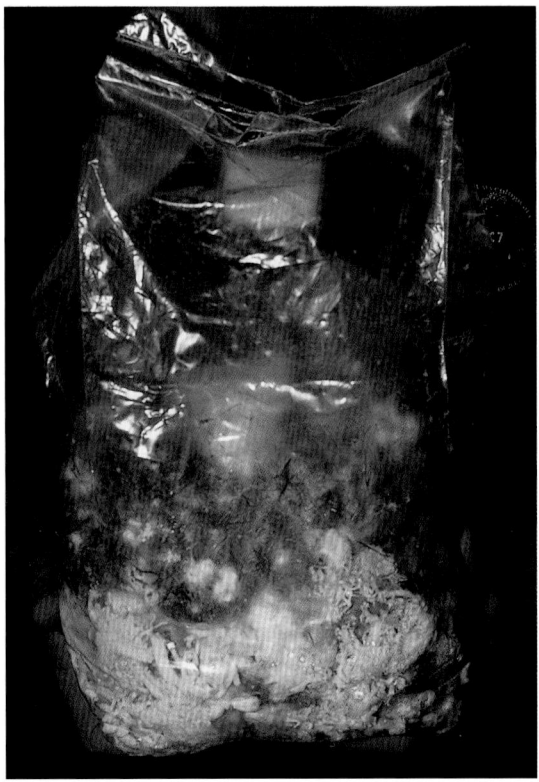

▲ **FIGURE 103**

Oyster mushroom primordia fruiting from human hair. Since hair naturally soaks up petroleum, it can be used to absorb oil floating in water and then be saprophytized by mycelium.

(2003) found that when the mycelium of shiitake was exposed to heavy metals (cadmium, mercury, zinc, copper, and lead) in dye-enriched effluents, laccase production increased, decolorizing toxic dyes while absorbing these heavy metals. Such examples show that mycofiltration can be used to treat complex wastes contaminated with heavy metals and other toxins.

In our food chain, toxins including mercury, polychlorobiphenyls (PCBs), and dioxins are passed from many sources upward from one level to another and become more concentrated at each step. Mammals at the top of the food chain suffer by ingesting toxins consumed by organisms lower on the food chain. Mycelia can destroy these toxins in the soil before they enter our food supply.

Fertilizers, munitions, pesticides, herbicides, textile dyes, and estrogen-based pharmaceuticals are all susceptible to enzymes secreted by mushroom mycelia. Some species can degrade several of these synthetic compounds at once, while others are more selective. On the opposite page is a very simple chart matching classes of contaminants with the products that contain them, and on the following page is a chart matching mushrooms that are active against well-known toxins. For charts on mushrooms that target microorganisms and heavy metals, please see pages 62 and 106.

Chemical Warfare Agents and Biological Warfare Agents

In 1988, during the Iran-Iraq war, Saddam Hussein deployed an arsenal of chemical weapons, including VX, against the Kurds in Iraq, killing more than 12,000 people in just 3 days. On March 22, 1995, five sect members from the Aum Shinrikyo cult released sarin gas into a Tokyo subway station, killing a dozen people and wounding thousands. Stored nerve gas agents pose a serious threat to world safety. Neutralizing neurotoxins is an international priority in combating and disposing of chemical weapons. Other, more persistent neurotoxins—placed in a class of persistent

intended or not. Inadvertently, the bioremediator may be in violation of one of the several issued patents specifically granted for matching fungus against a toxin. (Please see page 99 for a short list of some of the more significant patents issued to date.) Unintentional patent violation does not protect you from patent infringement and lawsuits. This legal mess is a major stumbling block preventing wide-scale fungal cleanup of chemical toxins.

Mycoremediation works first by denaturing toxins such as petroleum products (discussed earlier in this chapter), and second by absorbing heavy metals. Many contaminated habitats contain both. Hatvani and Mecs organic pollutants (POPs) and ranging from DDT to dioxins and PCBs—are far more widespread and insidious, as they accumulate in our environment and enter our food chain, causing chronic diseases such as cancer and neuropathy.

The most appealing bioremediation method for destroying these toxins is to decompose them on-site. Extracting, shipping, storing, and destroying toxins off-site is more expensive than on-site mycoremediation strategies, which could cost as little as $50 per ton of toxin-laden soil, compared to nearly $1,000 per ton for incineration. The mycoremediation method is elegantly simple: overlay straw or wood chips infused with the right mycelium to create a living membrane of enzymes that rain down on the toxins in the topsoil. Replenish annually with additional mycelium-treated substrate. Several sequential applications may be the necessary norm to reduce toxins to acceptable levels.

Two years previous to the WSDOT diesel-contaminated maintenance yard experiment, I informed chemists and remediation specialists from Battelle Pacific Northwest Laboratories of the idea

Toxins, Their Primary Origins and Research Showing Efficacy of Their Fungal Degradation

Type of Toxin	Products or Processes That Emit Toxins	Supporting Research References
Anthracenes	Dyes, pesticides, and derivatives: benzo(a)pyrenes, wood preservatives, fluorene, naphthalene, acenaphthene, acenaphthylene, pyrenes, biphenylene	Johannes et al. 1996; Knapp et al. 2001
Anthraquinones	Dyes	Kasinath et al. 2003; Minussi et al. 2001; Novotny et al. 2001, 2003; Hatvani and Mecs 2003
Benzopyrenes (PAHs)	Incinerators	Qiu and McFarland 1991
Chlorinated aromatic compounds: pentachlorophenol (PCP), trichloro-phenol (TCP), polychlorinated biphenyls (PCBs), dioxins, chlorobenzenes	Transformers, lighting fixtures, paper products, chlorine bleaching, paints and coatings	Gadd 2001
Copper/chromium	Treated wood	Humar et al. 2004; Illman et al. 2003
Dimethyl methylphosphonates (DMMP)	Chemical warfare agents: VX, sarin, soman	Thomas et al. 1999; Word et al. 1997
Dioxins	Incineration of industrial wastes, forest fires/wood burning, coal-fired plants	Chiu et al. 1998
Pentachlorophenol	Pesticides, preservatives	Kondo et al. 2003
Pesticides	Alachlor, aldrin, chlordane, DDT, hep-tachlor, lindane, mirex, atrazine, benomyl	Gadd 2004
Petroleum hydrocarbons	Oil, coal, tar, gasoline, diesel	Bhatt et al. 2002; Cajthaml et al. 2002; Eggen and Sasek 2002; Sasek 2003; Thomas et al. 1999; Moder et al. 2002

that mycelia of oyster and other mushrooms can decompose hydrocarbons in petroleum products since lignin has similar hydrogen-carbon bonds. (A couple years earlier I had encouraged Enviros, a Seattle company, to do experiments that resulted, much to their surprise, in oyster mushrooms fruiting in a tray filled with Alaskan crude oil.) Soon my discussions with Battelle went beyond oil spills and turned to the subjects of national defense and the threat of biological and chemical warfare agents. I

Mushrooms with Activity against Chemical Toxins

More species and toxins will be added over time. Several of the species probably act upon more toxins than the ones listed above. I will update this chart on www.fungi.com as more research is published.

	Anthracenes	Benzopyrenes	Chromated Copper Arsenate	Chlorine	Dimethyl methyl phosphonate (VX, Soman, Sarin)	Dioxin	Persistent Organophosphates	Polycyclic Aromatic Hydrocarbons (PAHs)	Polychlorinated Biphenyls (PCBs)	Pentachlorophenols (PENTAs)	Trinitrotoluene (TNT)	Brown (B) or White (W) Rot?
Antrodia radiculosa			X							X		B
Armillaria ostoyae					X							W
Bjerkandera adusta		X						X				W
Gloeophyllum trabeum			X			X						B
Grifola frondosa									X			W
Irpex lacteus								X				W
Lentinula edodes								X	X	X		W
Meruliporia incrassata			X							X		B
Mycena alcalina				X								?
Naematoloma frowardii (=Hypholoma)								X			X	W
Phanerochaete chrysosporium		X								X	X	W
Pleurotus eryngii						X						W
Pleurotus ostreatus		X			X	X		X	X		X	W
Pleurotus pulmonarius						X					X	W
Psilocybe spp.					X		X					W
Serpula lacrymans			X					X				B
Trametes hirsuta											X	W
Trametes versicolor	X		X		X	X	X			X	X	W

learned that Battelle was also interested in decomposing neurotoxins. As an approved Department of Defense test facility, Battelle had well-established programs in place for researching defenses against chemical and biological weapons.

As we brainstormed ideas with their chemists, Battelle became interested in my model of mycoremediation and process of matching strains with nutrients. I trained them in tissue culture of mycelium, showing them how I introduced to a nonnutritive agar media a sample of my preferred wood substrate upon which I wanted mushrooms to eventually form. The idea was to familiarize mushrooms strains to a nonnative wood substrate so that the mushroom mycelium could adapt its enzymatic pathways early in its life cycle. This is one of the ways I have selected mycelial strains for growing shiitake or maitake mushrooms on alder, a wood that they don't inhabit in nature. Some strains adapted; some didn't. I selected those that grew faster, sending out diverging fans of running mycelium. We expanded this model, using toxins instead of wood as added nutrients in the agar media. I provided Battelle, at their request, a select library of 26 of my most aggressive mushroom strains for testing as antidotes to bacteria, petrochemicals, and other toxins. We further refined this model, incrementally increasing concentrations of various toxins used in chemical weapons and decreasing other natural nutrients until the starving mycelia digested toxins as food. After a series of dilutions wherein the carbon source was incrementally replaced by the selected surrogate neurotoxins, we found, much to my surprise and our delight, that some of my strains adapted and grew when the toxins became the sole source of nutrition!

By capitalizing on the enzymatic versatility and learning ability of fungus in this way, we were able to customize strains so that they neutralized toxic weapons and wastes. As part of this venture, I sought and found mushroom strains with enzymes for making and breaking phosphorus bonds, the critical bonds that held these nerve gas toxins together. Once the enzymes broke the phosphorus bonds, we theorized,

FIGURE 104

Cellulose plug saturated with bunker C crude oil absorbed by oyster mushroom mycelium. The mycelium darkens with absorption, metabolizes the oil, and becomes white again after digestion.

the toxin would decompose into nontoxic forms. This is, in fact, what happened with several of the provided strains. With the merits of this approach established, we all became excited about the possibilities of using mushrooms for mending war-torn environments.

Months later, Battelle sent me an unpublished internal report dated July 14, 1997, that listed me as a coauthor. The report, "Adaptation of mycofiltration phenomena for wide-area and point-source decontamination of chemical warfare and biological warfare agents," gave the results of tests using my strains against chemical warfare agents and *Escherichia coli*, a deadly endospore-forming bacterium that is used as a surrogate for testing as it can be weaponized just like anthrax *(Bacillus anthracis)*. Much of this research is confidential, but, suffice it to say, the results were encouraging and succeeded in a "proof of concept" report. Eventually, the work progressed without me.

Warned by my Battelle colleagues that patent law exempts the government from charges of intellectual property theft, they suggested I protect my strains by disguising them with code letters instead of species names. The strains were then loaned to them while I maintained my ownership rights. After the positive results appeared, I coauthored a patent with the Battelle researchers on mycoremediation of chemical and biological warfare agents: US Patent Application 09/259,077.

The patent application uses codes to designate the most promising strains I provided. One of these active mushrooms is a resident in the old-growth forest and my clone gave rise to mycelium that denatured VX. Another mushroom was *Mycena chlorophos*, a luminescent mushroom (see figure 105) that I originally selected because it emits light using luciferase, an enzyme that liberates photons. I also picked phosphorus-metabolizing strains because many neurotoxic chemical warfare agents are similar to insecticides and also contain phosphorylated compounds. When broken, a process called dephosphorylation, the molecules of these neurotoxins unravel.

▲ **FIGURE 105**

A bag of glowing *Mycena chlorophos* mycelium.

The most common group of toxic chemical nerve agents includes anticholinesterases, which interfere with the regulation of signals passing between neurons. Highly toxic and fast acting, many anticholinesterases resemble in their activity dimethyl methylphosphonates (DMMPs), which are core constituents and parallel precursors to chemical warfare agents. Sarin, soman, VX, and other chemical weapons all have similar phosphorus bonds. This is also true of DMMP, a surrogate for neurotoxins, which laboratory researchers use for safety reasons.

This joint study by Battelle and Fungi Perfecti suggests that remediation of sites polluted by chemical warfare agents using mushroom mycelium is more effective and less expensive than conventional methods. In one of many tests, 2 of my strains neutralized "very close surrogates of chemical weapons such as sarin, soman, and the VX family of compounds" (Jane's Information Group 1999). Since one of my active strains is native to the old-growth forests of the Pacific Northwest, saving our old-growth forests could help national defense. (See also page 38 for another mushroom species indigenous to the old-growth forest that may be important for our national defense.)

Magic Mushrooms versus Nerve Gas?

The *Psilocybe* mushrooms such as *Psilocybe azurescens* and *Psilocybe cyanescens* absorb phosphorus from their surroundings in order to synthesize the psilocybin molecule (0-phosphoryl-4-hydroxy-N, N-dimethyltryptamine), which can make up 2 percent of its mass. Hypothetically, with the right enzymes *Psilocybe* mycelium could extract phosphorus from DMMP. Nonfruiting strains could be used for cleaning up environments contaminated with DMMP-like neurotoxins. I suspect that mushrooms rich in psilocybin have metabolic pathways for taking phosphorus from organophosphates, a group that includes many chemical weapons, pesticides, and herbicides, and other industrial toxins. (For more information on the habitats *Psilocybe* species prefer, see *Psilocybin Mushrooms of the World* Stamets [1999b].)

This discussion brings up an interesting issue: If these "magic" mushroom species proved effective for breaking down VX, would we choose not to use them since they are controlled substances and illegal in many countries? Nature responds to catastrophes with apolitical measures. We often do not.

Using Fungi for Destroying Munitions

Munitions destruction is a serious issue haunting the communities around military depots. As microbes decompose any organic matter, including munitions, they release heat. The heat released by microbes in compost piles can cause spontaneous combustion, a phenomenon witnessed by many cultivators. This exothermic reaction could spell disaster when it comes to decomposing munitions.

One avenue of munitions-destruction research I recommend is "cold composting"—that is, using mushroom species that metabolize at very cold temperatures. The natural advantage of using cold-tolerant mushrooms is that they continue to secrete enzymes in near or even below-freezing conditions, and the temperatures they generate are below the threshold of dangerously overheating. Examples include the enoki, or winter, mushroom (*Flammulina velutipes*) and the waxy cap (*Hygrophorus camarophyllus*). Similar species of cold extremophiles should be tested. Please refer to page 95 for list of classes of toxins and supporting research citations, and the chart on page 96, matching mushroom species for decomposing toxins.

Several patents have been awarded on using fungi to break down toxic wastes. Here is an abbreviated list of some of the key patents I'm aware of. More patents may have been issued, or are pending, than are listed here. Also, some patents may overlap one another when applied to complex landscapes containing a plurality of toxins. U.S. patents have, on average, a 17-year lifespan before expiration.

- Aust, D., and J. A. Bumpus, 1990. "Methods for the degradation of environmentally persistent organic compounds using white rot fungi." U.S. Patent 4,891,320.
- Aust, D., D. P. Barr, T. A. Grover, M. M. Shah, and N. Chung, 1995. "Compounds and methods for generating oxygen free radicals used in general oxidation and reduction reactions." U.S. Patent 5,389,356.
- Aust, D., and J. A. Bumpus, 1995. "Process for the degradation of coal tar and its constituents by *Phanerochaete chrysosporium*." U.S. Patent 5,459,065.
- Aust, D., D. P. Barr, T. A. Grover, M. M. Shah, and N. Chung, 1995. "Compounds and methods for generating oxygen and free radicals used in general oxidation and reduction reactions." U.S. Patent 5,468,628.
- Aust, D., and J. A. Bumpus, 1997. "Process for the degradation of coal tar and its constituents by white rot fungi." U.S. Patent 5,597,730.
- Bennett, J. W., A. M. Childress, K. G. Wunch, and W. J. Connick Jr., 2001. "Fungal compositions for bioremediation." U.S. Patent 6,204,049.
- Illman, B., V. W. Yang, and L. A. Ferge, 2002. "Fungal strains for degradation and bioremediation of CCA-treated wood." U.S. Patent 6,495,134.
- Illman, B., V. W. Yang, and L. A. Ferge, 2004. "Fungal degradation and bioremediation system for pentachlorophenol-treated wood." U.S. Patent 6,727,087.
- Illman, B., V. W. Yang, and L. A. Ferge, 2003. "Fungal degradation and bioremediation system for creosote-treated wood." U.S. Patent 6,664,102.
- Illman, B., V. W. Yang, and L. A. Ferge, 2002. "Fungal degradation and bioremediation system for ACQ-treated wood." U.S. Patent 6,387,691.
- Khindaria, A., T. Grover, and S. D. Aust, 1996. "Compounds and methods useful for reductive dehalogenation of aliphatic halocarbons." U.S. Patent 5,556,779.
- Lamar, R. T., D. M. Dietrich, and J. A. Glaser, 1995. "Solid phase bioremediation methods using lignin-degrading fungi." U.S. Patent 5,476,788.
- Raghukumar, C., T. M. D'Souza, R. G. Thorn, and C. A. Reddy, 2002. "White rot-lignin-modifying fungus *Flavodon flavus* and a process for removing dye from dye containing water or soil using the fungus." U.S. Patent 6,395,534.
- Stamets, P., 2004. "Mycofiltration of silts and sediments within delivery systems for mycotechnologies, mycofiltration and mycoremediation." U.S. Patent Application Serial No. 10/852,948.

- Thomas, S., J. Word, M. Pinza, P. Becker, and P. Stamets, 1997. "Mycoremediation." U.S. Patent Application Serial No. 09/259,077.
- Yadav, J. S., C. A. Reddy, J. F. Quensen, and J. M. Tiedje, 2000. "Degradation of polychlorinated biphenyl mixtures in soil using *Phanerochaete chrysosporium* in nutrient rich, non-ligninolytic conditions." U.S. Patent 6,107,079.

Amassing Mycelia for Mycoremediation

Mycoremediation could become a large business. In the late 1980s, mycoremediation studies initially focused on the activities of just a few aggressive saprophytes: *Phanerochaete chrysosporium*, *Phanerochaete sordida*, *Gloeophyllum trabeum*, *Pleurotus ostreatus*, *Trametes versicolor*, and *Bjerkandera adusta*. More than 100 papers were published between 1994 and 2004 on enzymes from such wood-rotting mushrooms being used to break down synthetic toxins.

Wood's main structural fiber, lignin, is one of the most recalcitrant molecules manufactured by nature. It can resist decomposition because its long-chained carbon-hydrogen design rebuffs most enzymes. However, mycelial enzymes are uniquely equipped to degrade lignin into shorter-chained molecules. Man-made wood preservatives are more difficult to decompose than lignin, since the scientists developing these recalcitrant compounds specifically tried to resist the effects of most wood-rotting fungi. Mushroom mycelia secrete peculiar enzymes, such as quinone reductases from the brown rot *Gloeophyllum trabeum*, powerful enough to consume many wood preservatives, or manganese peroxidases from the white rot fungus *Phanerochaete sordida* used for digesting petrochemicals (Ruttman-Johnson et al. 1994). For excellent descriptions of the biochemical pathways mushroom enzymes use to degrade lignin, see *Fungi in Bioremediation* by G. M. Gadd (2001) and *The Fungi* by M. J. Carlile, S. Watkinson, and G. W. Gooday (2001).

A major factor in breaking down plant fibers or toxins using mycelium is the influence of hungry microbes that prefer certain types of nutrients. Knowing how to appease or redirect their appetite to a menu specific to your needs is part of the art of mycoremediation and mycofiltration. Nature loves communities. When one species is suddenly introduced, the population dynamics shift in response. Introduced mycelium can become a launching platform for bacteria. Oftentimes toxins stall the biology of an environment—afflicting the immune system of the landscape and effectively undermining biological communities—until organisms tolerant of the toxins are selected. These toxin-selected organisms often need a new nutrient source, and the mycelium plays a pivotal role in jump-starting the decomposition process, directly through its extracellular secretions, and indirectly by providing a nurturing food source for the toxin-tolerant organisms. This panoply of players often gives the best results, although the processes are more complex than science presently understands.

Mycelium fosters microbial communities. Upon its cellular architecture bacteria ride, held in abeyance by the selective influences of the mycelium's antibiotics. As the mycelium declines in vigor, resident and competing bacterial populations bloom and use the mycelium and the fruiting body as staging platforms for explosive growth. The resident bacteria harbored by the mycelium stifle the growth of competing bacteria. Bacteria are better than mycelium at breaking down plants without lignin, such as most nonfibrous garden vegetables. Most bacteria are also better than mycelium at degrading toxins with smaller molecular weights, and many partner well with fungi after the mycelium is exhausted in its efforts in decomposing recalcitrant organic molecules of high molecular mass. After fungi break down large molecules, bacteria feast on these newly available nutrients. Our tandem approach of using both fungi and bacteria holds great promise for synergistic remediation. Fungi and bacteria are the biological pumps of the carbon-nitrogen cycle. With carbon dioxide and water as by-products, fungal metabolism is the reverse

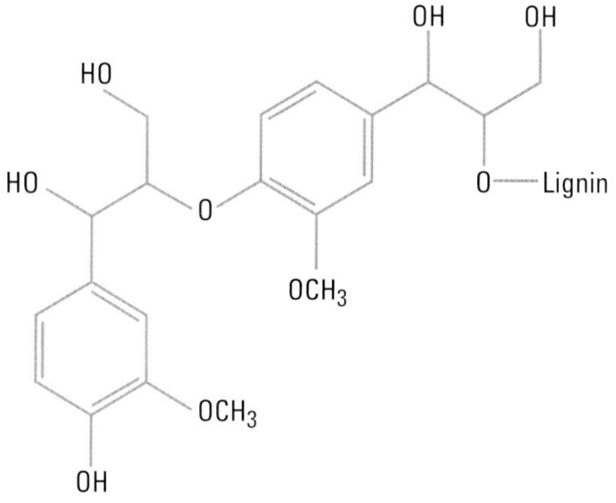

▲ FIGURE 106

Molecular structure of lignin.

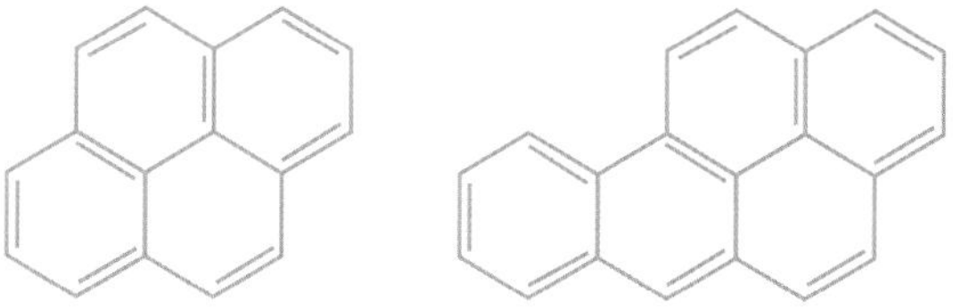

▲ FIGURE 107

Molecular structure of polycyclic aromatic hydrocarbons.

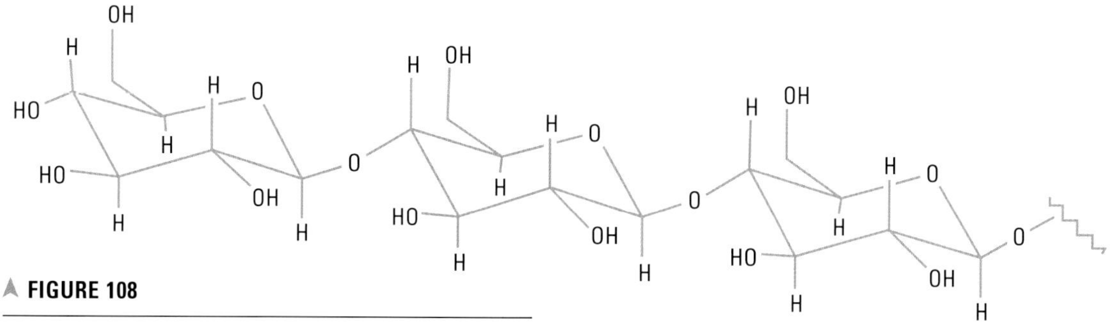

of photosynthesis, reducing cellulose and lignin to simple forms and remanufacturing them into chitins, polysaccharides, proteins, enzymes, and alcohols. The interactions of fungi and bacteria have evolved complex biochemical pathways for nutrient streaming, constituting the microbial foundation of our ecosystems.

As mycoremediators, we can deploy diverse mycelial systems to benefit the ecosystem. As the work of Tornberg, Baath, and Olsson (2003) has shown, I believe future research will affirm that most saprophytic fungi predestine and steer subsequent biological communities through the bacterial communities that mushrooms select for, influencing all other organisms in the food web, particularly plants. Mushrooms have a vested interest in the developing plant communities that will fuel their future life cycles with the debris they generate.

Mycoremediators must hone their skills so they can successfully inoculate landscapes teeming with potentially hostile microbes. Unless these microbial dangers can be overcome, competing microorganisms will thwart the beneficial mycelium. Native mushrooms already familiar with the micro biota have natural advantages. Native mushrooms' history of exposure to endemic hostile and beneficial organisms, acclimation to weather patterns, and existing niche within the ecosystem favor them over foreign imports. Some of the most powerful mycoremediating mushrooms, fortunately, are ubiquitous and grow on wood

▲ FIGURE 108

Molecular structure of cellulose.

throughout the world. Oyster *(Pleurotus ostreatus)* and turkey tail *(Trametes versicolor)* mushrooms can be found in virtually every woodland, and indigenous strains of these species can be manipulated by mycoremediators. But whether or not native strains are used, mycoremediators must amass and deploy populations large enough to have the desired effect.

In a cubic foot of spawn, there can be than 14,000 miles of fine, threadlike mycelia. A 1-inch-long rhizomorph has the tensile strength of more than 30,000 times its mass (see figure 71). When the spawn is broken up, the hyphae fork and refork, greatly extending their range, provided more food is available as fuel for these fungal cells. Each fragment of mycelium is a mycelial island, which as it grows seeks to join with other islands of itself, to eventually form a contiguous mycelial mat. Once you amass and distribute spawn, you not only have "mycelial mileage," meaning the spawn has enormous potential for further growth, but you enlist a powerful ally for transforming habitats. This sudden influx of fungal cells inalterably changes the microbial landscape, creating a mycelial tsunami, and setting a new path in habitat evolution for micro and macro organisms. The consequences of spawning are dramatic in the microbial universe. Bulk substrates such as straw and sawdust colonized by mycelium are generally more effective than grain spawn for mycoremediation. (If left uncovered, bugs, birds and other animals can consume the grain.) Nevertheless, many forms of mycelium can be used.

Mycelium for mycoremediation can be obtained from the following sources:

- commercial grain or sawdust spawn
- "spent" compost from a mushroom farm (compost spawn)
- transplanted mycelium from wild patches (wild spawn)
- stem butt spawn
- cardboard sheet spawn
- plug spawn
- bunker or burlap sack spawn
- spore mass spawn

Mycoremediation projects are natural side activities for mushroom growers who generate tons of "spent" compost annually. Spent but fresh mushroom substrate—what some call "mushroom compost"—is likely to produce the best results in mycoremediation projects. Spent compost from a mushroom farm, the myceliated straw medium after it has flushed mushrooms to a point of diminished returns, is far less expensive and as effective as, or more effective than, pure culture spawn used by mushroom growers to inoculate bulk substrates. "Spent compost" is not the same as the "mushroom compost" gardeners purchase from stores. If you acquire spent compost, make sure it has *not* been steam pasteurized ("cooked off"), a standard practice many farms use to kill bugs, molds, and microbes before they release their compost for sale to the public. In every case for remediating toxins that I know of, all but one—in the sequestering of mercuric ions (see Arica et al. 2003 and pages 105 and 301)—the mycelium must be alive and viable to work for mycoremediation.

Currently, oyster mushroom farmers have little or no market for their spent compost once it is has flushed mushrooms. They would probably be eager to find someone to take this material off their hands. Spent compost offers a fortuitous opportunity to dramatically heal sick landscapes. The sheer tonnage of enzyme reserves within this myceliated substrate is staggering. Not using this mycomass of by-products is to waste this resource. Mushroom composts of species that eliminate toxins should be valued as a form of natural capital, an idea that fits well with the concepts proposed in the book *Natural Capitalism*, by Paul Hawken, Amory Lovins, and L. Hunter Lovins (1999). Mushroom farms are ideal platforms for regional mycoremediation projects.

Here is a model for how we can use the natural capital of spent fungal compost, in combination with farm waste and wood debris:

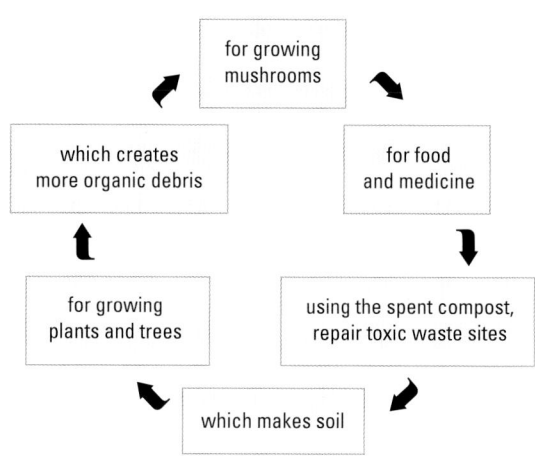

If a mushroom farm is not nearby, commercial spawn can be imported and used to make "bunker spawn" (burlap sacks filled with wood chips and mycelium), which is incubated outside and naturalizes to the outdoor climate during grow-out. If spawn is too expensive, a simple way to amass mycelium is to take stem butts (the bottom of the harvested mushroom stem) and expand them into bunker spawn (see chapter 9). Using a combination of spores and stems butts harvested during the fruiting season and inoculated into new habitats can be an effective method for expanding mycelium year to year. The stem butt method for bunker spawn generation will generally take a year longer than using commercial spawn, depending upon the strain and number of mushrooms that you collect the first year. Furthermore, the stem butt spawn method does not always succeed in generating usable mycelium because some strains die. Trial-and-error methods will be necessary until each cultivator perfects their skills. Alternatively, specimens can be collected and cloned in a laboratory to generate sawdust spawn. My company, Fungi Perfecti (www.fungi.com), offers a custom cloning service and provides spawn from mushrooms on a case-by-case basis. If you want to clone mushrooms and generate your own pure culture spawn, see *Growing Gourmet and Medicinal Mushrooms* (Stamets

2000a). I recommend planting spawn in early spring unless winter temperatures stay above freezing, in which case fall inoculations can give you a head start.

The implementation of mycoremediation has faced some serious hurdles. What mycoremediation trials have lacked to date is the participation of experienced mushroom cultivators in outdoor trials. Abundant research has proven the success of mycoremediation, but in situ (outdoor) trials have met with less success, primarily because mycelia failed to colonize the targeted substrate. Sasek (2003) summed it up when he stated that the skill of a mushroom cultivator is critical to the successful efforts of scientists attempting to use fungi for mycoremediation.

Mycelium, Mushrooms, and the Hyperaccumulation of Heavy Metals and Radioactive Elements

The Chernobyl reactor near Kiev, Ukraine, overheated on April 26, 1986, leading to a core meltdown that released a plume of radioactive debris for more than 10 days. Nearby residents absorbed levels of radiation 100 times greater than the levels absorbed by individuals at a similar proximity to ground zero at Hiroshima. The radioactive fallout dusted the surrounding terrain with toxic metals, causing marked increases in mutation and cancer rates. Months later in Germany, government health officials discovered that people were exhibiting pronounced increases in whole-body radioactivity after a single meal of wild mushrooms, contamination of which was caused by the Chernobyl disaster. Nearly all the radioactive cesium still detected in mushrooms in Sweden is traceable to Chernobyl. French officials confiscated mushrooms coming from Bulgaria because they contained 4 times the permitted levels of cesium. The *Japan Times* (2001) reported that a shipment of porcini mushrooms *(Boletus edulis)* from Italy was confiscated because it showed high cesium levels. The *New York Times* (Wines 2002) reported that Moscow officials passed Geiger counters over all foods for sale in marketplaces to test the safety of foods from radiation exposure. A single jar of dried mushrooms emitted radiation 20 times the admissible level. In all these cases, the radioactivity came from Chernobyl. This fallout will persist for decades, poisoning downwind ecosystems. However, this nuclear disaster stimulated new research on the accumulation of radioactive fallout in the food chain; of all the foods identified as bioaccumulators in this research, mushrooms top the list. Although mushrooms around Chernobyl are well known to be dangerous, for many, poor economic conditions, lax security, ignorance, and hunger override concerns about safety (Bellaby 2004).

Many mushrooms absorb radioactivity, but some are *hyperaccumulators*, with a peculiar ability to take in and concentrate elements such as cesium at thousands of times above levels in the surrounding area. What is most amazing to me about the discovery that specific mushrooms can superconcentrate radioactive metals is how species vary in their uptake levels, favoring a particular heavy metal over others. Often, a toxic waste site will have multiple contaminants, and if heavy metals are present some mushrooms will concentrate them and become too toxic to eat. (For this reason, mushroom consumers should know where their mushrooms come from—all the more the incentive for supporting organic growers and selecting mushrooms from pollution-free environments.)

This process of hyperaccumulation of heavy metals by fungi suggests a new, simple strategy for remediating sites contaminated by these metals: harvesting metal-laden mushrooms could gradually reduce cadmium, mercury, arsenic, lead, and even radioactive elements such as cesium-134 and cesium-137, by-products of the nuclear energy and weapons industries. Mushroom-forming fungi have a decided advantage over those that don't form mushrooms, since the heavy mass of compacted hyphae composing the mushroom's body makes it easy to collect the bioconcentrated toxins in solid form. In effect, the toxins move into a portable cellular vessel: a mushroom that, when removed, could reduce ambient levels of the toxins on the site. In addition,

numerous studies have found that the "coefficients of accumulation"—the levels at which the mushrooms concentrated heavy metals such as mercury, lead, arsenic, cesium, and cadmium from the background environment—ranged from 1 to 10,000 times. So the question becomes, how do we use mushrooms to remove or neutralize heavy metals? Here is one strategy, long-term and incremental, that could be put into practice.

In mycoremediating sites using this technique, mycorrhizal or saprophytic mushrooms could be grown through mulch and selectively harvested each year during the fruiting season, thus reducing the background contamination level. The metal-laden mushrooms could then be taken to a toxic waste site where they could be buried, stored, or incinerated, whereupon the residual metals could be resold to the metal recycling industry. To readers who usually collect mushrooms for the table, the idea of taking your harvest to a metal recycler may sound strange, but researchers in the field (Gadd 1993; Garaudee et al. 2002; Wasser et al. 2003; Sasek 2003; and Stijve et al. 2001) have suggested or have questioned that this may be the best mycoremedial path to remove toxic heavy metals from the environment.

In 2003, I visited Vashon Island, Washington, to address residents concerned about the environmental health of their community. After decades of airborne pollutants raining down on it from Tacoma's Asarco smelter, the island is heavily contaminated with arsenic, cadmium, and mercury (McLure 2002). For the first time, I advised people *not* to eat wild mushrooms picked near their homes, but to have them analyzed first to see which species might be accumulating metals from the smelter. Only after surveying the indigenous mushrooms and profiling the spread of contaminants could the community undertake appropriate mycoremedial responses. Although receptive and concerned, the community did not have a manual for mycorestoration, and to the best of my knowledge, no actions were taken. This is exactly why I wrote this book—so local communities can begin the process of healing their habitats. We must fund a library of mycological remedies so that future generations can learn, benefit from, and improve on these practices. Mycological societies throughout North America (most belong to the North American Mycological Association, www.namyco.org) are already well schooled in taxonomy; the task of further specifying how to use mushrooms in polluted zones is just an extension of their educational mission. In addition, many of these societies are associated with universities with which they share resources. I see these educational organizations as key in helping communities coordinate a mycological response to local toxic issues.

Here is another promising application of mycoremediation, this one for cleaning water leaching from heavy metal–contaminated soil: Arica and fellow researchers (2003) used turkey tail (*Trametes versicolor*) and phoenix oyster (*Pleurotus pulmonarius* var. *'sajorcaju'*) mycelia to remove 97 percent of mercuric ions from water. They did this by combining the water with small beads essentially composed of the mycelia, which selectively absorbs mercury, and a salt called alginic acid, which simply speeds the transfer of mercuric ions from the water to the mycelia. Subsequent to absorption, the heavier-than-water beads settle out of solution, and once removed, the water is decontaminated of the toxic mercury. Surprisingly, they noticed in lab studies that when dead mycelium was used, 73 to 81 percent of the mercuric ions were removed from the wastewater samples, suggesting that some physical feature resident on the cellular architecture of the dead mycelium bonded with the mercuric ions. Consequently, turkey tail and phoenix oyster mushrooms were identified by Arica and his fellow researchers as good candidates for mycofiltration of water high in mercury contamination. The live mycelium of some fungi produce organomercury lyases, enzymes that break down organomercury into HgO by mercuric reductase (Gadd 1993). Once freed, mercury can unimolecularily bond with selenium, another metal some mushrooms concentrate through *upchanneling*—moving the metal from the environment into the mycelial matrix. Using these types of reducing enzymes, fungi can precipitate many metals around their mycelia, such as silver, selenium, tellurite, cadmium, lead, and others.

I cannot help but wonder if the mushrooms that concentrate toxins are purposefully volunteering their services, absorbing heavy metals in order to protect other organisms from harm, in essence emerging to be picked and help the injured environment repair itself. Perhaps nature is more intelligent than we give her credit for. We may now be faced with a critical evolutionary decision, and an important lesson to learn: humans can use mushrooms to clean up heavy metals in our environment. But in order to do this, we must first discuss the cast of characters, both mushrooms and metals.

Mushrooms and Metals: Bioaccumulation of Arsenic, Cadmium, Cesium, Lead, and Mercury

Mushrooms concentrate many metals. Just as biologists do not know why *Psilocybe* mushrooms manufacture hallucinogenic phosphorus-based compounds, we do not know why other species concentrate heavy metals, such as arsenic, cadmium, cesium, lead, mercury, and copper. Determining the level of toxicity of heavy metal exposure is a work in progress. An excellent resource for information on environmental toxins is the Agency for Toxic Substances and Disease Registry (ATSDR) at the Centers for Disease Control (CDC; http://www.atsdr.cdc.gov). You can send them a dried mushroom for heavy metal analysis, at a cost of $35 to $150 per test, depending upon the toxins tested. To calculate whether you are ingesting unacceptably high levels of heavy metals in mushrooms, take the number of mg of heavy metals per kg in the mushrooms (dry weight—that is, the net weight of the mushrooms after they have been thoroughly dried) and divide that number by your body weight. That number should not exceed your daily allowable limit as reported by ATSDR.

So far, about 2 dozen species are known as hyperaccumulators as indicated in the chart on page 106.

Mushrooms versus Heavy Metals

This chart gives a general, preliminary guide to the bioaccumulation coefficients—concentration factors—of a mushroom species' ability to upchannel heavy metals from its myceliated habitat. This chart is a work in progress. Please consult the scientific literature cited in the text for more information.

	Arsenic	Cadmium	Radioactive Cesium	Lead	Mercury	Copper
Agaricus arvensis		X			150X	
Agaricus bisporus		X			X	
Agaricus bitorquis		X		23X	165X	
Agaricus brasiliensis		X			X	
Agaricus brunnescens	X	X			X	
Agaricus campestris		X		10X	10X	
Amanita muscaria		X			X	
Amanita rubescens		X				
Boletus badius			X			
Boletus edulis		10X	X	X	250X	X
Cantharellus cibarius			2X			
Cantharellus tubaeformis (Craterellus tubaeformis)			X			
Chlorophyllum rachodes	X			X	X	X
Clitocybe inversa	X	X				
Coprinus comatus	21X	8X			27X	
Coprinus spp.		X				
Flammulina velutipes	X					
Gomphidius glutinosus			10000X			
Laccaria amethystina	X		X			
Lactarius helvus			X			
Lactarius turpis			X			
Leccinum scabrum			X		X	
Lepista nebularis	X					
Lepista nuda					100+X	X
Lycoperdon perlatum			X	2X	100X	X
Marasmius oreades					X	
Macrolepiota procera					230X	
Morchella spp.				70–100X		
Morchella atretomentosa				X	X	
Paxillus atrotomentosus			1180X			
Pleurotus ostreatus		X			65–140X	
Pleurotus pulmonarius		X			X	X
Rozites caperata			X			
Suillus tomentosus				67X	6X	
Trametes versicolor					X	
Tricholoma magnivelare	22X					

When it is known, I indicate the mushroom's *bioaccumulation factor*, or how many times more concentrated a metal is compared to the background level. The mushrooms with an X but no coefficient multiplier accumulate, but evidence thus far does not clearly tell us how much. Several factors affect bioaccumulation of heavy metals. This preliminary and admittedly simplistic chart samples research and presents a guide for further comparisons.

One word of caution regarding interpreting the above tables: considerable variation has been found, and is to be expected, between samples. Mycorrhizal fungi, in particular, may show great variation depending upon the age of the colony tested. The longer the mycelium is in direct contact with the soil containing heavy metals, the greater the absorption into the mycelia. Younger mycorrhizal colonies are likely to bioaccumulate less than will colonies of longer residence and larger size. This is partly due to the fact that mycorrhizal fungi are plant symbionts living in close association with roots, and they penetrate deeper below the surface as time passes.

Saprophytic fungi, on the other hand, grow primarily on or above the ground. These species feed on debris fields and thrive rather briefly compared to mycorrhizal fungi, which can live on a site for decades. Saprophytic mushrooms may be more useful for cleaning up recently deposited or surface contaminants, while mycorrhizal species offer a transport system from areas deeper underground. Once the mycelium up-channels heavy metals into the mushrooms, they can be picked and transported out of the area. If the heavy metal–laden mushrooms are not removed, then bacteria and other fungi will cause them to decompose and return to the soil.

In theory, if these tainted mushrooms are collected, their metals can be concentrated further and recycled. But as long as we pollute, heavy metal contaminants will continuously accumulate in the food web. When this effect is compounded over time, biological populations intolerant to heavy metals will die. The fact that mushrooms selectively concentrate heavy metals may be good news for those seeking organic methods for restoring habitats in situ. Few other avenues of treatment are available. And mushrooms' affinity for absorbing metals promises new areas of research. As we better understand hyperaccumulation rates and selectivity factors, we can put these mushrooms to work mining for metals.

What follows is a short list of heavy metals and some comments related to their sources, mushrooms, and health. Many of the measures of toxicity of metals fall into parts per million (ppm) or micrograms (mcg).

ARSENIC

Used as a wood preservative, arsenic stops most fungi from growing on lumber. So much arsenic contaminates soil and groundwater that the Environmental Protection Agency (EPA) banned its use in treated lumber for home construction and renovation. Additionally, with the support of the National Academy of Sciences, the EPA lowered acceptable levels of this known poison and carcinogen. In 2001, the EPA ruled that the maximum level of arsenic permitted in drinking water is 10 mcg per liter (=ppb), assuming an average consumption of 2 liters of water per day. The previous level was 50 mcg per liter. We ingest most of our arsenic through drinking water. Excessive exposure can cause lung, bladder, and other cancers.

Mushrooms that accumulate arsenic include the prized North American matsutake (*Tricholoma magnivelare*), a mycorrhizal species, and shaggy mane (*Coprinus comatus*), a saprophytic species. A Canadian study done in the Northwest Territories measured an alarming 494 mcg/g (dry weight) of arsenic in shaggy manes growing near a mine. If you ate 100 g of these fresh shaggy manes (about 10 g dry weight), less than 1/4 pound, you would ingest 4,940 mcg of arsenic, about 250 times the allowed EPA limit in drinking water which is now 10 ppb.

Shaggy manes may be a bioindicator species for arsenic. I often find particular shaggy mane strains in a couple of notable habitats: the fertile soil of lawns, rich in organic debris, and the soil of recovering habitats, especially those with upturned, mineral-rich, subsurface soil that is quite low in organic material. Prolific fruitings in these disturbed settings, number in the thousands per locale, and they may decontaminate land poisoned with arsenic if the bioaccumulating mushrooms were removed.

One should not be unduly concerned about eating shaggy manes unless they occur on or near arsenic-contaminated land. I advocate avoiding all mushrooms found along roadsides, where exhaust fumes and asphalt supply arsenic to fungi. Jochin Obst and colleagues surveyed wild mushrooms around the North Great Slave Lake region in the Northwest Territories of Canada. The authors "recommend that responsible authorities advise the public not to harvest mushrooms from areas exposed to emissions and from contaminated sites such as locations near mine sites, tailings, crushed waste rock fill, landfills, roadsides, parking lots, disturbed areas, communities and cities…. Mushrooms from the Agaricaceae [the family including the *Agaricus* genus] should be avoided completely because of their high bioaccumulation properties of heavy metals" (Obst et al. 2001). Similar species-specific bioaccumulations of heavy metals have been found in mushrooms elsewhere, especially in Eastern Europe, China, and India.

CADMIUM

Of all the mushrooms mentioned in this book, *Agaricus* stands out as the genus with the most species concentrating acutely toxic cadmium. Cadmium hyperaccumulation has been a concern for the button mushroom *(Agaricus bisporus)* industry for many years. The maximum acceptable level for cadmium intake, according to the ATSDR, is 2 parts per billion (ppb) per kg of your body weight per day. Cadmium intakes as low as 1 mcg per day can harm the neurological system and the immune system and could cause cancer. *Agaricus* mushrooms grown in cadmium-laden lands, especially regions near steel smelters, could be harmful. Stijve and others (2003)

analyzed several *Agaricus blazei* sensu Heinemann (now known as *Agaricus brasiliensis*), comparing mushrooms from various suppliers. Specimens from China had 2.75 mg per kg of cadmium, whereas specimens from California contained 0.90 mg per kg and those from Washington had 0.53 mg per kg. The highest levels, in mushrooms from China, is likely due to fallout from industry—in other words, air pollution.

If the commercial mushroom industry uses cadmium-contaminated substrate, such as high-nitrogen chicken manure, their produce will concentrate cadmium. Nevertheless, cadmium levels vary widely among cultivated button mushrooms: some analyses show no cadmium at all (Lelley and Vetter 2004). Regardless, I recommend monitoring metal concentrations in all edible mushrooms, especially species known for hyperaccumulating toxic metals like cadmium. One complicating issue is that many analytical laboratories use tests that are not sensitive to occurrences below 1 ppm and hence cadmium, although possibly present, is not detected. This is problematic because cadmium is toxic at less than 1 ppm.

The mushrooms that absorb the least cadmium are oysters, which can have twice the concentration of cadmium in their flesh compared to the soil mass around them (Favero et al. 1990). Yet oyster mushrooms can flourish in environments heavily contaminated with cadmium and mercury, typically where petroleum products have spilled, while still degrading polycyclic aromatic hydrocarbons. Baldrian and others (2000) found that only when levels of cadmium reached 100 to 500 ppm and mercury 50 to 100 ppm did oyster mycelia die back. In a more recent study, Baldrian and Gabriel (2003) found that the cellulose- and hemicellulose-degrading abilities of oyster mushrooms were limited by cadmium levels. Purkayastha and fellow researchers (1994) reported that the mycelium of oyster mushrooms (*Pleurotus sajor-caju*, or *Pleurotus pulmonarius*) concentrate cadmium but not lead.

CESIUM

Cesium-134 and cesium-137 are by-products of nuclear fission. When the Chernobyl nuclear power plant melted down in 1986, the radioactive plume it emitted contaminated much of Europe with radioactive cesium. Afterward, some mushrooms highly valued for their edibility, such as *Boletus edulis*, commonly known as the porcini, or cèpe, contained unacceptably high levels of cesium. Wasser and others (2003) found that the coefficient of accumulation ranged from 1,180 times in *Paxillus atrotomentosus* to more than 10,000 times in *Gomphidius glutinosus*, both mycorrhizal mushrooms. Other surveys of cesium-contaminated mushrooms (Fielitz 2001; Epik and Yaprak 2003) found that fresh mushrooms variously concentrated cesium measured in Becquerels per kilo (Bq per kg), an alternative term for curies, as follows: *Boletus badius*, 3,030 Bq per kg; *Cortinarius hercynicus*, 6,750 Bq per kg; and *Elaphomyces granulatus*, 25,660 Bq per kg. The following mushrooms concentrated very little: *Armillaria mellea*, *Boletus subtomentosus*, *Cantharellus cibarius*, and *Boletus edulis*, all approximately 400 Bq per kg. *Boletus chrysenteron* contained 1,000 Bq per kg.

In 2003, the European Commission established new safety limits for acceptable cesium levels in mushrooms sold in the market—600 Bq per kg—in a report entitled "EC Recommendation Sets New Limits to Cesium Concentration in Wild Food Products on the Market," made by Finland's Radiation and Nuclear Safety Authority (STUK 2003).

The chart on page 106 is a sampling of species based on reports regarding wild mushrooms. However, radioactive fallout can be uneven, as was the case with Chernobyl. Pockets of high toxicity may have skewed the data and affected averages. Nevertheless, the list of species on page 106 could be helpful in assessing strategies for the gradual detoxification of an environment by trained mushroom harvesters. Although cesium-137 has a half-life of 30 years, during which it naturally decays into nonradioactive forms, collecting and disposing of specific mushrooms might be a viable method for extracting cesium from soil. Garaudee and others (2002) postulated that bay boletus mushrooms (*Boletus badius*) could mop up cesium, because they found that its isotopes bind to the mushroom pigment norbadione A. In much the same way that hemoglobin carries oxygen in the blood, the catalytic reaction from cesium's contact with norbadione could potentially lead to a new mechanism for scavenging—in this case—radioactive contaminants.

▲ **FIGURE 109**

In this mycelial mat the elm oyster mushroom *(Hypsizygus ulmarius)*, a brown rot fungus, fruits among grasses that feed on decomposing coconut fiber. Such mycelial mats can be used not only to absorb a toxic spill but as a pedestal for mycelial growth leading to ecological recovery. Mycoremediators can use toxin-specific mycomats to carpet a toxin-laden landscape in a dual attempt to destroy the underlying toxins and to give rise to customized descendant plant communities. Mycelium leads the way to habitat restoration. Bacteria, plants, and animals follow, fueling the food chain with nutrients and renewing life cycles.

LEAD

Most mushrooms do not actively concentrate lead above the level of concentration in the environment, with a few exceptions. One such exception, the morel, proliferates in burned habitats where fire has reduced the bulk of organic matter, an event that increases baseline metals in the "soil." Lead is made less soluble, and hence less extractable, in soils where the pH is near neutral; adding lime (calcium carbonate) significantly reduces the solubility of lead, cadmium, and other heavy metals, thus locking them up and reducing infiltration into water and/or living organisms. In contrast, acidic soils allow lead to be easily absorbed by plants and mushrooms. According to Garcia and others (1998), saprophytic mushrooms absorb more lead than mycorrhizal ones, and shaggy manes (*Coprinus comatus*), especially those found in cities, could be bioindicators of lead contamination. Naturally, one wonders if the very presence of shaggy manes could indicate lead contamination? I don't know, but I am suspicious when I find shaggy manes near industrial sites and avoid eating them.

The limit for lead in food set by the European Community is 2 mg per kg in dry weight. The U.S. standard for drinking water is <15 mcg per liter, assuming consumption of 2 L per day. Mushrooms growing along roadsides, particularly where there is exposure to exhaust from leaded gasoline, should not be eaten. Otherwise, the main environmental source of lead, now that lead-based paints are no longer in use, is air pollution from factories, which becomes concentrated downwind. I don't see any practical mycomethod of removing lead from polluted lands since most mushrooms, according to research I have read, do not hyperaccumulate this metal.

MERCURY (AND SELENIUM)

Compared to lead, for instance, mercury is a highly toxic metal that dramatically impairs human and environmental health. The maximum amount allowed in drinking water is 1 ppb or .001 mg per liter of water. The minimum risk level for ingestion of methyl mercury is 0.12 mcg per kg of body weight per day. Although we accumulate toxic levels of mercury by eating particular types of fish, mushrooms can be a source too. Arica et al. (2003) discovered that fresh or heat-killed mycelia of turkey tail, *Trametes versicolor*, and oyster, *Pleurotus pulmonarius*, removed mercuric ions from aquatic systems.

Selenium, a mineral important for cellular metabolism, can lower toxic mercury levels in the body. Selenium bonds with mercury in an unimolecular, biologically inactive form, making it nearly non-toxic. However, if too much selenium bonds with mercury, which could mean there is too much mercury in your body, then the body's selenium levels drop, affecting the immune system.

The *Journal of the American Medical Association* published a placebo-controlled clinical study of selenium and reported that 200 mcg of selenium per day reduced lung, colorectal, and, most dramatically, prostate cancer (Clark et al. 1996). Selenium activates the antioxidant enzyme glutathione peroxidase, which helps the body purge itself of environmental toxins. This enzyme is thought to mitigate the effects of aging, and its absence impairs the metabolizing of vitamin E, itself an important antioxidant. A lack of selenium and vitamin E impairs overall cellular vitality and system-wide immunity. The National Cancer Institute states that diets deficient in selenium may also increase mercury toxicity, further impairing immunological function (National Cancer Institute 2004).

Few foods are rich in selenium. Fish, Brazil nuts, meat, and mushrooms are notable sources. Mushrooms that are especially efficient at incorporating selenium from the environment could be considered health foods for this feature alone—at least when levels fall within a safe range. In a survey of 126 species, Alfthan (2003) found that porcini (*Boletus edulis*) contained the most selenium. The levels varied by a factor of 7 between samples, up to 51.6 mg per kg (dry weight). On the other hand, *Agaricus* species accumulated mercury up to 42.5 mg per kg, but the levels in these samples varied by a factor of 10. Blewits (*Lepista nuda*) and parasols (*Lepiota rachodes*) are also known to accumulate mercury (Kalac et al. 1991).

Some mushrooms are almost too good at concentrating selenium. The cultivated button mushroom's ability to concentrate this element initially drew interest in it as a health food. Researchers at Pennsylvania State University found a linear relationship between selenium uptake by mushrooms and selenium supplemented into composts (Spolar et al. 1998). The uptake is so efficient that some button mushrooms, if eaten daily, would cause the consumer to exceed the recommended daily intake and overdose. The recommended daily intake of selenium for an adult, according to the NIH, is in the range of 50 to 77 mcg, but a daily intake of 200 mcg is recommended by Dr. Andrew Weil, director of the Program for Integrative Medicine at the University of Arizona Medical School, Tucson (Weil 2004). This is in also agreement with research by Clark and others (1996). More than that can be deleterious: at levels of 1,000 mcg per day, selenium may cause skin rashes, loss of fingernails, and nervous disorders (Medsafe 2000). One sample of button mushrooms in the Penn State study concentrated enough selenium to make them potentially dangerous. This finding, that *Agaricus* mushrooms could accumulate too much selenium, caused concern on the part of those promoting selenium-enriched mushrooms, who had floated the idea that such mushrooms could be marketed as health supplements.

Indeed, excessive selenium can be toxic. Minute amounts of mercury, however, are more dangerous than large quantities of selenium. Selenium from mushrooms has a positive protective effect against mercury poisoning, until the balance of effects from the diversion of selenium impairs selenium's immunological role as it diverted and bonds with mercury. The key is to help bind mercury into a nontoxic form while sufficiently serving the immune system by supplementing the diet with enough selenium from foodlike mushrooms. Ostensibly, the addition of selenium to a mercury-contaminated substrate is one way to neutralizing toxicity. However, navigating through the proper combinations, the applications, and the simple mechanics presents its own problems.

The table on page 106 shows mushroom species and the heavy metals they concentrate. The bioaccumulation factors—the coefficients of concentration—are listed for some mushroom species. This listing is only an approximation. Strains vary within a species and regional conditions also vary. Nevertheless, mushrooms concentrating heavy metals offer an advantage to habitat remediators. Focusing on mushrooms as bioindicators, environmental scientists can test them to obtain a reading of heavy metals in a polluted habitat.

After Chernobyl, media reports stated that mushrooms became unusually large. I wonder if this is nature's way of purging heavy metal contamination from the environment. Sampling mushrooms to identify "hot zones" could direct mycoremediation efforts. When mushrooms over-concentrate metals, they become toxic to eat, but furnish a new opportunity. By removing these mushrooms we reduce metals in the immediate environment. People living in industrialized areas—on reclaimed landfills, near smelters, military installations, highways, or polluted rivers—should be most concerned. Much pollution is spread by wind and water in concentrations decreasing by an inverse square of distance from the source.

Thoughts on Strategies for Mycoremediation

What's the first step you can take toward mycoremediation? Look at the history of the use of your land. Take soil samples and have them analyzed for targeted toxins. If possible, match mushroom species from the above list, and when they grow on contaminated land, have them analyzed, prioritizing for the metals known to be concentrated within them. If the toxic substance in the mushrooms exceeds the minimum risk levels in mg consumed per kg per day, then the land may be polluted (Chou 2004). Medicinal or gourmet mushroom products, no matter where they originate,

◀ **FIGURE 110**

Gomphidius glutinosus, a mycorrhizal mushroom, can concentrate radioactive cesium-137 to more than 10,000 times the background level.

should not be ingested if they contain dangerous levels of heavy metals.

When choosing mushroom species to mycoremediate a toxic site, select species that naturally grow in that landscape. These native strains can be enhanced as the primary remediating fungi. See chapters 9 and 10 for techniques for creating spawn from wild mushrooms; also see *Growing Gourmet and Medicinal Mushrooms* (Stamets 2000a). For example, as previously mentioned, the slippery *Gomphidius glutinosus* concentrates cesium-137 more than any other species yet surveyed, 17,117,000 Bq per kg (dry weight), tens of thousands of times the background levels (Wasser et al. 2003). The mycelium of this mycorrhizal mushroom remains in contact with its host, a conifer tree, allowing for the long-term concentration of cesium before the fungus fruits. Of the species I have checked, the garden giant *(Stropharia rugoso annulata)* appears to be the best at consuming bad bacteria and stimulating plant growth, and it resists absorption of heavy metals from its immediate environment. This is good news because this is also one of the most versatile mushrooms for habitat restoration. Oyster mushrooms continue to degrade PAHs, the core molecules

of petroleum products, even in the presence of high cadmium and mercury (Baldrian et al. 2000). They also concentrate mercury up to 140 times over background levels in the soil (Bressa et al. 1988), but generally resist cadmium uptake (Favero et al. 1990). These are just a few examples showing how different species of mushrooms react differently with heavy metals in their habitats.

Examples of Mycoremediation Strategies for Toxic Waste Sites

Contaminated habitats can be simple or complex in their toxic profiles. In aged industrial sites, many toxins reside. Determining the best mycoremedial strategies for a particular site is a challenge. The chart on page 106 shows matchings of mushroom species versus primary toxins. Each toxic habitat is distinct and demands a localized approach by a skilled remediator. In many cases, mycelium is implanted to begin the sequence of biological community building, which then completes the process of restoration. Mycoremediation is the destruction of life-limiting toxins that enables other ecological restoration strategies. This is a gateway technology, and once implemented, a

Contaminated Habitat Scenarios, Their Toxins, and the Mushrooms That May Heal Them

Contaminated Habitat Scenario	Recommended mushrooms
Petroleum products (oil, diesel, gasoline, petrochemicals)	*Pleurotus ostreatus*
Chemical dyes	*Ganoderma* and *Trametes* species
Industrial metals (lead, cadmium, arsenic, mercury, selenium, radioactive cesium-137 and cesium-134)	Large *Agaricus, Lepiota,* and mycorrhizal species
Munitions (TNT)	*Hypholoma* and *Flammulinas* species
Organophosphates, chemical weapons (VX, sarin)	Polypores, oysters, and *Psilocybe* species
Biologicals (*Escherichia coli, Bacillus* sp.)	*Calvatia gigantea, Coprinus comatus, Fomes fomentarius, Ganoderma* species, *Piptoporus betulinus, Pleurotus* species, *Polyporus umbellatus,* and *Stropharia rugoso annulata*
Nitrates and phosphorus-bound toxins	*Agaricus bernardii, Agaricus silvicola* and allies, *Coprinus comatus,* and *Psilocybe* species

domino effect comes into play. The following guidelines are meant to aid habitat restoration and will be adapted, revised, and expanded in the future. This book will be updated in future editions that will further elaborate on these strategies. Because mycoremediation is an infant technology, many experiments and proof-of-concept trials need to be conducted before commercialization. Once more data becomes available, more precise methods may be discovered.

The scenarios in the above chart each include a different class of toxins. In reality, few landscapes are affected by just a single type of toxin. In many cases, overlapping and sequential mycelial mats are recommended over the long term in order to reduce multiple types of toxins. In some cases, both mycofiltration and mycoremediation may be used simultaneously. Additionally, mycoremediation strategies are best integrated into habitat restoration programs that also utilize the bioremediative properties of plant, bacteria, and algal communities.

Can Fungi Consume Radiation?

Researchers at Einstein College of Medicine of Yeshiva University, led by Dr. Ekaterina Dadachova (2007), have discovered that fungi having melanin (the pigment in your skin that darkens upon exposure to sunlight) use radiation to help their cell growth, analogous to how plants use chlorophyll to generate chemical energy from sunlight. This first came to their attention when they read that robotic cameras detected dark pigmented fungi growing on the walls within the highly radioactive Chernobyl reactor vessels. In experiments, the authors found that ionizing radiation approximately five hundred times higher than background levels significantly enhanced the rate of growth in several species of melanin-rich fungi. The implications of these findings are huge: opening up the possibility of generating fungal foods for space travel, and that fungi could exist on other planets lacking sunlight but exposed to ionizing radiation that permeates the universe.

CHAPTER 8

Mycopesticides

My house, built in 1910 in the middle of an old-growth forest in Washington, was a natural disaster. Thousand-year-old trees created so much darkness in this shadowed woodland that settlers, seeking daylight, cleared the land so a homestead could be built. When I bought this small waterfront farm in 1984, the remnant of an old-growth stump adjacent to the house sported an artists conk, *Ganoderma applanatum* (see figure 20). As I would soon discover, the mycelium from this stump flowed underneath the house in fungal fans that channeled water, rotting the floor joists and setting the stage for insect invasion.

When I moved in, I saw that my small house, which rested upon timbers, had soft spots in the floor where fungi were emerging through the carpets. Upon investigating, I was both fascinated and disturbed to find huge waves of brown and white mycelium coursing through the floors and walls of my house. The house was constantly infused with a sickeningly sweet, musty scent; perhaps a perfect abode for a budding mycologist.

▲ **FIGURE 112**

Our house was destroyed: first by fungus; then by carpenter ants; and then by humans with the help of this machine. The ants followed the path of mycelium and pulverized this house. The author, desperate for a solution, tried something new and received a patent for the idea in 2003: U.S. Patent 6,660,290.

◀ **FIGURE 111**

Dusty Yao inspects an aerial termite nest attached to a palm tree in the Bahamas. From these nests, termites march far and wide, consuming wooden structures.

Unfortunately, the spread of wood-digesting fungi such as the artist conk is well known to precede or coincide with termite and carpenter ant invasions; the fungi soften and moisten the wood, setting the table for these munching insects. And so I discovered that the slumping house was under attack by carpenter ants (*Camponotus modoc*). A friend doing remodeling for me quipped, "If all the carpenter ants stopped holding hands, your house would fall down." Each day, I awoke to find piles of sawdust on the floor near the walls; overnight more of my house had been milled into sawdust by ants. Visitors showed displeasure at the sawdust raining down upon them, despite my best attempts to make light of the situation.

I searched the Internet to find information about fungi that naturally parasitize insects, and I began to read about *Metarhizium* molds, especially *Metarhizium anisopliae*, which had been used to kill termites. (Since *Metarhizium* does not harm mammals, does not cause human allergies, and is limited in colony size, this fungus does not lead to "sick house" syndrome, caused by some allergenic fungi.)

I decided to give it a try and ordered a strain from a commercial tissue-culture library. When the culture arrived via overnight delivery service, I opened the package to find a test tube containing a green mold similar to the *Penicillium* types one finds on rotting oranges. While all houses have populations of molds and their spores, spawn laboratories such as ours are super clean so that spores are not free flying, and cultures are kept captive in petri dishes or other sealed containers. I was horrified at the prospect of purposely letting a mold culture loose in my clean rooms, which could contaminate my mushroom cultures. With the utmost care, I used a needle to pull out a fragment of tissue, placed it quickly and gingerly into a malt-enriched agar-filled petri dish, and slapped the cover on tight. In a few days I noticed a whitish circle of growth emanating from the point of inoculation, which was soon followed by green spore zones. After transferring the mold several times, in effect creating downstream "generations," I saw something unusual—

a white wedge of growth fanning out from the green mold colonies. "I'll chase that white sector and see what happens," I thought, hoping that the white wedge of mycelium would have delayed sporulation. The result was a discovery that might retool the entire pesticide industry.

From this divergent whitish growth sector I grew some presporulating *Metarhizium* mycelium on sterilized rice—a standard spawn medium. I took a teaspoon of mycelium on rice from the laboratory, placed it into a plastic bag, and took it home. I exclaimed to my then-17-year-old daughter, La Dena, that we were going to trick the carpenter ants into spreading my stealthy spawn. Not wanting to place the mycelium directly on the carpet, I requisitioned one of the dishes from her childhood dollhouse. I placed the whitish *Metarhizium* mycelium on the dish and put it next to the ants' biggest sawdust pile, assuming that this was a nightly path used by the foraging carpenter ants.

Around midnight, en route to the bathroom, my daughter looked down at the dish and gasped. She spotted a swarm of ants all over the little glass dish. "Wake up, wake up! You have *got* to see this," she exclaimed. We snuck up to the doll dish and witnessed the ants picking up the kernels of myceliated rice, one at a time, and retreating into the dark, damp walls of our incrementally collapsing house. Each ant became a distribution vector for the lethal mycelium, a point of inoculation, promptly infecting the nest. A couple of weeks later, my old, decomposing farmhouse was free of carpenter ants and was never reinvaded. I had discovered something amazing: the mycelium of parasitic fungi prior to sporulation acts as a Trojan horse, attracting its insect victims. My mind raced with ideas about the implications.

By good luck, I told a friend, a bemushroomed patent attorney, about my carpenter ant experiment. I asked him if I could patent my discovery. With a twinkle in his eye, he answered, "I think you may have something patentable here. Let me look into it."

I hypothesize that the house, once treated, has repelled subsequent invasions because the parasitized insect carcasses became moldy from the repellent spores, thus warding off other carpenter ants. Once a wood structure is treated this way, the moldy fungi provide a resident protective shield.

The Emergence of Mycopesticides

Swarming over the planet are legions of insects whose diversity boggles the imagination. Current estimates of insect diversity range from 4 to 6 million species (Kirby 2002), while fungi hover between 1 and 2 million species (Rossman 1994). According to Dr. Roger Gold, an entomologist at Texas A&M University, about 5 percent of insects are economic pests.

However, biodiversity across the planet is in serious jeopardy. According to an article in the journal *Nature*, 15 to 37 percent of current species could die out by the year 2050 (De Siqueira et al. 2004). Human activity is at the root of this emergency. Cutting down forests and burning fossil fuels cause global warming. Pesticides, herbicides, heavy metals, estrogenlike synthetics, and radioactive waste inject doses of toxins into the biosphere. At what point in this crisis of declining diversity will the survival of the human species be in doubt? Will the ecosystem sustaining human life collapse when we lose half of the world's species? Or 90 percent? We are falling down a slippery slope, with the interests of short-term politics and commerce muffling the voices of alarm echoing from the scientific communities.

Pesticides were invented to fight destructive insects and protect crops and structures. However, many of the chemicals used in pesticides, especially the organophosphates, harm nontargeted organisms, pollute water, and impair human health. The toxicity of certain chemical pesticides was not widely known until tons had been dispersed into the environment. Chemical pesticides permeate the food chain: most Americans, and indeed most people on the planet, accumulate these compounds and their derivatives in their fatty tissues and bloodstream. The need for alternative, nontoxic pesticides is critical, since the med-

ical and ecological impact of toxic pesticides poses a cascading health hazard and a global threat to our biosphere. Fortunately, the EPA has recently banned many of these, and as of February 2004, more than 50 countries—signatories of the Rotterdam Convention treaty—also limited toxic pesticides. Finding a replacement for environmentally dangerous pesticides will bring us one step closer to preserving biodiversity and saving the ecosystem.

The search for ecologically rational methods to control insects has focused on biopesticides—nature-based remedies that cause only negligible collateral damage, or none at all, to other organisms. Biopesticides, especially select fungi, do not pose a persistent threat to the environment after use, in contrast to many conventional chemical treatments that cause long-term damage.

The interrelationships between fungi and insects are complex and are only now being understood. Practically all insects engage, consume, or succumb to fungi. However, not all fungi are insect friendly. We know that thousands of species of fungi (called *entomopathogenic* fungi) attack thousands of insect species. This dance between predator and prey has led to the evolution of survival strategies for both, some of which border on the bizarre. By exploring these relationships, we may be able to adapt new techniques for controlling insects without harming the environment.

In 1834, Agustino Bassi noticed that spores of the fungus *Beauveria bassiana* were causing the disease muscardine, a plague that imperiled the international silk trade. He is credited with conceiving "germ theory," a major tenet of modern medicine, well before Louis Pasteur discovered the role of microbes in 1858. As more entomopathogenic fungi were observed, often found on the moldy carcasses of dead insects, the pesticide industry explored the use of fungal spores as natural insecticides.

Since the 1990s, several patents have been awarded exploiting these mold fungi, raising expectations for treatments in the emerging field of entomopathogenic mycology. Most of these mycopesticidal patents target an insect species using a fungus strain, with or without an effective delivery system. Species in the genus *Metarhizium* (especially *Metarhizium anisopliae*), a green mold fungus highly prevalent in soils; in the genus *Beauveria* (especially *Beauveria bassiana*); and some species in the genus *Paecilomyces*, hosting many white mold fungi, have earned the most attention from researchers. When insects come into contact with spores of these entomopathogenic fungi, the spores attach to the insects and germinate, boring hyphal pegs through the insects' exoskeletons using chitin-dissolving enzymes. Other portals of entry include the respiratory tract, anus, and mouth. Once inside, the mycelium forks and runs through the internal organs, interfering with the creature's metabolism and causing malaise, necrosis, and death in a few days. The insects, looking mummified with fuzzy mycelium, then become a launching platform for further sporulation (see figure 113). With some species of *Metarhizium*, *Beauveria*, *Hirsutella*, and *Paecilomyces*, a tiny club-shaped mushroom classified in the genus *Cordyceps* can sprout from the dead insect carcass (see figures 114 and 115). This is an example of fungal dimorphism: this organism can express itself in either of 2 forms, as a mold or as a mushroom.

One species of *Cordyceps* (*Cordyceps lloydii*), when it infects a carpenter ant, compels the insect to climb

▲ **FIGURE 114**

Cordyceps myrmecophilia mushroom fruiting from the carcass of carpenter ants of the genus *Camponotus*. Some species of *Cordyceps* can express a mold state such as in the green mold, *Metarhizium anisopliae,* which produces spores highly pathogenic to ants, termites, locusts, mosquitos, and mites.

▲ **FIGURE 115**

Cordyceps lloydii fruiting from a carpenter ant *(Camponotus* sp.) in Costa Rica. The ant, once infected, has the irresistible impulse to climb to the top of the jungle canopy. Once it has ascended and locks its mandible into a leaf, it dies, whereupon a mushroom sprouts from its carcass. This behavior ensures that the spores of the mushroom will be spread far and wide by the winds.

◄ **FIGURE 113**

Mycelium emerging from a carpenter ant, *Camponotus modoc.* This ant ingested the pre-sporulating mycelium and became mummified. The infection here came from within.

to the canopy of the Costa Rican cloud forest where it resides. Once it has ascended, the ant locks its mandible into a leaf and dies. A *Cordyceps* mushroom erupts out of its head (see figure 115). By releasing as-yet-unidentified chemical compounds and compelling the insect to ascend to the top of the forest canopy, the mushroom can release spores that fly far and wide, ensuring wider distribution than would be possible nearer the ground. Many such fungi use insects as vehicles for distribution and, later, their carcasses as centers for spore production.

With social insects—those that have a queen, such as ants and termites—the partitioning of fungi-infected

▼ **FIGURE 116**

The *Cordyceps-Metarhizium* life cycle.

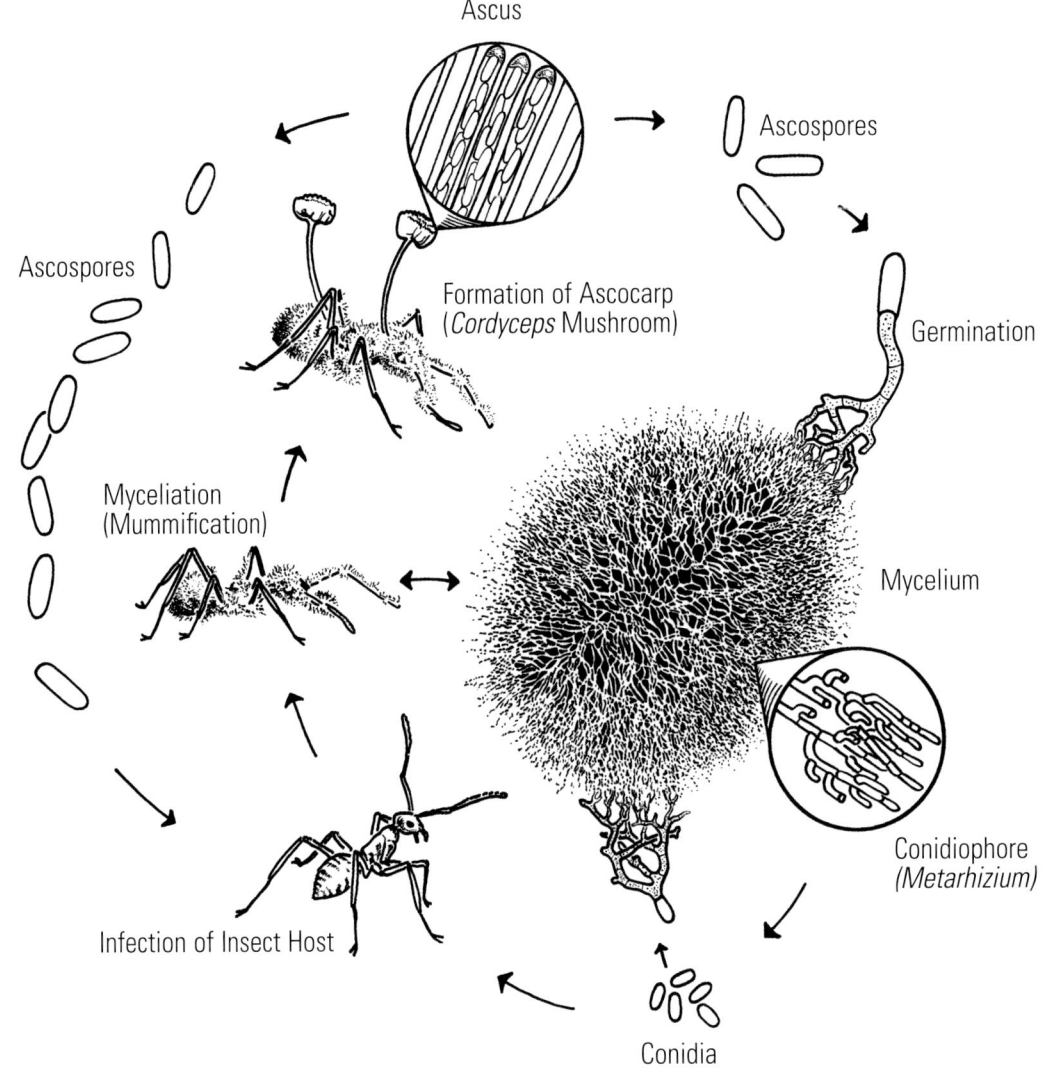

Ascus

Ascospores

Ascospores

Formation of Ascocarp (*Cordyceps* Mushroom)

Germination

Myceliation (Mummification)

Mycelium

Conidiophore (*Metarhizium*)

Infection of Insect Host

Conidia

members is critical for the nest's survival. The populations of some termite and ant nests number into the millions. In order to protect these large nests, the insect communities set up a sentry system guarding the queen and her brood. Some entomologists have termed these nests "factory fortresses," with soldiers throughout the nest on high alert to prevent entry from enemies, including infected members. If an infected individual is recognized, it is promptly killed and placed in a graveyard away from the nest. In other words, the insects know that the plague is nearby when they sense or smell it, and they immediately mobilize to prevent infection of their colony. Millions of years of common evolution have allowed fungi and insects to learn to recognize each other and engage in a constant biochemical dialogue.

Worldwide, termite species number at least 2,500, of which about 150 species can attack buildings, but within those 150 species, only a few cause widespread economic losses. There are nearly 600 species of carpenter ants, of which about a dozen attack buildings. Carpenter ants are nature's way of recycling wood. Carpenter ants and termites dance with fungi in debris fields—and wooden buildings. Aboveground colonies of Formosan termites (*Coptotermes formosanus*) can contain 500,000 to 3.5 million individuals; this species does not need to forage outside the infested structure, unlike its subterranean cousin *Reticulitermes flavipes*.

The damage, primarily to buildings, caused by native subterranean termites in the United States alone exceeds $10 billion per year, while the damage from the imported Formosan termite exceeds $1 billion annually. Whole neighborhoods, and even entire cities such as New Orleans, are being consumed. That city loses $300 million per year to termites, with forecasts of greater damage as the colonies penetrate deeper into the city's wooden infrastructure.

On an individual level, the threat of losing one's house—for most people their largest single investment—can drive a homeowner to spend whatever is necessary to prevent damage by insects. This drive to protect one's home has been a major economic stimulus for the pesticide industry, and a global market for

insect-defense measures and cures continues to spur the invention of new pesticides.

The environmental persistence and inherent toxicity of conventional pesticides have only recently been recognized. The market value of the pesticide industry in the United States has swelled to $9 billion; of that, 5 percent or $450 million is from biopesticides. The biopesticide market share is increasing 15 percent per year.

The pesticide industry has traditionally employed monitoring stations to capture and identify the insects infesting wooden structures. Once the type of invasive insect is identified, a treatment strategy is chosen. A current practice, declining in popularity, is to tent and fumigate a wooden building with a dose of toxic gas, or to heat it with portable propane-fueled heaters. These types of techniques succeed temporarily at best and are generally disappointing since the bugs usually come marching back. Plus, chemical treatments can be toxic to humans, other animals, and the environment.

The use of entomopathogenic fungi as a biological weapon against insects has many advantages but also faces obstacles. Several companies have expended a great deal of money to develop spore-delivery systems, especially using the entomopathogenic green mold *Metarhizium anisopliae* and the white mold *Beauveria bassiana*. However, spraying an entire building with spores is impractical and unlikely to assassinate the queen of the colony. Although captive insects are easily infected with spores in the laboratory, insect communities in nature have developed sophisticated strategies for defending themselves against contamination. In groups of social insects, spore-carrying workers are refused entry to the nest, with several tiers of guards preventing any disease-bearing insects from entering.

Throughout the late 1980s and 1990s, pesticide researchers tried to place spores in bait or treatment stations—fine-mesh brushes at portals—where entering insects would pick them up. This way of killing insects looked promising at first, since entomopathogenic spores quickly penetrate the exoskeletons of termites or ants upon contact. But the colony's aversion to pathogenic spores and the insects carrying them stymied commercialization of these insect control stations.

Although incidental contact with free-flying or soil-borne spores may succeed to infect individual insects, natural selection has given a decided advantage to insects that can detect entomopathogenic spores and avoid them. A fungus that would always alert its desired target insect with smelly spores would be selected against unless there was an alternative strategy for tricking the insects to engage the fungus. I surmised that since the insects can detect infectious spores, essentially "knowing" a mold plague when they sense, or possibly smell, one, the fungus also "knows" that the insects "know." I realized that I could create a temporary sporeless state in *Metarhizium* or *Beauveria* that not only lacks the repellent smell but also actually contains an attractant.

In the course of growing *Cordyceps sinensis*, a fungus traditionally used in Chinese medicine, I was intrigued by the idea that it could also serve as an insect pathogen and that it expressed different forms of mycelium. My curiosity led me down a path of discovery not taken by other researchers in the field. When I first obtained cultures of the green mold *Metarhizium anisopliae* for my experiments with carpenter ants, I selected white-wedge sectors of mycelium that lacked or had delayed spore formation, and I made descendant cultures (see figure 117). After successive transfers of the selected presporulating mycelium, its whitish form soon dominated. In culturing these mold fungi, and through further generations of transfers, I isolated a nonsporulating mycelial phenotype. These sectors initially appear as white V-shaped wedges of growth that gradually lose the ability to produce spores. I pursued these presporulating growth forms in vitro and soon discovered that the sporeless mycelium, still lethal to insects, emits attractants and feeding stimulants.

Although it is possible that ants and termites might recognize, over time, that the preconidial mycelium is pathogenic, its occurrence in nature is rare compared to the sporulating state. That we have many entomopathogenic fungi to tap into, essentially

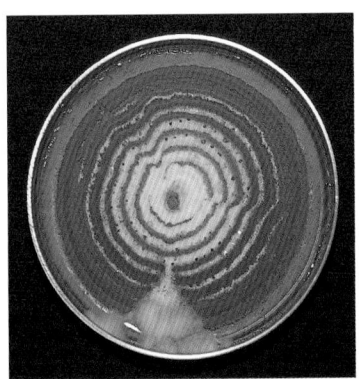

▲ FIGURE 117

This strain of *Metarhizium anisopliae* shows the emergence of a white wedge sector, which when subcultured, leads to strains that are delayed in sporulation. The resultant presporulating mycelium attracts many species of pest insects that unsuspectingly consume and spread the infection.

▲ FIGURE 118

After subculturing, the mycelium enters into a presporulating state from its parental green mold form, opening up reservoirs of enticing feeding stimulants and attractants to ants and termites. These two culture dishes are in fact the same strain. With many cultures, the white forms peter out as they are subcultured over time.

meaning that moldy insects give us a constant reservoir of new strains, makes recognition by the insects difficult because we can constantly switch strains.

Since insects avoid known lethal spores, attracting them to presporulating insecticidal fungi has obvious advantages, including that the insects themselves could become agents for dispersing this insecticidal fungus, which could later sporulate, throughout a colony.

Soon after my pilot test on my crumbling wood-framed house, I initiated a series of research trials at Texas A&M University's Entomology Department under the guidance of Dr. Roger Gold. These trials showed that Formosan termites *(Coptotermes formosanus)*, eastern subterranean termites *(Reticulitermes flavipes)*, and fire ants *(Solenopsis invicta)* were attracted to the mycelium, fed upon it, and carried fragments back to the nest. In 2 to 3 weeks, they died from fungal infection. In choice tests, termites preferred the mycelium to wood, and in one alarming example fire ant workers enthroned the queen on a bed of mycelium, where she soon expired along with her subjects. With termites, it appears that the mycelium kills the large protozoa in their digestive tract, which thwarts their ability to metabolize cellulose and starves them. The insects sicken and become susceptible to infection by the fungus, both internally and externally.

Furthermore, we found that even water and ethanol extracts of the presporulating mycelium on rice yielded a powerful attractant and a feeding stimulant. Through serial dilutions, we discovered that our baseline extract worked best after being diluted with water, thus reducing production costs. And I hypothesize that native-born fungi from an already infected termite colony are more attractive to that species of termite than to others. In many cases, *Metarhizium* strains from one insect species' colony did not attract other insect species. We also determined that only some strains of *Metarhizium anisopliae* produce species-specific attractants, which may lead to the design of products to target specific insect pests.

This discovery—that the extracts made from presporulating mycelium of entomopathogenic fungi—

▲ **FIGURE 119**

Ants swarming on presporulating mycelium of an entomopathogenic fungus. After contact, the ants become infected and spread the mycelium which subsequently regrows, resulting in their death.

▲ **FIGURE 120**

Here, when an extract of a presporulating entomopathogenic mycelium grown on rice is placed onto cellulose, termites tunnel to it, ignoring three nontreated controls.

may have greater implications. I also found that *Beauveria bassiana* produces attractants prior to sporulation. Many other genera of entomopathogenic fungi—*Hirsutella*, *Mucor*, *Paecilomyces*, *Aspergillus*, *Nomuraea*, and species in the order Entomophthorales, for example—are likely to produce similar attractants, feeding stimulants, and trail-following agents for the insects they target.

Follow-up research at Texas A&M determined that the presporulating mycelium of *Metarhizium* and *Beauveria* showed other unexpected properties. Termites, seeking wood, burrowed to the location of the mycelium and stopped. The entomologists viewed this "arrestant" behavior as a unique benefit of our natural pesticide. After contacting most other pesticides, the insects continue on their destructive path, but with our method they penetrated no further. Additionally, termites that contacted the mycelium then recruited other individuals to travel to the presporulating mycelium, who, in turn, solicited even more insects. The compounding effect of recruitment means that a little mycelium can go a long way, as opposed to contact poisons such as pyrethrins, which quickly kill the insects before they can return to the colony and thus must be used in large amounts.

This discovery of how to use the green mold *Metarhizium* for controlling insects is a splendid example of "green chemistry" at work: a fungus can entice an insect to carry and cache it as food before it is recognized as a pathogen. The active constituents causing attraction, feeding, dispersion, recruitment, and eventual death have not yet been molecularly characterized. Characterizing these behavior-modifying agents may not be as simple as discovering, for instance, a single antibiotic. We may eventually find that, rather than a single molecular compound causing such diverse effects, suites of synergistic compounds work in concert, like a combination to a safe. Time will tell.

The benefits of this discovery, the economic and ecological impact, may prove extraordinary. *Metarhizium anisopliae* does not infect plants, mammals, fish,

bees, or other beneficial insects. For this reason, the EPA has encouraged the study of *Metarhizium anisopliae* as a biopesticide. Other benefits include the following:

- replacing toxic pesticides with an effective natural method for eliminating termites, ants, and flies
- protecting groundwater and habitats from contamination by toxic pesticides
- recruiting the pestilent insects to implement their own destruction
- using the genome as an indigenous source of new strains, thus limiting tolerance buildup
- minimizing potential harm to nontargeted insects while precisely targeting unwanted insects
- providing attractants to the pesticide industry that allow insect contact with spore- and chemical-based treatments, lowering the amount required to be effective
- affording long-term protection of treated sites because, after treatment, spores repel future invasions
- allowing use of structural building materials otherwise susceptible to insect destruction, reducing the need for termite-resistant tropical hardwoods from rain forests, for example, while retaining jobs and helping the U.S. wood products industries
- saving diverse biological communities

On December 9, 2003, with the help of my attorney friend, I received approval from the U.S. Patent Office for a patent on this biotechnological breakthrough for attracting and killing Formosan termites (*Coptotermes formosanus*), eastern subterranean termites (*Reticulitermes flavipes*), and carpenter ants (*Camponotus* species). On October 17, 2006, I was awarded a second patent using these presporulating fungi for controlling all social insects, which are estimated to be nearly 200,000 species. I have pending an expanded set of divisional patents—basically derivative patents from my first patent—within the Continuation-in-Part ("CIP") application. Recent trials show that fungus gnats (phorid and sciarid species and allies) and blowflies (*Calliphora* species) are attracted to this mycelial extract, which strongly suggests that this technique may also be effective against non-social insects. In addition to *Metarhizium anisopliae*, several other entomopathogenic fungi show similar activity using the patent-pending techniques in the expanded Continuation-in-Part patent. The future of this green mycotechnology looks highly promising for treating insects that undermine structures, decimate crops, and carry diseases.

This type of green technology confronts the paradigm articulated by the motto of the 1950s and 1960s, "Better living through chemistry." Clearly, the chemical miracles worked in the mid-twentieth century had a devastating impact on environmental health. Big profits trumped good science, and as a result, future generations will have to deal with more pernicious forms of pollution.

Any revolutionary technology, such as the techniques described in this chapter, will be seen as a threat to conventional industry. Therefore, I follow these guiding principles when signing industry licenses for my mycopesticide patents:

- To develop effective and environmentally safe methods for controlling insect pests utilizing fungi. My philosophy is not to wage war against the insect kingdom but to enlist fungal allies for the intelligent, natural, and localized control of targeted insects when and where they threaten people, buildings, jobs, or the environment. We seek balance, not extinction.

- To license the mycopesticide patents to those who will best promote my nature-friendly alternative to toxic chemical treatments, and not to license this

technique to those who would suppress or delay its use in order to maintain consumer demand for toxic chemical treatments.

- To respect the intellectual property rights of all peoples, and to commercialize fungal strains traditionally used by native peoples in natural pest-control techniques, but only with their consent and participation. I am opposed to "biopiracy."

- To encourage the use of environmentally safe mycopesticides in impoverished countries to control previously uncontrolled pests, or to reduce the use of dangerous chemical pesticides; to provide financial support to nonprofit organizations that can manage and promote the use of mycopesticides in poverty-stricken communities and environments that are at risk from pests.

- To respect the sanctity of all species, to preserve and honor biodiversity, and to protect the environmental health of all inhabitants.

Since entomopathogenic fungi, especially in the form of mycelium, can be projected via many carriers— wood chips, sawdust, paper, cardboard, biodegradable fabrics, and many agricultural waste products—I envision landscapes customized with mycelial matrices. These would be populated by fungi that target specific insects and the diseases they carry, forestall or prevent pest outbreaks associated with livestock and farms, and stop beetle blights from spreading across forests—all while preventing and cleaning up pollution.

Part III

GROWING MYCELIA AND MUSHROOMS

Key to growing mushrooms, whether for deploying mycorestoration strategies or eating or using them medicinally, is to first grow mycelia—the subject of the final part of this book. Mycelium is the cellular fabric of our food web and although pervasive in nature, getting a mycelial mat to infuse through a virgin habitat is both an art and a science. As cultivators, it is our job to help mycelia navigate through the highly competitive microbial universe, one habitat at a time, always keeping hungry parasites at bay. Thankfully, all of the challenges that come up when growing mycelium, even those early hurdles (choice of strains, how to get the mushrooms into culture, and how and where to project the mushrooms), can be overcome using the techniques described in the following chapters.

Most important of the techniques to come is how to use "natural spawn," that is, how to transplant and nurture wild spawn for mycorestoration projects. Wild spawn has the major advantage of being already acclimated to habitats teeming with competitors. Conversely, pure culture spawn, bought from commercial spawn suppliers, must either be implanted into clean substrates or slowly adapted to the complex microbial ecosphere that surrounds us.

Natural spawn can be made by transplanting wild patches of mycelium, from germinating mushroom spores, and from regrowing stem butts. For many, buying and using pure culture commercial spawn can speed up the early steps of the process, but the mycelium ultimately fairs better if it is inherently fortified against the onslaught of wild microbes eager to consume it. In the end, taking the time to create natural spawn will prevent a lot of very time-consuming troubleshooting that goes hand in hand with starting from pure culture spawn. For the stubborn pure culture spawn devotees however, there exists a compromise: Commercial spawn can be naturalized through a process analogous to vaccines—it can be brought into contact with a less-than-virulent dose of the microbes found in the destined habitat. The mycelium's natural antibiotics preselect microbes that help it survive in the wild. A synergistic blend of microbes joining with the mushroom mycelium provides a defensive shield, much like fortifying an immune system, that not only forestalls competition but streams nutrients and mycelial stimulants. Once the pace of mycelial growth quickens—regardless of its beginnings—mats can become quite large; in the extreme, they can cover thousands of acres.

Remember our mycomotto: Move it or lose it! A mycelium grower is really a mycelial herdsman; no matter how successful you are in getting mycelium to adapt and grow in one habitat, that success is a temporary episode in the theater of life. Mycelium consumes its preferred habitat resources and then strategizes for transporting itself to new niches. Unless the mycelium is recharged with basic nutrients, it will move on as it transforms debris fields into soil.

Mycelium is, in essence, a digestive cellular membrane, a fusion between a stomach and a brain, a nutritional and informational sharing network. It is an archetype of matter and life: our universe is based upon these networking structures. Your job is to become embedded into the mind-set of this matrix and use its connections for running with mycelium.

CHAPTER 9

Inoculation Methods: Spores, Spawn, and Stem Butts

This chapter describes methods of inoculating by using spores, spawn, or stem butts. Although each method can generate mycelium and eventually mushrooms, my preferred method for outdoor colonization is to combine uses of spores, spawn, and stem butts, techniques I will explain as I progress through the chapter. The wisest method for generating mycelium is the one that works. My experiences of success and failure may differ from the experiences of someone in India, for instance. Climate makes a huge difference, as do the seasons.

Some Thoughts on Failing

Over the years, I have learned the value of attempting that which seems unlikely to succeed. Several times, I thought I had failed only to have the mycelium surge and fruit later. My failures oftentimes become my successes. But with every "failure," if I have paid attention, I hone my skills and sensitivity to the mycelium's needs. Hence, another of my mottos: Every failure is the price of tuition I have paid to learn a new lesson.

Your method of inoculation will depend upon your access to spawn or mushrooms. Buying pure culture spawn has its advantages, but in this chapter, which could be titled "Liberation Mycology," I'll show you how to use nature as your source for mycelium. The materials upon which mycelium is grown or inoculated are called *substrates* by mush-

room growers. Substrates usable for mushrooms are extraordinarily varied, although most come from by-products from forestry and agricultural practices (see chapters 11 and 12).

Several factors drive the decision-making process. The habitat you want to inoculate predetermines which mushrooms are suitable. Your ability to obtain mycelium—in the form of spores or commercial spawn, or by gathering naturally growing mycelium—also influences your decisions. Other factors—experience, labor, ease, and commercial or recreational intent—come into play. Because so many methods are now available, you have many alternatives to choose from. I have learned to make the wisest choices by listening to the mycelium.

Musings on Using Spores: The Fairy Dust of Mushrooms

Frank Herbert, the well-known author of the Dune books, told me his technique for using spores. When I met him in the early 1980s, Frank enjoyed collecting mushrooms on his property near Port Townsend, Washington. An avid mushroom collector, he felt that throwing his less-than-perfect wild chanterelles into the garbage or compost didn't make sense. Instead, he would put a few weathered chanterelles in a 5-gallon bucket of water, add some salt, and then, after 1 or 2 days, pour this spore-mass slurry on the ground at the base of newly planted firs. When he told me chanterelles were growing from trees not even 10 years old, I couldn't believe it. No one had previously reported chanterelles arising near such young trees, nor had anyone reported them growing as a result of using this method.* Of course, it did work for Frank, who was simply following nature's lead.

Frank's discovery has now been confirmed in the mushroom industry. It is now known that it's possible to grow many mushrooms using spore slurries from elder mushrooms. Many variables come into play, but in a sense this method is just a variation of what happens when it rains. Water dilutes spores from mushrooms and carries them to new environments. Our responsibility is to make that path easier. Such is the way of nature.

Frank went on to tell me that much of the premise of *Dune* — the magic spice (spores) that allowed the bending of space (tripping), the giant worms (maggots digesting mushrooms), the eyes of the Freman (the cerulean blue of *Psilocybe* mushrooms), the mysticism of the female spiritual warriors, the Bene Gesserits (influenced by tales of Maria Sabina and the sacred mushroom cults of Mexico) — came from his perception of the fungal life cycle, and his imagination was stimulated through his experiences with the use of magic mushrooms.

Spores: How Mushrooms Travel Distances

Mushrooms reproduce through spores that can travel great distances via water, by air, or by hitchhiking on other carriers. For instance, many species of mushrooms use insects to carry spores. The scent of the birch polypore (*Piptoporus betulinus*) attracts beetles that then burrow into the mushrooms, piercing the spore-rich underlayer to feast on the soft internal flesh. In doing so, the beetles become covered with spores.

---

* *Eric Danell and Francisco Camacho first reported the growth of chanterelles with a seedling of* Pinus sylvestris *in 1997.*

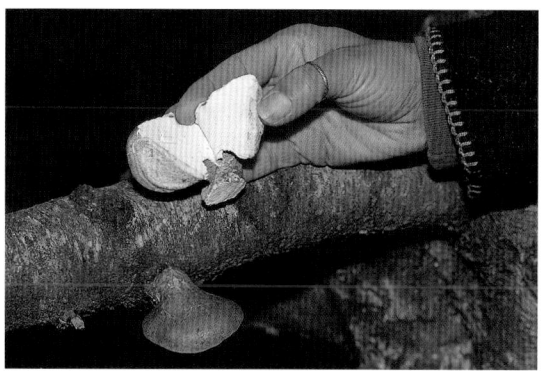

▲ **FIGURE 121**

The birch polypore *(Piptoporus betulinus)* growing on birch in Germany. This species is a potent medicinal mushroom. See pages 40, 45, 48, 275-277.

When the beetle travels to another tree and burrows into its bark to lay eggs, the tree is inoculated with spores of the birch polypore. The emerging fungus provides the beetle's developing larvae with food. The tree softens as mycelium grows into it, and soon woodpeckers arrive in search of beetle grubs and other insects that are attracted to the myceliated wood. Once the woodpeckers leave, carrying spores with them, other birds and insects take up residence. An entire ecosystem spirals from the pockmarked tree, which can become a launching platform for more fungi, insects, and birds. Many other polypore mushrooms coexist within such a tree. And not only do beetles and birds spread mushroom spores, but bears do too. I have seen scratchings on trees in the Hoh River Valley in Washington State where bears scratch trees to mark their territories and in the process, create apt habitats for spore entry (see figure 75). Mycelium and mushrooms sprout from these scratchings, attracting more beetles and bugs, and the process begins again. These intersecting cycles sustain biodiversity.

Mushroom spore casts can initiate satellite colonies of mycelium, forming from a few feet to several hundred feet from their parental sources. When spores are dispersed, density of spores decreases

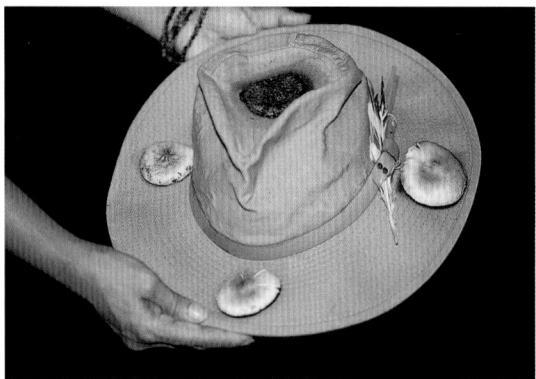

⋀ FIGURE 122

Mushroom spore prints on a hat, when you wear it you'll trail invisible spores behind you like fairy dust. From these spore trails mushrooms can emerge long after the traveler has moved on. Mushrooms use us as vehicles for transporting and dispersing spores.

⋀ FIGURE 123

Clothing is also a good fabric to make spore prints on. Such "mushroom wear" is increasing fashionable amongst a group of West Coast mycophiles.

exponentially—as an inverse square of the distance—from the parent mushroom. Since, in most cases, the mycelia from 2 compatible spores must merge before fertile mycelium can be created (see chapter 2), it becomes increasingly unlikely that 2 spores will join the further apart they land. This may be one of the reasons mushrooms produce so many spores—to guarantee the widest possible matchings distant from the parents.

One clever method mycologists have developed to help them determine whether a mushroom species is hiding in a habitat without actually finding a specimen is to germinate a single spore of the species in question in a petri dish on sterilized media. This monokaryon can be used as a "spore trap" to catch compatible spores in its cellular net (James and Vilgalys 2001). The mycelium germinating from a single spore is a monokaryon and typically infertile—meaning that it cannot form a sporulating parent mushroom until it sexually combines with another compatible spore of its own species. By placing mono-spore-generated mycelium-covered petri dishes in the woods and exposing the culture to the air, the research mycologist can wait to see whether compatible spores from difficult-to-find mushrooms land and mate with

⋀ FIGURE 124

These hats, being used as temporary vessels for carrying mushrooms (in this case *Psilocybe azurescens*) are impregnated with billions of spores, allowing for the spread of this species as far and wide as the wearer travels.

the spores in the dish. If mating has occurred, in a matter of days the formation of clamp connections, distinct elbowlike cell structures will signify this (see figure 73). This ingenious method can tell you if a species grows in the ecosystem even if you cannot find the mushroom!

Bears, birds, insects, and animals are not the only carriers of spores; humans carry them too. Many mushroom species turn up on my property after I carry specimens home from forays. When I was studying the taxonomy of psilocybin-containing mushrooms at the Evergreen State College in the 1970s, I was amused to find that many of the species I collected from miles away began appearing in the wood chips around the laboratory. I soon realized that boots, bikes, books, and backpacks are convenient vehicles for spore transport; millions of opportunistic spores trail like fairy dust behind us. Airplane travel and mail service spread spores throughout the world. Even handshakes spread spores.

We are attracted to mushrooms for their myriad flavors and uses, and through our use of them we spread their spores. Chopping wood, trimming trees, gardening, cutting grass, playing with your dogs, frolicking with your kids, and cooking gourmet mushrooms—all of these activities make us effective vectors, allies unwittingly recruited by the mushrooms. Whenever you handle mushrooms, their spores stick to you. The wind blowing the spore dust off our bodies adds to spore mobility. I guesstimate that in the course of a day the average human accumulates between 10 and 100 million fungal spores on his or her body and clothes. Close contact with others in densely populated cities allows for more points of distribution, although the ambient mushroom spore load may be less than that of suburbs or forests, whereas the ambient spore load of molds may be greater, depending upon the state of decay of the city's infrastructure. In fact, everyone is a mushroom grower, from Manhattan high-rise dwellers to Pacific Northwest loggers, whether they want to be or not. We are all Johnny Appleseeds, or rather Johnny Mushroom spores, in the service of the fungal kingdom. Once you are aware of how spores work, you can use this knowledge to start your own mushroom gardens.

Collecting Spores for Inoculation

The air is a sea filled with invisible spores of microorganisms. In my backyard, using a high-tech laser particle scanner I commonly detect concentrations of airborne particles exceeding 3 million per cubic foot. Of these particulates, about a third of them fall within in the size range of most fungal spores, which typically measure 1 to 30 microns. Many of these airborne fungal spores compete for the same habitat, which hampers any effort to cultivate a single species. Success is a numbers game: you need to find a strategy that favors your chosen mushrooms and discourages competitors. Several tried-and-true techniques work for gathering enough spores for successful inoculations. The simplest method is to take a spore print.

Spore Prints on Paper: Spores for Spawn or Mushroom Art

When you place a mushroom onto paper, falling spores collect in delicate patterns, graceful and beautiful. The collection of these invisible spores amasses as a visible dust. You never know how a print will look until you remove the mushroom that has cast its spore dust.

The mushrooms discussed in this book produce spores from gills, pores, teeth, or folds. The garden giant (*Stropharia rugoso annulata*) and parasol mushroom (*Macrolepiota procera*) are gilled mushrooms. Reishi (*Ganoderma lucidum*) is a polypore mushroom, with pores instead of gills for sporulation. Monkey's head (*Hericium erinaceus*) has long, cascading spines or "teeth." Morels (*Morchella esculenta* and *Morchella angusticeps*) have deeply convoluted folds.

Mushrooms increase in their spore release as they mature from adolescence to adulthood. The length of time of spore release in a mushroom's life cycle varies from species to species. Spores released in the wild are easy to see as a dust on the ground or wood directly beneath the mushrooms. Sometimes taller mushrooms will cast their spores onto the caps of other nearby mushrooms.

Nowhere else in botany is color a key feature for separating so many genera. Although genera vary so much in classic spore color, every spore print is mandala-like (see figures 125–128). As art from nature, spore prints unveil one of the hidden mysteries of the mushroom life cycle. And making spore prints is not only fun—it is essential for identification.

The classic method for capturing a spore print is to place the mushroom, spore layer down, on a piece of paper. The spores then descend like a fine mist. These same spores can be used for culturing in the laboratory (see my book *Growing Gourmet and Medicinal Mushrooms* [2000a]) and for growing mushrooms outdoors. Here are the steps for taking a spore print from a gilled mushroom, such as a portobello or garden giant.

MAKING A SPORE PRINT

(Caution: Wear a dust mask when handling masses of spores. Anyone whose immune system may be compromised, is allergic to fungus, or suffers from lung disease should be especially careful. Wash your hands afterward.)

1. Go mushroom hunting. Before going on a foray, learn about the species you want to grow by enrolling in a class or picking up a field guide. Being a West Coast mycophile, my two favorite field guides are David Arora's *All That the Rain Promises and More* (1991) and *Mushrooms Demystified* (1986). On your foray, choose a mushroom specimen, preferably one that is approaching maturity. With many gilled mushrooms, this means that the mushroom cap is convex in shape but not flat. Once mushrooms mature to a flattened state, spore production usually decreases, and in some cases the edges will upturn at full maturity. If you prefer, order a mushroom-growing kit from a mail-order company like mine, Fungi Perfecti (www.fungi.com). An oyster mushroom kit weighing 8 pounds will produce around 1 to 2 pounds of mushrooms and, correspondingly, 1 to 2 g of spores (approximately 1 to 2 billion).

2. If the gill color is not white, use a white sheet of paper. Typing or photocopy paper works well. If the gill color is white, as is the case with shiitake or oyster mushrooms, use a colored piece of paper.

3. Sever the cap from the stem, and place the cap, gills or pores down, on the piece of paper. Caps of larger mushrooms do not need to be covered. Smaller mushrooms may dry out, hindering spore release, so it is best to cover with a cup or bowl to lessen the rate of evaporation.

4. Allow the sporulating mushroom to sit for at least 12 hours. When you carefully lift up the cap, you

▲ **FIGURE 125**

Spore printing *Agaricus brasiliensis,* the almond portobello, on paper. Note the spore trails, which follow air currents.

▲ **FIGURE 126**

Spore prints can be made in a journal when traveling.

will discover the spore pattern that has collected on the paper. The spores fall in a pattern mirroring that of the gills or pores and, in the case of gilled mushrooms, pile up in the form of ridges. You can use these for cultivation or as decorative art. If you wish to keep your spore prints as artwork, spray them with the same aerosol fixative that artists use to protect chalk drawings. After being sprayed, these fun-to-make nature portraits are unusable for cultivation and are not allergenic.

With most species, you can create additional prints by moving the cap to a clean space on the paper and waiting for another 12 hours or so. Some caps will yield up to 6 distinct prints over a week before the mushrooms dry up or succumb to bacteria, other mold fungi, or maggots. If the mushroom seems wet, replace the paper daily to remove the excess moisture. Or you can place the mushroom on a bottle cap so it stays elevated off the paper. These spore prints tend to be smudgy and less distinct. If you want to grow mushrooms from the spores on the print, then place the dried print in a plastic resealable bag until needed. Write down the date, mushroom species, geographical location where collected, and any other notes. Keeping

a specimen for future identification is always advisable, since identification of mushrooms by spores alone is very difficult unless their DNA can be analyzed.

Alternatively, you can make a spore print (with spores of any color) on a clean pane of glass—my favorite method. Once spores have fallen (step 4, above), carefully remove the cap, overlay with another pane of glass, and tape the edges all the way around. The result is a glass-enclosed "spore gallery" that can be stored at room temperature for years. Storing spores this way allows them to be easily observed without the risk of contamination by extraneous airborne spores, while the glass protects against damaging UV waves. To retrieve the spores for future use, just cut the binding tape, open the spore gallery, and use a razor blade to scrape the spores into a pile.

Spore prints made either on paper or glass, once sealed, can also be sent through the mail with few ill effects.

Spore prints come in all of the colors of the rainbow. Within most genera of mushrooms, the color is consistent or varies only slightly. For instance, oyster mushrooms have a whitish or sometimes a light lilac color. Most field guides for identification are organized using spore color as a primary distinguishing feature.

⋀ FIGURE 127

Spore printing is fun!

⋀ FIGURE 128

Making spore prints of the parasol mushroom (*Chlorophyllum rachodes*) on clean glass.

Among the better field guides are David Arora's *All That the Rain Promises and More* (1991) and *Mushrooms Demystified* (1986), Gary Lincoff's *The National Audubon Society Field Guide to North American Mushrooms* (1981), and Alan and Arleen Bessette's *Mushrooms of Northeastern North America* (1997). Among European guides, I like Rose Marie Dahncke's book, *1200 Pilze* (1993), which is currently available in German. *Edible and Poisonous Mushrooms of the World*, by Ian Hall, Stephen Stephenson, Peter Buchanan, Wang Yun, and Tony Cole (2003), is a good overall guide.

The Bag Technique of Collecting Spores

Some mushrooms cannot easily spore print on paper. In particular, morels and lion's manes are architecturally different from classic button-style mushrooms and can broadcast spores by expelling them in all directions. Collect spores from these mushrooms by enclosing them in a paper, wax paper, or plastic bag. For best results, keep the mushroom upright, as it is in nature, tenting it an upside-down bag so insects and other debris will fall out. A paper bag has the advantage of being biodegradable, whereas a plastic bag allows for a cleaner collection surface when you're amassing larger quantities of spores. However, plastic bags sometimes collect so much condensation that spores are rinsed down the sides in an unusable sludge. I prefer paper sacks, especially for morels. (The sacks can be sandwiched into cardboard to create cardboard spawn that can then be placed outdoors.)

Collecting Spores from Airstreams

Air passing over mushrooms carries spores. Studies by E. R. Badham (1982) show that some *Psilocybe* mushrooms, like many others, orient themselves into the wind, a response known as *aerotropism*. A pressure drop forms under the cap, allowing spores to be sucked into the ambient airstreams. And if you channel air past a sporulating mushroom into a chamber, you can collect many spores. Here are 2 ways of accomplishing this; there are no doubt more, so feel free to improvise:

1. Place a nylon air sock around the exhaust tube coming from a forced-air mushroom drier. As the mushrooms dry, spores from the drying mushrooms flow into the sock. Make sure your drier is set at the lowest possible heat setting so that drying does not occur too quickly.

2. Use an aquarium pump or a small peristaltic air pump to suck air from the spore-collection chamber with the sporulating mushrooms gills up, and aim the spore-enriched air directly into a paper bag, onto a pane of glass, or into water if you want to make a spore slurry—which in most cases would have to be used within 48 hours or frozen until needed. This method channels spores to a specific target. If the tubes clog with spores, use a probe to free the airflow. "Gazillions" of spores can be amassed with these methods.

If you want to gather spores from thousands of mushrooms, particularly oysters, shiitake, reishi, maitake, and other cultivars, I have an elegantly simple method: Contact your local grower of exotic mushrooms and ask for spores from their air ducts in their growing rooms or from their spore-laden air filters. Most cultivators, small or large, use plastic ductwork and coarse air filters in their ventilation systems. As the air is recirculated, the spores from mushrooms get sucked into the ducting and collect en masse. The ductwork can be taken down and the spores can be shaken into an airtight gallon jar; up to a liter of spores can be gathered from a 1,000-square-foot growing room every few months. Store the spores in a cool, dark place. (Note: Using a desiccant during storage will keep your spores dry, preventing premature germination, and reduce competitor molds that feed on the nutrient-packed spores. In fact, the spores of most temperate species, after a period of drying, can be frozen for many years. However, I have found that the spores of most tropical species, like their mycelia, die when frozen.)

Although the vast majority of the spores in the ductwork will be from the mushrooms being cultivated in the growing room, bear in mind that spores of outside fungi and contaminant mold fungi will inevitably be in the mix. The cleaner the crop inside, and the cleaner the incoming air, the more pure the collected spore mass will be. If more than one species is being grown, then obviously the spore mass will reflect the type of mushrooms sporulating in the growing room.

Harvesting Spores from Dried Mushrooms

This technique is more difficult than most others listed. Occasionally, I want to start a culture from a dried mushroom specimen. Dried specimens have plenty of spores on their gills, but many times the spores are not easily cultured when in contact with the gill flesh, soon contaminating when rehydrated. Using high heat in the drying process may lessen viability. Nevertheless, to release the spores, allow the gills to float in water, which will rinse the spores off. They will temporarily float before becoming saturated with water and sinking. Before they sink, you can skim the water's surface with a fine sieve to remove the spores.

▲ **FIGURE 129**

These two jars contain spores. One is full of reishi *(Ganoderma lucidum)* and the other has a mixture of maitake *(Grifola frondosa)*, shiitake *(Lentinula edodes)*, oysters *(Pleurotus ostreatus)*, nameko *(Pholiota nameko)*, and pioppinos *(Agrocybe aegerita)* collected from the plastic ductwork in a growing room every few months. The more than 2 pounds of spores depicted here are roughly equivalent to 1,000,000,000,000 (1 trillion!) spores—enough to inoculate tens of thousands of stumps! This is an undervalued product most mushroom farmers discard, not realizing its utility.

Collecting Spores with Variable Mesh Screens and Centrifuges

This is a fun technique for gathering spores, especially from subterranean trufflelike mushrooms like the mycorrhizal *Rhizopogon* and *Glomus* species, which live above the root zones as well as underground. Gather the mushrooms at maturity. Dry brush the outer layer using a piece of mattress foam and allow it to dry, a process that make take hours or days. The spores often mature as a defensive response to being plucked. Pulverize the dried mushrooms in a household food blender and place them on a stack of shaker screens with decreasing mesh fineness from top to bottom. Since most mushroom spores measure between 3 and 20 microns, finer screens will eliminate all but the smallest nonspore particles. Most of the time, 3 or 4 screens will suffice, ideally with the finest screen openings measuring .0007 inch or 18 microns. If this type of screen is not available, using a .007-inch screen (180 microns) will work, although some impurities will fall through. The coarsest screen can be of the window variety, but it should be rinsed with water and dried before use. For making ultra-pure samples, place the spore mixture into distilled water and place it in a centrifuge. Since spores have a fairly consistent color and specific gravity, distinct populations will soon settle out and form bands of discolorations. If your mushroom has a distinct spore color, you can remove the undesirable layers with a turkey baster or syringe until you come to the spore stratum. (Rinse well before drawing from within the spore-producing layer.) Although color will be a good indicator of which strata has the spores you desire, should there be any question, a quick look under a microscope can tell you if the shape and size matches your mushrooms. (Please consult one of the previously mentioned recommended mushroom field guides that have good microscopic descriptions.)

Isolating Spores by Electrostatic Fields

Mushroom spores tend to be electrically charged and so adhere to opposing charged surfaces. A coating of spores is most noticed on the walls in a room or on other dry, well grounded, positively charged materials. Electrostatic filters, widely used in homes, can be set up downwind of a mushroom kit to collect 99 percent of the spores produced from the mushrooms. Before trapping the spores, thoroughly clean the filter screens. After the filter is unplugged, the screen can be cleaned to collect the spores using a brush or, better yet, pressurized air from, for instance, an air compressor. Blow the spores into a suitable vessel such as a large plastic or paper bag. Wear a mask so you don't inhale spores.

How Many Spores Can a Mushroom Produce?

Oyster mushrooms have been reported to convert 50 percent of their mass into spores, a figure that seems high to me. Nevertheless, if this were true, a pound of fresh oyster mushrooms from a single mushroom kit would yield about 45 g of dry mushrooms. If the maturing mushroom released 50 percent of its mass in spores, or 22.5 g, then the output would be approximately 25 billion spores. Even if this estimate is off by a factor of 10 (in other words, if a kit produces only 2.3 g or 2.5 billion, a more realistic number), since there are an estimated 1 billion spores per g, measuring about 10 microns in length, then they would span 25,000 meters or 15.5 miles if laid end to end!

Perennial mushrooms produce many more spores than annual species like oysters. Arora (1986) estimated that a large artist conk (*Ganoderma applanatum*) can produce 5 trillion spores annually (see figure 20)! With spores averaging 8 microns in length, this means that a string of these spores would be 40 trillion microns, or 40,000 kilometers, or approximately 24,854 miles—just about enough to encircle the Earth.

This abundance of sporulation of just one species, when placed in the context of total fungal biodiversity, shows how infused our biome is with fungal DNA. Mushrooms are mycological geysers spewing spores by the billions and competing with each other to dominate an ecological niche.

A new astromycological unit, the amount of spores equivalent to encircle the earth, the Spore-Earth Unit (SEU), could be used, for instance, to calculate the outflow of fungal spores into space. The Earth is sporulating, seeding the galaxy with its germplasm. How many SEUs are leaving the Earth's atmosphere every day? Traditionally, scientists have thought about panspermia from outer space that could infect the Earth. Equally likely, the Earth is inoculating the heavens.

Germinating Spores

With most saprophytic mushroom species, germinating spores is not difficult. Spore germination depends upon the species, the material upon which they are grown, and the temperature. Each species reacts differently to different substrates. The genetic variability in spores allows for options in growth media and methods and is part of fungi's adaptive strength. This is nature's way of guaranteeing a mushroom's survival should its natural habitats suddenly change. Some mushrooms are highly adaptive to a wide range of substrates. Oysters, turkey tails, and *Hypholoma* fungi are amazingly aggressive. Their spores germinate with ease. On the other hand, in my experience, shiitake spores must dry first and then rehydrate before they can germinate en masse.

Spores may not germinate for any number of reasons. One problem with spore prints is that when the protein-rich spores concentrate, they become a fertile breeding ground for bacteria. Spores need to be diluted in order to tip the balance in favor of fungal growth and against competitors. In the laboratory, the common practice is to germinate spores under sterile conditions on nutrient-enriched agar media. See *Growing Gourmet and Medicinal Mushrooms* (Stamets 2000a) and *The Mushroom Cultivator* (Stamets and Chilton 1983).

Most mushroom spores love to grow on the moist surfaces of dead plants, especially when they're scattered like checkers on a checkerboard. Spores germinate quickly in water. The problem is that most life on this planet also loves water. So when you immerse spores in water, other organisms, especially bacteria and protozoa, consume the fungi as food. The art of cultivation is to give the mushroom spores a head start, in advance of competition, to initiate the ever-increasing spiral of germination. Ideal spore dispersal is a balance between the conduciveness of the habitat and its background population of competitors. Once a few spores germinate, adjacent spores are triggered into germination. The following descriptions outline a variety of successful techniques. You'll discover the best one for you and your mushroom through trial and error.

Spore Germination in Nutrified Water: A Sugar and Salt Broth

In most instances, fresh spores of saprophytic mushrooms contain enough nutrition to germinate in water without added nutrients. However, as they get older, age becomes a barrier to germination. Soaking spores in a sugar and salt broth often causes them to germinate more quickly than competitor spores. After 1 to 2 days, the actively germinating spore-mass slurry will be ready to transfer, so be ready. If you incubate it too long, the decline in spore viability will provide an opportunity for hostile competitors to grow and eclipse your chosen mushroom spores. The activity in this broth is a microbial race, and your coaching goes far toward determining who wins.

Spore prints from paper and glass are the easiest to use with this germination method. By scraping the spores off the glass or paper with a razor blade, you can collect about .5 to 1 g of spores (approximately 500 million to 1 billion spores) for use in your solution. The salt limits bacteria growth without stifling spore germination. To make a sugar and salt broth:

1. Add $1/4$ teaspoon noniodized salt and 1 tablespoon sugar or light molasses to 1 gallon of water in a large pot, and boil for 10 minutes. Remove from the heat.
2. As it cools, pour the broth into a clean, well-rinsed container made of plastic, stainless steel, or glass. Make sure the container has not held chemical products and/or milk.

3. Once the broth is cooled to room temperature, add 1 teaspoon (1 g, or $1/28$ ounce) spores. *(Use a dust mask and do not inhale while working with the spores! Wash your hands thoroughly. Avoid contact with spores if you have fungal allergies.)*

Alternatively, you can use the water left over from immersing 2 or 3 mature 4- to 8-inch-wide mushrooms, for example. However, spore broths made from spore prints tend to have fewer bacteria than those made from the spores from submerged mushrooms. Or, you can rinse spores off a mushroom's gills, allowing the spore concentrate to collect in a gallon of water. Never use souring mushrooms. Use only those that are pristine, firm but fairly mature, and growing rapidly. (Many mushrooms continue to grow after you pick them, as their flesh is transformed into a spore-generating material in a last heroic effort to produce offspring.)

4. Cover the broth immediately, and incubate it for 24 to 48 hours in a shaded, place with even temperatures between 50 and 80°F. Temperature preferences are species specific. (To zero in on the best temperature for germination, check a field guide that identifies the range of temperatures at which the mushroom fruits in the wild. Often, but not always, this will be in the temperature window for germination.)

5. During incubation, shake the broth vigorously twice a day. Depending upon the species and circumstances, a hyperactive state of germination is triggered and fine threads of mycelium, barely visible, float in the media. If the spores are old, fewer will germinate, which may mean incubation in water should be extended, but not more than 5 days (otherwise bacteria growth may overwhelm spore growth). Generally, you should use spores not more than 2 years old.

You have just created a gallon of spore broth—a liquid spore-mass slurry—useful for inoculations. Once the spores have sat in the water for 1 or 2 days, they can be broadcast onto a substrate. This is the simplest method of cultivation, but it is not the most successful in every circumstance. Whether you are successful in growing mycelium, and then mushrooms, depends upon many factors, some obvious and some mysterious.

USING THE LIQUID SPORE-MASS SLURRY FOR INOCULATION

When spores are collected under sterile conditions and then put into a sterilized nutrient slurry on a laboratory stir plate, a more technically demanding technique than the simple technique described above, spore germination usually occurs in about 3 to 4 days; by day 10 the mycelium is so thick it collects on the top in the form of a mat the size of a pancake. In nature, spores face more challenges, especially from other microbes. To limit the challenges to your spores, choose the cleanest and freshest possible substrate. If the substrate houses smaller populations of microbial competitors, and if enough spore mass is placed on the substrate, your attempts have a better chance of succeeding. Spore-mass slurries can be used for inoculating coarse substrates like straw; wood chips; grasses; and human-made materials like pellets for wood stoves, kitty litter, rugs, textiles, and cardboard. Spores have the advantage of quantity—many millions of points of inoculation. Their disadvantage is their fickleness—sometimes they work well, and other times not, depending on the species and the substrate. There are other ways of using spores.

Spores in Oils

Spores can be immersed in canola, corn, or safflower oil, which can be used as a lubricant for chain saws or other cutting equipment. As trees, brush, or plants are cut, the spore-infused oil distributes spores to the newly cut surfaces, an efficient method of transfer. Another advantage of using oils is that they help the spores stick to the surfaces upon contact and have less chance of being washed or blown away. I filed a patent on spores in chain-saw oil, but another inventor from Europe had already come up with a similar idea first; his idea was to introduce a benign fungus that would compete with a parasitic mold that infected shrubs after cutting. So the patent office rejected my patent application, but I'm reapplying in the hope of getting it approved, as my approach is to speed up decomposition and habitat restoration, a different but complementary goal to preventing competition.

We did a simple test using 10 million, 100 million, and 1 billion spores of oyster mushrooms in 1000 ml of canola oil. With 100 million spores in a liter, we had the best results in getting good mycelial growth using an estimated 10 ml of spored oils when cutting through an alder tree (*Alnus rubra*) about 6 inches in diameter. The solution of 10 million spores also did well, but it took another 2 weeks to catch up to wood slices that were inoculated with the richer, 100 million spores/liter. I am sure that lesser concentrations would also be effective, but probably would require additional time to achieve the same mycelial mat density. We plan to conduct more tests in the future.

▲ **FIGURE 130**

One of these spored oils was made especially for Ken Kesey and the Merry Pranksters and contains hundreds of millions of spores of *Psilocybe azurescens*. See also figure 77, showing a mycelial colony emanating from point of contact with spored oil.

The solution of 1 billion spores did not do well because when spores are too concentrated in one space, they become a fertile substratum for bacteria before they can germinate and mobilize a defense.

Germinating Spores on Cardboard

Germinating spores on corrugated cardboard is a good method for creating cultures. Corrugated cardboard, with its ridges and valleys, favorably selects mushroom spores to the disadvantage of many other fungi. The sweet wood-based glues used in cardboard provide a boost to mycelial growth. I really do not know why, but contaminating green molds (like *Trichoderma* species) do not grow as well on cardboard as many gourmet and medicinal mushrooms do.

When you tear apart a box and soak a 1 by 2-foot piece of cardboard in hot water for 1 hour, the fibers soften, making a friendly bed for spore germination. Place a mushroom upon the moistened cardboard and let it sit until its spores have fallen en masse, usually

▲ FIGURE 131

Mushroom spores germinating on corrugated cardboard. In this case, the cardboard was exposed to the mushroom for a few hours and then covered. A week later, island colonies of germinating spores appeared.

overnight. Remove the mushroom and incubate the spores on the cardboard for a period of a few weeks in a container set in a cool, dark place. I prefer plastic storage containers with snap-on lids. After 1 week, and periodically thereafter, open the storage container to inspect for germination. The cardboard should remain moist; usually a small volume of excess water will remain on the bottom of the container and will keep the humidity high and the cardboard continuously moist. Once colonies are visible, transfer this mother colony of cardboard to a sheet of soaked corrugated cardboard about 4 times as large for further expansion. This can be done repeatedly every 2 to 3 months until you have enough cardboard spawn for starting a mushroom patch (see pages 142 to 145).

Germinating Spores on Straw

One method that is useful to those living in colder climates is the transfer of spores onto moist, untreated straw. Here I use the term "colder" loosely, meaning when temperatures hover between 35 to 55°F. Spores can be gathered from a spore print on glass or from a spore sock (or by other methods you prefer). Immerse .1 g of spores in a liter of water, and then spread the mixture over 10 pounds of straw. Once inoculated with spores, the straw is placed in a suitable container (a perforated bag, burlap sack, cardboard box, or wooden crate) and thereby protected from rapid water loss.

In the Pacific Northwest, wet straw does not become contaminated with mold fungi for several months when stored at low temperatures. We have found that if oyster mushroom spores are placed on moist, untreated wheat straw and left outside at temperatures of 35 to 50°F in January, mushrooms begin fruiting in late March. I call this method "cold incubation." Although the process is slower than pasteurization methods, it works with little effort, underscoring the power of spores. Once the mushrooms fruit, the spore mass generated from the new fruitings provides exponentially more mass than the starting spores. What this means is that, under certain conditions—cool temperatures, the right substrate, and an aggressive species like those in the genus *Pleurotus*—the spiral of generation of mycelial mass is potentially infinite. Given the ability of *Pleurotus* species to break down toxic wastes, this method is one even rural communities without electricity can put into practice. As with all low-tech methods, every attempt will not be successful, but your experiences can be used to hone your skills over time.

Germinating Spores on Burlap

Like the spores of many other similarly aggressive saprophytes, oyster mushroom spores germinate easily on fabric cloth, especially burlap. Choose burlap (usually made from jute or hemp) that has not been treated with fungicides, preservatives, or insecticides. (Burlap imported from Southeast Asia often smells of kerosene or diesel, used as an insect repellent, but the samples we have received from Canadian and U.S. burlap manufacturers appear to be free of preservatives.) This technique can be refined as you learn the preferences of each mushroom.

Soak a few burlap sacks in water for an hour or more and then spread them across a flat surface, such as a picnic bench. Place a 4 to 6-inch maturing mushroom for every square foot on the burlap fabric. Cover with newspaper or cardboard to prevent other airborne spores from raining down upon the exposed burlap, to protect the mushrooms from wind, and to prevent dehydration. Allow the spores to drop overnight. Once the spores are discharged, remove the mushrooms. The burlap can be rolled and tied with twine (each sack separately), and stuffed into cardboard mailing tubes. Pour on a gallon or so of water to thoroughly soak the spore-infused burlap sack. To further offset evaporation, put these moistened burlap-stuffed tubes into a cardboard box and store the box in a shady location. The gradual release of water will help limit competitors and give the spores a head start. After 2 to 4 months of incubation, remove the burlap from the tubes, unroll, and inspect for signs of growth. Reroll the burlap and return it to the tube if growth is not thorough. Wait another month, then re-inspect. After

several months, depending upon the species, rate of growth, and numerous other factors, mycelium will appear as island colonies. The scent of the mycelium— the species' signature fragrance—will be a key feature allowing you to recognize whether or not the mycelium is resident. Oyster mushrooms (*Pleurotus* species), for example, emit a unique fragrance that smells mildly but pleasingly of anise. Without the use of a microscope or experience handling the mycelium of the species you want to propagate, you may not know if you have mushroom mycelium until fruitings occur. Your only choice, then, is to run with the mycelium you see.

Once the burlap is at least 25 percent colonized, meaning a fourth of the burlap is fuzzy white with mycelium, gently unroll the sacks. Stuff the sacks with wood chips, preferably wood chips from a species of tree on which this mushroom species naturally grows. They will look like sandbags. Stack the stuffed burlap sacks outside, three high, preferably atop plywood, a pallet, or thick cardboard in a shady location out of the wind. Cover with shade cloth and incubate for several months. The burlap bags, with luck and a little help from nature, will begin producing mushrooms with the onset of conducive weather conditions, usually during the rainy months in the spring and fall.

Spawn

Spores germinate into mycelium. When this mycelium is used to inoculate more material, it is called *spawn*. Spawn can be sourced from the wild or from spawn laboratories (see www.fungi.com). Wild spawn will not be as "clean" as pure culture spawn, but wild spawn has the advantage of having been in contact with microbes and so may be better able to acclimate to wild habitats. Spawn can come in many forms. The most common forms of commercial spawn are made with sterilized grain, sawdust, dowels (or plugs), or wood chips. Methods for making spawn are described extensively in my earlier books *Growing Gourmet and Medicinal Mushrooms* (2000a) and *The Mushroom Cultivator* (1983), coauthored with Jeff Chilton.

⌃ FIGURE 132

Two bags of commercial spawn: the yellowish one is maitake
(Grifola frondosa) sawdust spawn; the other is shiitake
(Lentinula edodes) grain spawn.

I like to use laboratory-grown spawn to first create "mother patches" that incubate for 6 months to a year outdoors, acclimating to the fluctuations of weather and the wild microbial world. I then expand the mother patches from 2 to 20 times their mass using the strategies described in this book. If successful in growing mycelium from spores, a similar expansion schedule can be adopted using spore-generated mycelium. The key is to achieve enough mycelial mass to get the job done. For most people, buying spawn from a reputable laboratory is the most direct and quickest method. I recommend buying commercial spawn for those who are impatient, or who do not have ready access to mushrooms or natural mycelium (see Resources on page 309).

First, let's look at how to gather wild mycelium from nature.

Transplanting Wild Mycelium: Mycelial Footprints on the Path to Mycological Paradise

Fans of mycelium race everywhere. You can easily find mycelium beneath a downed log, in among stacks of aging firewood, or in piles of leaves or wood chips. Practically any fallen tree or piece of wood that has been lying on the ground for a few months will host mycelium on its underside. The mycelium pulps the wood over time, slowly digesting its primary components, lignin and cellulose.

Identifying mycelium without its mushroom is difficult, since it has few characteristics to help you identify it. Some examples that are easily identifiable are the blewit *(Lepista nuda)* and the garden giant *(Stropharia rugoso annulata)*. The blewit's brilliant purplish mycelium grows underneath some of my fir trees. The garden giant's mycelium has a ropy quality and a rich and uniquely sweet fragrance, which it also emits when grown in the laboratory.

In a sense, the mycelium "breathes," emitting a fragrance that is a valuable feature to the experienced mushroom gatherer. The outgassing from the mycelium carries scents that are often species specific and recognizable. While walking through the woods, I have often detected the anise-rich fragrance of *Clitocybe odora* or the pinelike fragrance of matsutake. My nose knows the smell, memories resurface, and my brain is titillated by the encounter. I then begin to search for the mushrooms of those species, and if I don't find them that day I make a mental note to check the site again during the mushroom season. Mushroom scents, even in the absence of mushrooms, inform us that the mycelium is alive and thriving. Though in most cases scent alone cannot be relied upon as an identification tool (you need the mushroom), mushroom scents can reveal the presence of otherwise hidden mycelial mats. Once you find a natural patch of mycelium, which I prefer to call a *mycelial lens* (or if it is exceptionally large, a "mother lode"), you have a golden opportunity to dance through life with a fungal friend. Here is how.

▲ FIGURE 133

We inoculated these wood chips the year before and created a mycelial lens. We use this as a mother patch and transplant mycelium from it to other locations.

▲ FIGURE 134

Scraping away the surface chips, we usually discover islands of mycelium which are then scooped up and transplanted. (By disturbing the wood chips in the mother patch, its mycelium is spurred into vigorous regrowth.) Usually a concentration of inoculation from 1:20 to 1:4 works best. A 1:20 inoculation rate means that 1 gallon of naturalized mycelium can inoculate 20 gallons of substrate, in this case more wood chips. Each year the new descendant patches can be amplified at a similar rate. In a few years, the amount of mycomass that can be generated is impressive: one 4 by 4-foot patch can be expanded to over 128,000 square feet, over 3 acres, in 3 years. But, to be successful, you must learn how to run with mycelium.

The simplest way of transplanting mycelium is to scoop it up and move it to a new location. The transplanted mycelium, *virgin spawn*, will be used to make a *mycelial footprint* at another location.

When virgin spawn is transplanted and placed into contact with the right mixture of materials, the mycelium will regrow, expanding the colony. Like all living beings, each individual will have preferences for different combinations of food. When providing materials for the mycelium to consume, the obvious choices are to try to replicate the native ecosystem from which it grew, or to find new materials for the mycelium to sample. This task is simpler with the saprophytic mushrooms than with the mycorrhizal ones. This book concentrates on these easier-to-grow decomposers. Check chapter 14 for species' individual preferences.

Many early cultivators simply gathered mycelium-rich logs having, for instance, shiitake mushrooms fruiting wildly upon them, and then placed these fruiting logs adjacent to newly cut logs. Logs hosting competitor fungi are removed before the competing

➤ FIGURE 135

When selecting natural spawn from a wild mushroom patch, choose mycelium that tenaciously grips the wood chips. One test for vigor of the mycelium is whether or not the rhizomorphs can suspend the chips in the air.

mushrooms can sporulate. The developing shiitake mushrooms then cast spore dust onto the new logs, effectively inoculating them. This method can be repeated for many years, but the population must be constantly culled so the selected mushroom predominates. Ultimately, however, other mushroom competitors are likely to take hold, with turkey tails (*Trametes versicolor*) being the prime if not the most benign competitor. Turkey tails are empowered with excellent medicinal, mycoforestry, and mycoremediation properties, thus, it's a serendipitous quirk of nature that the most aggressive competitor is one of our strongest allies. So if turkey tail is the invader, it should be welcomed and put to use.

Saprophytic mushrooms growing in wood chips are some of the easiest to transplant. To find this type of mycelium, look for areas where the chips are held together in a clump. Healthy mycelium grips the wood chips so firmly that some strength is required in order to tear them apart. This tenacity is a reflection of the strength and vigor of the mycelial mat. Once you find

mycelium, you need to give it a new, friendly environment. Mycelium needs moisture, air, and darkness. A loosely folded cardboard box or paper sack is a fungus-friendly container. Once mycelium is placed into a cardboard box or paper sack, it should be exposed immediately to the outdoors after being transported to your home, unless temperatures are below freezing or too hot. If it is blazing hot, the mycelium is best planted at dusk or early morning before the sun is high in the sky. If you are concerned that it is too hot or cold, the transplanted mycelium can be placed near a cool cement foundation of a house, for instance, in a basement up against a wall. The mycelium can be stored until you are ready to move it outdoors. In this case, I lay another panel of cardboard on top of the natural mycelium as an extra buffer against evaporation. Apply water every 2 to 4 weeks as needed, pouring about

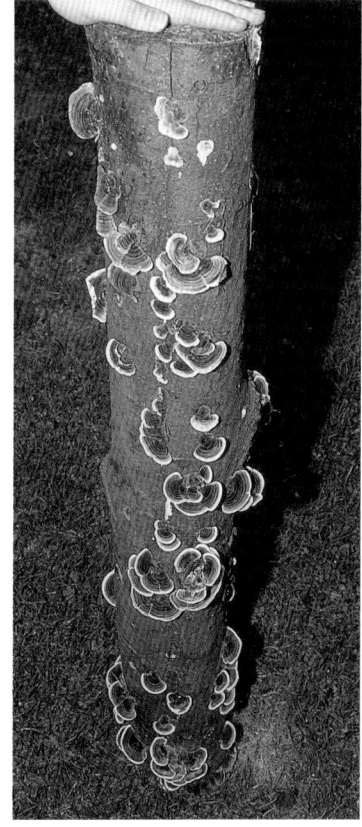

> **FIGURE 137**

Turkey tail fruiting from a log inoculated with pure culture shiitake spawn. The wild turkey tail got there first and prevailed. In this case, the competing fungus is also a beneficial medicinal.

▲ **FIGURE 136**

Before the invention of pure culture spawn, Japanese shiitake growers collected logs from the woods and then used the "soak and strike" method, which means to immerse the logs in water and then violently hit them to induce fruiting. By interspersing newly cut logs among the fruiting logs (see background), the growers encouraged the spores from the shiitake mushrooms to inoculate the neighboring logs.

1/2 gallon in a 2 by 3-foot box. Store your transplanted mycelium in the shade if you incubate it outdoors until you prepare your mushroom bed.

In some cases, the mycelium can stay healthy for months after leaving its native habitat, especially in cold temperatures, since growth is slowed as temperatures drop. (See page 161 for a discussion of cold incubation.) When spring returns, the mycelium can be planted outdoors in your newly prepared mushroom bed. In my experience, I have found that the ideal time for transplantation is from March to June.

Mycelial Footprints and the Wonders of Cardboard Spawn

One method I favor for expanding naturalized mycelium is to use cardboard. The idea for this method came to me when I observed an elderly couple who were suppressing weed growth in their rhododendron garden by placing cardboard over the mulch around their rhododendrons. Seeing my wife use cardboard in the garden for a similar purpose reinforced the idea. Although the husband and wife team were seeking to minimize weed growth without the use of chemicals, they unexpectedly grew mushrooms. Covering wood chips and mulch with cardboard promoted mushroom growth. Since that initial observation, I have adapted this method, improved upon it, and used it for running mycelium outdoors. The elegance of this method is in its simplicity. Another variation of this method is described on pages 146 and 147.

To use this method, first flatten 2 corrugated cardboard boxes. (Cardboard from the United States, Canada, and Europe tends to be free of dioxins, thanks to environmental regulations, and is not toxic. Cardboard from other countries may have some residues.) Tear the cardboard panels so that the corrugations are exposed. Wet the cardboard until it is saturated. Take the torn panels and place them with the exposed-corrugation side in contact with the mycelium recently implanted. You can encourage mycelium to grow upward through the wood chips or mulch and surface just beneath the cardboard. If the mycelium already

resides just below the surface, the patch is maturing: scrape away the top layer of wood chips or mulch to expose the mycelium to the cardboard. The sooner the terrestrial mycelium surfaces and makes contact with the wet cardboard, the better. (For this method, I prefer aerial mycelium, with uplifting fans of growth.) If the air is dry, first saturate the bed of mycelium with water before you cover with it with moist corrugated cardboard. Covering the cardboard with a loose layer of straw reduces evaporation and provides shade. Incubate for several months, periodically watering and checking to see if the mycelium has attached to the cardboard.

Once the mycelium covers 25 to 50 percent of the cardboard, it can be transferred. If it is overincubated, the mycelium will die back. For this reason, it is best to move the mycelium at the crest of its growth—what I call "surfing the mycelial wave." Among the saprophytes listed in this book, mycelial waves may continue for 4 months to 2 years before the mycelium consumes the cardboard, is consumed, and/or dies back (see figure 99).

Once the underside of the cardboard is colonized, with the growth running through the corrugations, pick up the myceliated cardboard and do either of the following:

- Place it on top of a new bed of fresh wood chips, mycelial face down (see figure 139). I call these *mycelial faces*. When placed onto a new substrate, the face becomes a *footprint*.
- Use it to make bunker spawn by sandwiching it between 2 burlap sacks filled with wood chips.
- Use the myceliated cardboard to inoculate more sheets of corrugated cardboard. Simply strip the smooth paper to expose the corrugations, soak the cardboard, and sandwich it between the new cardboard panels. By alternating panels with and without mycelium, you can build a mycelium cardboard tower.

Depending upon how thoroughly the mycelium has colonized the paper and the number of walls in the corrugated cardboard, several sheets of mycelium can be delaminated—separated into individual panels—

A successful transplantation strategy: taking a mycelium-covered fragment of wood and placing it onto wet cardboard. This method is one way to create natural spawn. Note the forking of the mycelium as it runs.

You can use an outdoor master patch to make several contact mycelial prints by periodically replacing the colonized corrugated cardboard atop the mushroom bed over the course of a year. Each one of these myceliated sheets can be used as a faceprint, laid upon wood chips, or used as a footprint with wood chips placed upon it. Or, the myceliated cardboard can be sandwiched between burlap bags filled with wood chips to make burlap bag spawn. (The patch featured here was originally grown from dowel spawn inoculated with stem butts. See page 148 and figures 152 to 156.)

substantially increasing the surface area of mycelium for use as cardboard sheet spawn. When placing cardboard with an exposed mycelial face upon wood chips, make sure the substrate is moist before contact. If living mycelium touches dry wood, the wood will suck the moisture from the mycelium, causing cells walls to collapse and harming them. If you place the sensitive mycelium onto a wet surface, it can grab onto the surface without having to struggle to manage moisture within its own cells. This strategy allows the mycelium to leap off. "Leap-off" is a term growers use to describe the mycelium's recovery from the shock of inoculation and the ensuing growth surge. Generally, the better the leap-off, the better chance the mycelium will win the microbial race.

Humans have 5 or 6 layers of skin, whereas the mycelium has just 1 layer. Our thick skin protects our cells from infection and dehydration. Mycelium does not have such protection. The mycelium is surrounded by millions of hungry microbes in the soil, many of which will quickly consume it if it weakens. Young mycelial cell walls are not callous, but sticky; they form a moist membrane through which metabolites are continually secreted. The mycelium is hypersensitive to water loss, and growth stops when it contacts dry surfaces.

I try to imagine what the mycelium feels when it is exposed to dry air. I imagine that the feeling is like that of an open wound exposed to blowing wind, a burning pain. In contrast, I think the mycelium must experience a soothing feeling in rain or fog. Of course, we may not ever know what the mycelium feels, but I do know the end result of mycelium's exposure to dry surfaces or arid air. My recommendation is to transfer mycelium only under moist conditions onto wet materials. By easing the transition and providing healing water, you can help foster cell growth while protecting cell viability.

Once you place the cardboard spawn on your new mushroom bed, the mycelium runs downward. If you lay down 2 layers of myceliated cardboard with wood in-between, the cardboard barrier helps to prevent organisms from attacking from below or above. In most instances, the lower level of cardboard is placed mycelial face up, with several inches of wood chips piled on top. Cover the top surface with the other myceliated panel face down. Placing 2 layers of mycelial cardboard has been one of my more effective methods for colonizing outdoor habitats.

If you are so lucky as to have mushrooms in hand that you want to cultivate, then you can have lots of fun making spawn from stem butts. Stem butts have some remarkable, unexpected properties.

Stem Butt Spawn and the 1-Dowel Revolution

This method may revolutionize the outdoor cultivation of many mushrooms. Perhaps you collected mushrooms while on a walk, or you have mushrooms fruiting. This is the time to make use of one of nature's most fortuitous fungal opportunities. Natural spawn can be generated from the basal rhizomorphs radiating around the stem's connection to the nurturing mycelium. If you cut the base away from the stem, carefully protecting its rhizomorphs, the tissue stays alive. A few stem butts can inoculate a bed of wood chips or wooden furniture dowels ("plugs") or be sandwiched into corrugated cardboard. For years thereafter, you can create dozens of satellite colonies from the stem butts of mushrooms from your mother patches, and each colony can be continuously expanded into constellations of colonies.

The effectiveness of the stem butt inoculation technique is a quirk about mushrooms that is unknown to most people, coming as a surprise even to many seasoned mycologists. I discovered this method by accident and am still perplexed by the chi power, the life-force energy, emanating from the stem base. Practically all connoisseurs of mushrooms trim and discard stem butts before cooking. But, surprisingly, with many saprophytic mushrooms, stem butts, but not the areas above the stem base, regrow with astonishing vigor when transplanted into wood chips, cardboard, or wooden dowels. This method of regrowth is another smart evolutionary advantage of the mushroom, since foraging animals, including humans, prefer to eat the softer flesh of the upper fruiting body rather than the tough, often woody

flesh at the stem base. When we throw stem butts away instead of eating them, we inoculate new habitats.

To use this method, harvest mushrooms with stem bases intact and ideally with their radiating rhizomorphs—dangling stringlike mycelial strands—attached and infused with substrate particles (see figures 142 and 147). The rhizomorphs by themselves do not grow as well as they do when attached to a stem butt. The stem butt acts as nutrient source for the rhizomorphs as they grow rapidly onto new materials.

The following 2 methods for growing spawn from stem butts use cardboard and dowels, respectively. Both methods allow growers to expand mycelium.

MAKING CARDBOARD SPAWN FROM STEM BUTTS

To use stem butts to make cardboard spawn:

1. Gather fresh mushrooms of varieties known to have stem butt regrowing capacity (see page 151).

2. Using a knife or scissors, clip off the base of the stem just above where it narrows, keeping the rootlike rhizomorphs intact and attached to the stem.

3. Soak cardboard in water until it is fully saturated. Tear off sections to expose corrugations.

4. Place 1 stem butt on the cardboard roughly every 16 square inches, and sandwich them between panels of corrugation.

5. Soak the stem butts and cardboard in water and place in a cardboard box, trunk, old bathtub, sink, or trough, or simply on the ground, and cover with a shallow layer of wood chips.

6. Keep in the shade, with the incubating container on the ground to limit temperature fluctuation.

7. Incubate for 4 to 8 months before transplanting.

At this point, you can use each sheet of cardboard as a mycelial footprint to inoculate an outdoor bed, stacked burlap sacks (bunker spawn, see figure 140), or straw bales. Or you can place sheets interspersed within a stack of moist newspapers and let them become myceliated over several months. Your imagination is the only limit. However, the mycelium must be moved (transplanted) or it will die. Hence, our motto with mycelium: "move it or lose it."

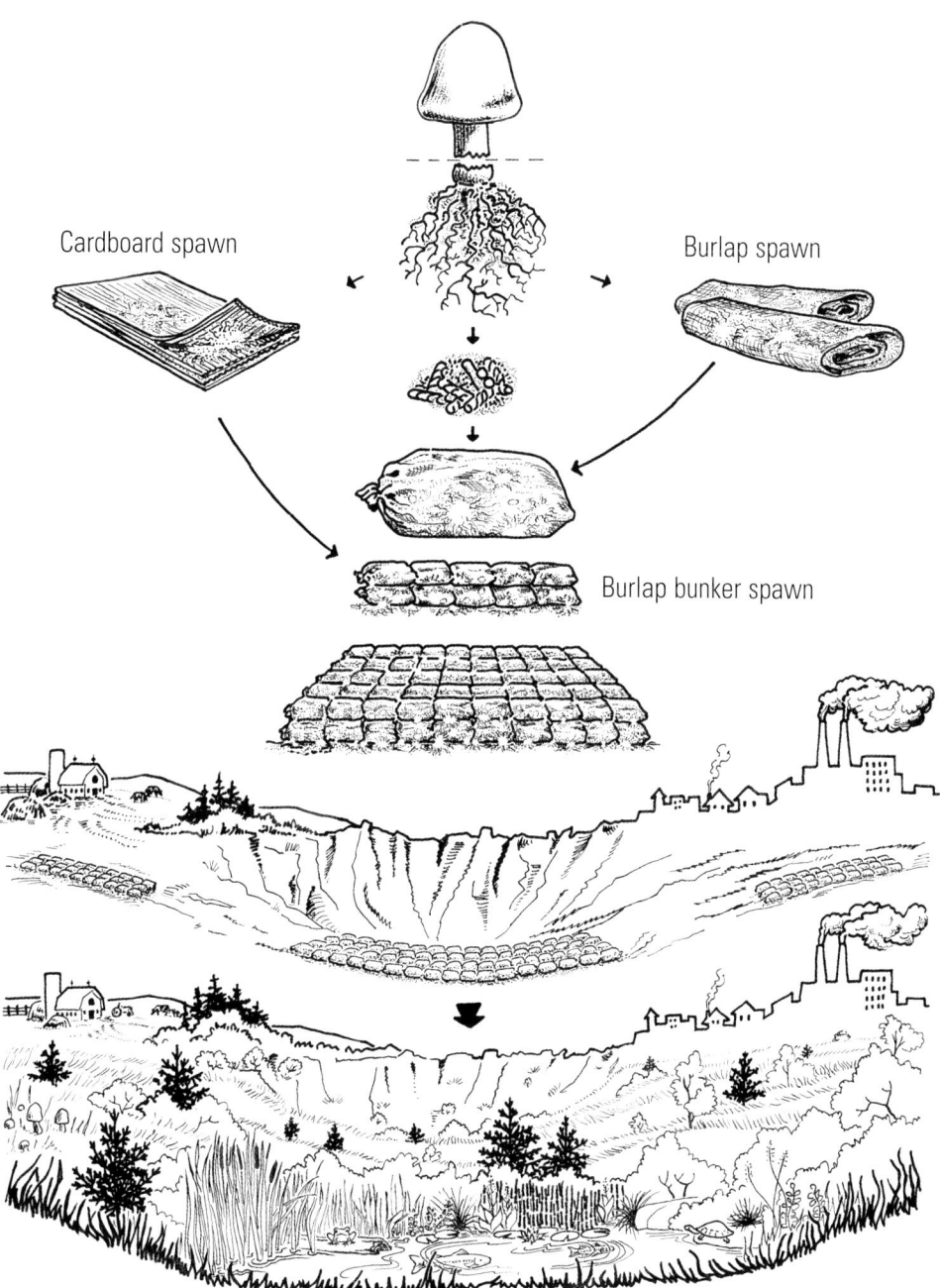

Cardboard spawn

Burlap spawn

Burlap bunker spawn

⋀ **FIGURE 140**

Overview of using stem butts to generate mycelium for mycorestoration.

▲ **FIGURE 141**

First, gather fresh mushrooms, moist corrugated cardboard, and an incubation container.

▲ **FIGURE 142**

This fluffy stem base regrows with vigor after being cut.

◄ **FIGURE 143**

Cut the stem butt from the mushroom.

➤ **FIGURE 144**

Place the stem butt pieces onto the moist cardboard and fold.

▼ **FIGURE 145**

Three weeks later, mycelium surges from the cut stem butt and is well on its way to creating cardboard spawn.

⋀ **FIGURE 146**

Dusty Yao plucks a garden giant *(Stropharia rugoso annulata)* from her garden, being careful to keep the stem butt and its nurturing rhizomorphs intact.

⋀ **FIGURE 147**

Rhizomorphs attached to the stem butt can regrow—if handled carefully. Be especially careful that they don't dry out before use.

⋀ **FIGURE 148**

Cutting the stem butt from the mushroom.

⋀ **FIGURE 149**

Placing cut stem butts, with rhizomorphs attached, into moistened, folded, or sandwiched corrugated cardboard. This folded cardboard, once inoculated, is placed into a cardboard box or plastic tub with a loose lid and drainage holes, and incubated outdoors in the shade and on the ground.

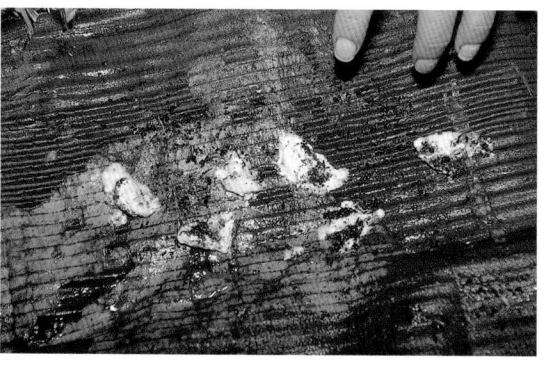

◄ **FIGURE 150**

Regrowth after 4 days.

▲ FIGURE 151

After a month, the mycelial colonies from each portion of the cut stem butts merge to form a contiguous colony. This cardboard can now be used as spawn for inoculating more wood chips by placing it mycelial face down onto wood chips or putting several inches of wood chips on top of it to make a mycelial footprint. Alternatively, this sheet of myceliated cardboard spawn can be sandwiched between two moist straw bales, burlap sacks filled with wood chips, or more cardboard. Like all spawn, its life span is limited: move it or lose it. Mycelium of many mushroom species can be grown in a similar fashion.

▲ FIGURE 152

Stem butts are placed into 20 pounds of soaked birch dowels (10,000), pushed just beneath the surface.

MAKING DOWEL SPAWN
FROM STEM BUTTS

Another permutation of the stem butt technique is to use stem butts to inoculate unsterilized wood dowels. The dowels are first submerged in water for a few days or weeks, drained, and then inoculated with stem butts from fresh mushrooms. In the example depicted in figures 156 to 162, 10,000 dowels half fill a box measuring 12 by 20 by 10 inches and weighing about 20 pounds. Into this box, I placed 4 to 8 stem butts each about 4 inches apart and pushed them underneath the surface to a depth of 2 to 4 inches. I covered the inoculated dowels with several layers of exposed corrugated cardboard and placed it in a blackberry patch. After 6 months, I harvested the several large mycelial islands that had grown and used them for inoculating stumps, logs, cardboard, chips, and more dowels. After removing a third of the dowels, I mixed up the remaining ones to help colonization, since at this step in the process mycelial growth benefits from disturbance. The broken rhizomorphs then regrew with astonishing vigor, soon covering the remaining dowels. These colonized dowels are powerful platforms for launching mycelium onto cardboard, coconut fiber, rope, or burlap bags filled with wood chips, or they could be used to plug logs and stumps or even to inoculate more dowels. What I like about this method is that the mycelium is acclimated to microbes and is tenacious and feels pure; it produces high quality spawn that can be expanded many times over (see figures 152 and 344). From this stage, you now have many options for traveling down the continuously diverging mycelial path.

This method works so well that I call it "the 1-dowel revolution." How much dowel spawn could you make from a single mycelium-covered dowel? Using my experience as a basis, I calculate that the first year you can make 10,000 mycelium-covered dowels from a single stem butt. The second year the mycelium on each dowel could also be multiplied 1,000 to 10,000 times, for a total of 10 to 100 million dowels, weighing about 2,000 to 20,000 pounds, after just 2 years. If you use the dowels to inoculate burlap bags filled with wood chips,

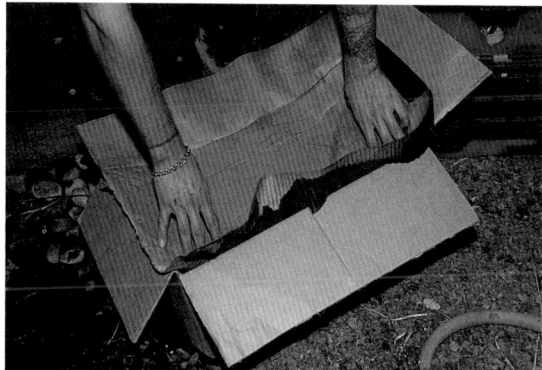

⚠ **FIGURE 153**

A layer of cardboard is placed on top of the stem butt–inoculated dowels. Alternatively, a woven mat (coconut fiber, hemp, bamboo, etc.) can be used.

⚠ **FIGURE 154**

The cardboard box holding 20,000 dowels, now inoculated with a few stem butts, is left outside to cold incubate through the winter and spring. The cardboard incubation box is soaked by natural rainfall and exposed to freezing temperatures, but recovers. Only mushroom species native to freezing ecosystems survive overwintering. The breathability of the cardboard box helps the mycelium grow.

⚠ **FIGURE 155**

Half the dowels have been colonized 6 months later. Large islands of mycelium can be harvested. Many of the nonharvested loose dowels are mixed through the remaining dowels, which, a month later, are also covered with whitish rhizomorphic mycelia. Once mycelium is at this stage, this box of "mother spawn" is explosive in its growth potential and can pump out waves of mycelium if you repeatedly harvest mycelial islands and then replenish with more dowels.

⚠ **FIGURE 156**

Ethan Schaffer enjoys the scent from plug spawn made from stem butts.

△ FIGURE 157

A dowel laced with whitish rhizomorphs. Each dowel becomes a platform for projecting rhizomorphs. The mycelium on this single dowel could potentially remediate many acres of polluted land. When mycelial colonies springing forth from the dowels touch, they share nutrition and grow stronger.

△ FIGURE 159

On November 9, mushrooms appeared.

△ FIGURE 158

On March 20, myceliated dowels were spread over wood chips and then covered with another layer of fresh wood chips 1 to 2 inches deep.

growth may be necessary. Even if only 25 percent of the burlap bags fully colonize, a reasonable expectation, this low-tech method can be a dramatic tool for restoring polluted environments, and for building life-sustaining soils. If grasses or trees are added into the mix, each bag becomes a pedestal of ecological rebirth. Diverse, complementary biological populations can use each burlap bag as a springboard for growth. (See figure 165.)

For mycoremediation or for growing mushrooms, this technique is revolutionary in its simplicity and large-scale production possibilities. It works remarkably well for generating a large mass of spawn without having to depend upon a spawn laboratory, provided you have patience, have the right strain, and can run with mycelium, planning several years ahead. Mycelial masters could artfully employ this mycelial expansion to craft wide-scale restoration from stem butts.

WHICH MUSHROOMS' STEM BUTTS WILL REGROW?

Undoubtedly, hundreds of species fit into this category. My rule of thumb is this: If the mushroom is a saprophyte with rhizomorphs, its stem butt will likely regrow when replanted. However, I encourage you to experiment with the stems of likely and unlikely mushrooms, since new knowledge about species' regrowth abilities could be put to a positive use.

by the end of the third year you could have generated enormous mycomass given an inoculation rate of 10 to 100 myceliated dowels for each bag. The number of burlap bags that could be inoculated is mind-boggling, from 100,000 to 1,000,000. When a burlap bag is laid upon the ground, it covers 2 to 3 square feet, meaning that enough mycelium can be generated to cover 200,000 to 3,000,000 square feet, equivalent to 4.5 to 69 acres of coverage. Culling bags that fail to show good

The chart on this page lists of some of the more interesting species (all saprophytes) known to regrow. I expect that in the years to come this list will be greatly expanded as a result of my research and yours.

Species Known to Regrow from Stem Butts

Genus/species	Common name
Agaricus augustus	Prince
Agaricus brasiliensis	Brazilian blazei, himematsutake
Agaricus subrufescens	Almond mushroom
Agrocybe aegerita	Pioppinno
Agrocybe species	Ground dwellers
Chlorophyllum rachodes	Shaggy parasol
Hypholoma capnoides	Clustered woodlover
Hypholoma sublatertium	Kuritake or brick top
Macrolepiota procera	Parasol
Morchella angusticeps	Black morel
Morchella esculenta	True morel
Morchella species	Morels
Pholiota nameko	Nameko
Pholiota species	Scaly caps
Pleurotus ostreatus	Oyster
Pleurotus pulmonarius	Phoenix oyster
Psilocybe azurescens	A-z's
Psilocybe cubensis	San Isidros, cubies
Psilocybe cyanescens	Wavy caps
Psilocybe cyanofibrillosa	Cyclone psilocybe
Psilocybe subaeruginosa	Australian psilocybe
Sparassis crispa	Cauliflower
Stropharia rugoso annulata	Garden giant or king Stropharia
Stropharia species	Swordbelt mushrooms
Trametes versicolor	Turkey tail or yun zhi

Mycorrhizal species can also be transferred to new habitats using stem butts—a fact that may come as a surprise to some mycorrhizologists. Differing from those of saprophytes, these stem butts (such as those of chanterelles, matsutakes, porcini, and *Amanita* spp.) must be placed directly into the rhizosphere of a young tree or plant host, especially ones lacking mycorrhizae like many of those coming from indoor nurseries, which are often grown in pasteurized soils. The soil in which the stem is embedded is concentrated with spores, making this region rich in regenerative potential. Whether the mushrooms are saprophytic or mycorrhizal, I encourage you to take the stem butts of mushrooms collected on forays and cast them about your property to create satellite mycological communities.

Burlap Bag Spawn: Amassing Mycelium, or Preparations for Creating a Mycelial Tsunami

Burlap bag spawn can be created from inoculations—individually or in combination—of dowel spawn, cardboard spawn, stem butts, spore slurries, and pure culture spawn. When spawn is mixed into materials such as wood chips and stuffed into burlap sacks (made of jute, hemp, cotton, or wood fibers), a living pod-with-a-purpose is made. Using burlap bag spawn, homeowners can create mycological landscapes in their backyards. On a larger scale, this type of spawn can help repair damage to ecosystems and create buffers between sensitive ecosystems and toxic environments.

Immediately upon construction, these spawn bags filter biological or chemical wastes and prevent downstream contamination due to the uncolonized wood chips' extensive surface areas and absorptive abilities. As the mycelium grows, its netlike cells increasingly trap silt and bacteria, and the "sweat" of the mycelium denatures many toxins, both chemical and biological. Upon colonization, a burlap bag of spawn (now weighing 30 to 60 pounds) becomes a mycofiltration vessel that can also be put to work for habitat restoration or can be staged for further expansion of mycelium into more wood chip–filled burlap sacks (see figure 140).

These bags, at a glance, resembles sandbags used for preventing swollen rivers from overflowing and hence are sometimes called "bunker spawn." In a sense, bunker spawn serves a similar function: preventing shallow sheet flows of contaminant-laden sediments and controlling erosion. Vertical walls or horizontal plateaus of myceliated burlap sacks or bunker spawn can be built according to the needed ecological applications and other design criteria. The portability of bunker spawn allows for a variety of applications. I filed a patent on this idea.*

---

* *U.S. Patent Application Serial No. 10/081,562 for delivery systems for mycotechnologies, mycofiltration, and mycoremediation, filed February 18, 2002.*

⋀ **FIGURE 160**

A handful of plug spawn made from stem butts (see figure 152) inoculates untreated wood chips in a burlap sack.

⋀ **FIGURE 161**

Incubation of inoculated burlap bags filled with wood chips, which I call "bunker spawn."

⋀ **FIGURE 162**

Mycelium penetrates the burlap, which decolorizes as it's decomposed.

⋀ **FIGURE 163**

Sometimes, the best growth occurs where the burlap bags make contact with native soil.

If uninoculated wood chip–filled bags are stacked in contact with myceliated bags in a checkerboard pattern but no more than 2 bags high, leap-off of mycelium onto the virgin bags is rapid and direct. From my experience, stacking the bags 4 or higher results in the growth of competitor fungi. The densest growth takes place near the ground (see figure 163). This simple technology—creating bunker spawn—may be a keystone event leading to better habitat restoration methods.

If the bags are left in place on the ground, oftentimes red worms move in to consume the mycelium, and, in turn, a nurturing soil forms, creating a springboard for vast communities of other organisms. The following methods for creating bunker spawn are easy for anyone to put into practice.

There are several effective sources of mycelium and suitable materials for creating bunker spawn. Burlap sacks filled with wood chips can be inoculated with mycelium coming from these sources:

- spores
- grain, sawdust (wood chips), or dowels
- stem butts
- myceliated cardboard sheets

But regardless of the materials you use, the mycelia that result have one thing in common: a preference for fabrics. The architecture of fabric, specifically that of burlap bags, is well suited to rapid colonization by mycelium. The mycelium, being a fabric itself, recognizes this structure and grows through it with a surprising velocity. The breathability and biodegradability of the fabric are essential for success. This biodegradable membrane allows for rapid colonization by the mycelium, encompassing its contents like a living pillowcase. Inside the fabric membrane the mycelium grows outward as well as inward. This mycelial conversion grips the substrate to the disadvantage of competitor fungi, which fail to compete with the ever-advancing avalanche of mushroom mycelium.

In effect, the surface fabric of burlap sacks becomes the contact point for the mycelium to leap off into the contents. By first inoculating the sack, we allow the mycelium to run quickly through the membrane and encase the contents of the bag with a sheath of mycelium on burlap. Since the fabric's tensile strength deteriorates as a result, the bag will degrade in about 6 to 12 months, falling apart and spilling its mycelial package.

Here are some of the diverse ways you launch mycelium into burlap sacks filled with wood chips. Some methods will be more appropriate than others depending upon your access to natural or commercial sources of spawn or mushrooms.

▲ **FIGURE 164**

This bag was inoculated with oyster mushroom mycelium and then submerged in a waterway to capture bacteria from a pond. Upon retrieval, oyster mushrooms popped out.

⋀ FIGURE I

To reduce coliform bacteria from an upland farm, two rows of woodchip-filled burlap sacks, inoculated with oyster mushroom mycelium, catch surface water before entering a sensitive salt-water estuary in Mason County, Washington, USA. Rows of myceliated bags are added annually as a best management practice. I prefer using woodchips from storm-created debris and inoculating with a native mushroom with proven antimicrobial properties. Similarly, such an arrangement can be used to catch and destroy chemical effluents.

◄ FIGURE 165

Another application I favor is using myceliated burlap sacks as an ecosphere for habitat restoration. Not only does this "mycopod" catch silt, trap bacteria, and encourage insects and worms, it also acts like a nurse log to help a Douglas fir seedling get started. The grass soon climaxes, dies, and further fuels the fungal carbon life cycle. After two years, the burlap sack disintegrates. Complex communities can be created, customized to each particular location and needs.

SPORE PRINTS AND SPORE MASS SLURRIES FOR INOCULATIONS OF BURLAP SACKS

Create a spore-mass slurry (see page 135) using 1 g of spores (from, for instance, 6 oyster mushrooms, 4 to 6 inches in diameter) per gallon. The typical size of burlap sacks that have worked well range from 12 by 24 inches to 18 by 30 inches. (Larger bags become unwieldy due to their weight.) This method works best if the burlap is hot-water pasteurized, sterilized, or chemically treated using hydrogen peroxide to reduce spores of competitor fungi in the fabric. If burlap is soaked using hydrogen peroxide, make sure it's thoroughly rinsed with clean water after treatment, lest the residual peroxide harm your implanted mushrooms spores. (For more information on the use of hydrogen peroxide, see page 162.) Once spores germinate into mycelium, however, residual peroxide levels are usually not detrimental to the implanted fungus, as the mycelium's digestive enzymes contain natural peroxides. Once the bags have been treated and inoculated with the spore-mass slurry, fill them with wood chips and stack them into a wall 3 tiers high (see figure 160).

After the bags are saturated with the spore-mass slurry, or after adding water once prints have been made upon them, roll up the bags and stuff them into cardboard mailing tubes. Incubate for a few months outdoors and then remove the rolled burlap from the mailing tubes. Oyster mushroom mycelium will produce whitish zones of growth in 2 months to 6 months after contact. Select the best colonized bags. Unroll the bags with the best myceliation, stuff them with wood chips, tie them closed, and stack them 3 high on a pallet outdoors. Try to get the mycelial faces to contact uncolonized areas of the bags stacked below or above them. This method, although it can work, is not nearly as predictable or successful as the methods described below. *Caution: When using spores or when using these low-tech methods, wear a filter mask and avoid black, yellow, or brightly colored molds. Some of these can be allergenic or even pathogenic.*

▲ **FIGURE 166**

Pure cultured dowel or plug spawn.

▲ **FIGURE 167**

Aggressive leap-off of mycelium from acclimated, natural-culture dowel spawn of *Psilocybe cyanescens* onto crude, raw, contaminated, dirty, nasty, untreated wood debris. The ability of naturalized mycelium to project from wooden dowels and compete is impressive compared to that from dowels inoculated with pure culture mycelium, which often dies or is eaten upon contact with this same material. Pure culture spawn, if not immediately parasitized, slowly adapts its immune system to newly contacted microbial populations, eventually naturalizing to the resident microflora and then rebounding with spurts of growth.

USE OF COMMERCIAL SPAWN
(GRAIN, SAWDUST, WOOD CHIPS OR DOWEL)
FOR THE INOCULATION OF BURLAP SACKS

Pure culture spawn is available from a number of commercial laboratories (see the list of spawn companies in the Resources section.) The success of using pure culture spawn to inoculate raw wood or straw is limited because of competition from the millions of spores already resident in that material. Having been raised in a sterile laboratory and never before exposed to competing microbes, the mycelium's immune system must work to overcome its competitors. To give pure culture spawn a chance to devise defensive strategies in advance, a few days before use, introduce a small sample (a few grams, for example) of the raw substrate to the spawn (5 pounds, for example). In effect, you are purposefully contaminating the pure culture spawn with a sample of the microbially rich habitat in which it is destined to reside. As is the case with vaccinations and homeopathy, exposure at low levels fortifies the immune system, preparing it for its impending contact with the many populations resident in the natural world.

Typically, pure culture spawn added to a grain substrate in burlap bags does not work well unless the substrate has had its population of undesirables knocked down using heat treatment (sterilization or pasteurization), hydrogen peroxide, or any means that renders competitor spores ineffective without damaging the receptivity of the substrate. Pure culture spawn added to a sawdust substrate tends to work better, in many situations, but "leap-off"—the ability of the mycelium to recover from the shock of being introduced to a new substrate—is considerably slower.

A clever method we have developed for growing both grain and sawdust bunker spawn is what I call "cold incubation," the inoculation of raw substrates during cold outdoor temperatures (35 to 45°F). At these temperatures, most major competitors are dormant or have such a slow rate of growth that they are unable to compete with your introduced spawn, which must be a cold-weather variety in order for this strategy to work.

After several months, as temperatures rise with the advent of spring, the mycelium zooms to the forefront ahead of competing microbes, and is better fortified to defend itself from microbial attack. This procedure of using spawn in combination with a cold incubation strategy is useful for mycorestoration strategies.

One variation on the strategy for creating bunker spawn stands out: inoculating bags of wood chip–filled burlap sacks with wooden dowels, made either from pure culture spawn or from stem butts. Surprisingly, each dowel, or "plug," by itself has enough nutritional mass to allow mycelium to project deeply into the surrounding substrate. Dowel spawn, traditionally used for inoculating logs, can also be used for creating outdoor beds. Cold incubation works particularly well with dowel inoculations of temperate mushroom species of burlap sacks filled with wood chips. Experimenting with combinations of strains and materials will give you valuable insights regarding the best methods.

STEM BUTT INOCULATIONS OF BURLAP SACKS

Stem butts can be deployed in either of 2 ways: interspersing them throughout the wood chips in a burlap bag, or sandwiching the stem butts between 2 bags after they've been filled with wood chips. For both of these methods, the workable rate of inoculation is 5 to 10 butts per 40-pound bag of wood chips.

If you want to create many sheaths of mycelium from stem butts, first soak similar sizes of flattened burlap and torn corrugated cardboard in cold, clean water. Place at least 2 stem butts per square foot onto the grooves of the corrugated cardboard, cover with a flattened burlap sack, add another layer of moistened corrugated cardboard, add more stem-butts and so on. This lasagna-building approach will result in a tower of mycelium, which should be stored outdoors on a wooden pallet, slightly above ground, in a cool, shady place. This mycelial tower needs to be broken down, separating the cardboard from the burlap, within 3 to 6 months, since the bags will fall apart if the tower is overincubated. Optimal timing balances colonization by the mycelium, tensile strength of the bags and cardboard, and separability of the layers. After the tower is taken down, the mycelial bags can be filled with substrate, while the cardboard can be inserted between newly filled bags of wood chips.

CARDBOARD SPAWN INOCULATION OF BURLAP SACKS

Cardboard spawn, which can be created using spores, spawn, stem butts, or transplanted face or footprints from overlaying mycelium (see figure 151) is an effective tool for inoculating the interface between 2 burlap sacks. The sacks are first filled with wood chips and soaked with water, and then, while they're still moist, the cardboard is delaminated to expose the mycelium. Immediately after exposure, the mycelial face is placed directly on the soaked burlap. Another soaked burlap bag filled with wood chips is laid on top. Together, the bags are covered with either loose cardboard or shade cloth and incubated outdoors until colonization is substantial. Don't wait for 100 percent colonization, since this spawn should be moved at the peak of its surging growth (see figure 163). By inoculating a population of bags and spot-checking their growth, you can get a good sense of whether the mycelium's growth rate is climaxing or descending. It's always best to run with mycelium when its pace is the briskest, projecting it at its optimum rate of growth. Once it peaks, vigor declines and the mycelium loses its momentum. I follow this rule at each step in the expansion of mycelium.

Bunker Spawn for Habitat Restoration

Bunker spawn is an ideal tool for habitat restoration. It can be used on a commercial scale and in many different ways. Bunker spawn can be placed on the peripheries of watersheds to protect them from the surface flow of contaminants (see figure 139). Bunker spawn can also filter contaminants emanating from chicken, pig, and cow farms, in the process protecting downstream environments from bacteria, viruses, protozoa, pesticides, nitrates, and many other pollutants, including the so-called inert ingredients commonly used in fertilizers. Adding bunker spawn to riparian buffers creates a mycofiltration layer even on the first day of application: the wood chips, with their large surface area, catch substantial colonies of bacteria.

The benefits of inoculating wood chips are that you can be selective about the fungi you grow, and that colonization is far faster than it is in the wild. Uninoculated fresh wood chips are vulnerable to "spore fall"—wild spores raining down from the air and coming up from the ground. Spore fall typically produces island colonies of several competing species, usually 2 years after wood chips have been manufactured. Essentially, the window of opportunity for get-

➤ **FIGURE 168**

A habitat is reborn. David Sumerlin stands in a planted grove of poplars rapidly growing from a forest floor built of burlap sacks filled with wood chips permeated by *Psilocybe cyanescens* mycelium. Bags of wood chips were placed on clay-mineral earth around transplanted trees. Leaf and twig fall fuels the carbon cycle: as trees grow, mycelium thrives, soils thicken, and mushrooms form. From what was once barren soil, a whole ecosystem flourishes in about 5 years.

ting a species you desire to grow on recently chopped, undiseased wood chips usually closes in 2 years, when native fungi surge and dominate. In contrast, inoculated wood chips colonize with mycelium in 6 to 12 months, and the mycelium, secreting antibiotics and free radical enzymes, destroys many airborne and embedded spores of competing fungi.

Selecting mushroom species with unique filtration or remediation properties allows the landowner to customize a mycological response specific to the environmental threat. When creating bunker spawn, selecting native mushrooms is recommended over importing foreign species; for instance, I recommend this method for growing garden giants, clustered woodlovers, turkey tails, cauliflowers, and some woodland *Agaricus* mushrooms. No doubt dozens if not hundreds of species of mushrooms could be used in bunker spawn production, whether or not the substrate is inoculated with grain spawn, plug spawn, stem butts, or laboratory-grown sawdust or chip spawn. I also recommend the bunker spawn method for getting extra crops of shiitake, turkey tail, or reishi mushrooms by recycling spent substrate from a gourmet mushroom farm. If not for the extra mushrooms that can be produced, the best use of this form of mycological currency may be for their pollution-destroying enzymes, placing them in key locations in and around toxic waste sites.

If bunker spawn is being incubated in a staging area before placement, the burlap sacks should be incubated near to but off of the ground—for instance on wooden shipping pallets or sheets of cardboard. If incubation is to occur on the site where you want your mycelium to reside, then bunker spawn bags can be laid side by side to create a 4- to 6-inch deep contiguous layer of wood chips directly upon bare ground. (I prefer to place the mycelial face down rather than up in the air.) The mycelium digests the wood chips and the bag over time, beginning the process of soil building.

Not every bag of bunker spawn will colonize. Often, spots of dark green and black mold can appear, but they are weak competitors to those mushroom species that project ropy rhizomorphs. The art of using the bunker spawn method is in knowing which bunker bags of mycelium to expand, and when. Sacks that are poorly colonized should be removed. On the other hand, the sacks that become fused together by thousands of interconnecting rhizomorphs have enough chi power to leap off into fresh wood chips, allowing deep penetration and overcoming ecological barriers that weaker mycelium could not. These are the bunker spawn bags I select for expansion. One strange twist in this method is that often the mycelium from the burlap surfaces penetrates more effectively to the depths than does the mycelium from within the bags. I think the mycelial mat that forms in the burlap surface becomes a platform of strength, a membranous sheet of cells that helps the mycelium project its tendrils deep into the depths of the internal wood chips

Once a naturalized bag of bunker spawn has matured, in many cases becoming snow-white in color, this spawn can inoculate more wood chip–filled burlap sacks for maximum mycelial mileage, or be mixed into wood chip beds to create mycological landscapes.

Bunker spawn generated by any method of inoculation (stem butt, pure spawn, native mycelium, cardboard sheet spawn, or spore-mass slurry) creates a mycelial footprint of substantial mass. The methods described here could generate hundreds of tons of mycelium. The sacks are easily used in conjunction with commercial filling systems, chippers, and bark blowers. Mulching equipment already in use by restoration and landscape companies would require only simple modifications in order to serve mycological needs. Naturalized bunker spawn works superbly and is less expensive than pure culture spawn. This simple technique may be the bridge of knowledge that so many mycorestoration and bioremediation projects have needed in order to achieve success.

COMBINATION METHODS FOR CREATING BUNKER SPAWN

Mushrooms are often mysterious in their ways. Sometimes, they inexplicably fail to grow. If the substrate is suitable and none of the above methods has worked, then I recommend that combinations of the previously mentioned strategies be used. Combining spores, stem butts, and mycelium to create a complex inoculum often leads to development that surges past barriers that would ordinarily stop the mycelium if only one method was used. The bottom line: If you want the best chance of culturing your mycelium, then inoculate the burlaps sacks using 3 methods: spore printing, stem butt inoculation, and mycelium transplantation. If the conditions are right, then the surge of regrowth is nearly unstoppable.

PREFERRED PATHS OF EXPANSION FOR BUNKER SPAWN

Now that I have described several methods of getting the mycelium to run using bunker spawn, what paths are likely to lead to the greatest success? Methods that are most effective in the Pacific Northwest may not work as well in Thailand. Nevertheless, the principles are the same; it's the techniques that may differ. As a cultivator, you will face critical decisions as you follow the path toward mycelium expansion. The good news is that several paths lead to success. Which one is best for you depends upon your experiences.

Here are 3 paths that I prefer in my bioregion.

- Pure culture naturalized plug spawn mixed into fresh, raw wood chips in burlap bags (100 to 200 plugs per sack).
- Sawdust spawn or spawn from a recycled mushroom kit stuffed into wood chip–filled burlap sacks.
- Stem butts used in 2 ways: (a) to make wood dowel spawn, which is in turn mixed into wood chips placed in burlap sacks, and (b) to make wood dowel spawn, which is in turn mixed into wood chips to make a mushroom garden as shown in figures 157 to 159.

The first uses laboratory-grown spawn, preferably acclimated. The second takes advantage of natural spawn. The third uses stem butts as a source for mycelium. All 3 can create bunker spawn, which can subsequently be expanded. However, the third method is particularly favored, since once the mycelium naturalizes, meaning that its immune system adapts to the complex microbial environment in the wild, it really runs, and it is aggressive. Pure culture spawn and recycled mushroom kits, on the other hand, must overcome adverse microbial hurdles before the mycelium can move.

➤ **FIGURE 169**

La Dena Stamets holding oyster mushrooms fruiting from a bag of trimmings of U.S. paper currency, which is made from hemp and cotton. (You can make money growing mushrooms on money!) The U.S. Treasury trucks its paper-money trimmings to recycling centers. Nearly 2 pounds of oyster mushrooms fruited from this 6-pound bag of trimmings—the most efficient yield I have seen from a first flush.

Subsets of this second-generation naturalized spawn can be selectively expanded by at least 2 orders of magnitude. In effect, each bag of bunker spawn can generate 10 to 100 more bags. Once mushrooms begin fruiting, stem butts from the new crops can inoculate dowels or corrugated cardboard panels, thus renewing the cycle of expansion. Growers who can successfully expand the mycelium by incorporating a variety of propagation techniques heighten the art of cultivation. Like an orchestra conductor, a good grower conducts a strategy that makes the best use of the instruments in the theater and is sensitive to peculiarities of the players

Alternative Materials for Growing Mycelium and Mushrooms

Most new cultivators are astonished, as I was, by the variety of waste products mushrooms can decompose. For example, most of the major debris that comes from farming or forestry can be used in some fashion by mycologists. Many other materials, listed below, can also be used. Mycelium grows best on most of these materials when they have been broken into smaller pieces, $1/4$ inch to 2 inches, and are 60 to 80 percent saturated with water. Mixing fine and coarse materials together often creates a friendly matrix for the mycelium, since it loves complex mixtures with fractal-like microstructures. However, with most of the species in this book, consistently using fresh material that is known to be hospitable to your mycelium can make the learning curve less steep.

Many of these materials are discussed in more detail in *Growing Gourmet and Medicinal Mushrooms* (Stamets 2000a) and *The Mushroom Cultivator* (Stamets and Chilton 1983). The excellent website www.mushworld.com also provides extensive lists and details recent cultivators' successes. MushWorld has published 2 excellent mushroom grower's handbooks, *Oyster Mushroom Cultivation* (Mushworld 2004) and *Shiitake Mushroom Cultivation* (Mushworld 2005), distributed for free to developing countries in an effort to help fight poverty.

The list of alternative materials below, compiled from the collective experiences of cultivators worldwide, will expand as cultivators like you continue to experiment. These materials can be processed using the methods described in this book for making spawn, expanding natural mycelium, and creating mycelial mats:

- bamboo
- brewery waste
- cacao shells
- cacti
- coconut and coir (coconut husk fiber)
- coffee beans, grounds, hulls, and debris
- corn, corncobs, and cornstalks
- cotton and cotton waste
- fabrics
- garden waste, grass clippings, and yard debris
- hair
- hemp
- leaves
- manure
- nut casings and seed hulls
- oils (vegetable and petroleum)
- paper products (newspapers, cardboard, money, and books)
- soybean roughage
- straws (wheat, rye, rice, oats, barley, etc.)
- sugarcane and sugarcane bagasse
- tea, tea waste, leaves, and trimmings
- textiles
- tobacco and tobacco stalks
- trees, shrubs, brush, and wooden construction waste
- water hyacinth

If you are uncertain which material to choose for running mycelium, choose a few that are readily available and run a preference test by placing mycelium on each one. By inoculating several materials on the same day, you start a mycelial race. The materials most quickly colonized, as a general rule, are usually the best. Experimentation using mini-trials will give you ideas that can make larger projects less tedious and more likely to succeed. Have fun with your fungus!

CHAPTER 10

Cultivating Mushrooms on Straw and Leached Cow Manure

Despite the fact that only a few fungi are native to straw, many mushrooms can be grown on cereal straws due to the mushrooms' powerful enzymes that break down plant fibers. Straws, especially wheat, barley, rye, and rice, are preferred. In order for mushroom mycelium to grow on straw, the mycelium must be given an advantage over competing microbes, which often prevent the mycelium from growing. The role of the cultivator is to set the stage so that mycelium predominates and forms a contiguous colony, which then consolidates into a nutritional platform sufficient for supporting mushroom growth.

We can treat straw with a variety of methods so that mycelium will grow upon it. My other books *Growing Gourmet and Medicinal Mushrooms* (2000a) and *The Mushroom Cultivator*, coauthored with Jeff Chilton (1983), discuss commercial methods. What I describe here, however, may be put into practice by small-scale cultivators, people who live in rural regions, and those who are simply interested in techniques that are less expensive than commercial methods.

I recommend the following methods, which will allow you to grow mycelium without having to spend a lot of money:

- cold incubation
- hydrogen peroxide (peroxidases; enzymes) treatment
- heat pasteurization

Other commonly used methods, which I do *not* recommend due to their environmental toxicity, include the use of formaldehyde, chlorine, methyl bromide, and gamma radiation. Although effective against molds and insect pests, these processes, or their residual effects, pose a health risk to people, nontargeted organisms, and the environment. Unfortunately, the use of these toxic chemicals is common in developing countries such as China, whose environmental laws do not yet reflect our modern understanding of chemical toxins. In growing mushrooms to benefit human and environmental health, it makes no sense to use toxic methods when alternatives are available.

Cold Incubation

Cold incubation is a method I've developed that takes advantage of the fact that some cold-weather mushroom strains will grow well at low temperatures; often well below the minimum threshold for growth of competitors. In this method, the straw (or wood chips, in some cases) is kept at 35 to 50°F for several weeks after it has been inoculated with mycelium from spawn laboratories or natural patches. Incubating at lower temperatures slows the principal competitors and gives the mycelium an advantage.

The first step on the path to success is choosing the right strain of a cold-loving mushroom such as an oyster, enoki (*Flammulina populicola* or *Flammulina* velutipes), or woodlover (*Hypholoma* species) to place into untreated straw. Use the most vigorous mycelium possible. Simply broadcast the mycelium onto wet straw, inoculating at a 10 to 30 percent rate, meaning the mass of the spawn is 10 to 30 percent the total mass of the receiving substrate. Mix thoroughly and then place this mixture into your incubation vessel: a plastic bag, a box, a tray, an old bathtub—whatever you think is suitable. Laying parallel layers of spawn can help speed up colonization in some circumstances. Incubate outdoors.

While cold incubating mycelium on straw, cover with a shade cloth, cardboard, or plastic (if the straw is placed in a shady spot). Incubate until fully colonized, when the brown straw has become white with mycelium. Once colonized, the mycelial straw is brought to a suitable temperature for fruiting. (See chapter 14 for the ideal temperatures for fruiting for each species.) This method, although the simplest, incurs the highest risk of failure; but less so for cold-tolerant species of *Hypholoma*, *Stropharia*, *Psilocybe*, *Agrocybe*, and similar wood chip–loving terrestrial species that also favor straw substrates.

Peroxide Treatment

Mushroom mycelium sweats peroxidases. These digestive enzymes are free radicals that break down plant fibers and also kill foreign spores. The natural secretions are the mycelium's way of neutralizing competing spores surrounding it. Fortunately for us, hydrogen peroxide is readily available, relatively inexpensive, and effective for treating small batches of straw or sawdust, even burlap.

Peroxide treatment is great for giving the mycelium a head start on competing microbes. The expense of hydrogen peroxide is less than that of the equipment and fuel required for heat pasteurization, which has a similar effect. Another advantage is that peroxide is fairly harmless—H_2O_2 (hydrogen dioxide, another name for hydrogen peroxide) creates only water and oxygen as by-products. Its main limitation is

that this method is more convenient for small-scale applications than for large-scale ones.

The peroxide available for use by consumers at retail stores is approximately 3 percent concentration. Other products with higher concentrations are available for commercial use and are sold under various brand names, including Oxidate, for instance, which contains 27 percent hydrogen dioxide.

To treat straw with hydrogen peroxide:

1. Acquire 1 liter (about 1 quart) of hydrogen peroxide diluted in water to a concentration of 3 percent. (Caution: Higher concentrations can cause burning!)

2. Soak straw with water until thoroughly wet, drain, rinse again with water, and drain. Gather 5 to 10 pounds of straw, and place in a waterproof container (plastic bag, bucket, or other container).

3. Pour 1 liter of hydrogen peroxide into 1 gallon of fresh water, then pour onto the straw, thoroughly mixing and, preferably, submerging the straw into the peroxide bath. Cover and soak for 24 hours.

4. Rinse the peroxide-saturated straw with fresh, clean water, submerging the straw, and then drain. Rinse and drain again.

5. After rinsing the second time with clean water, immediately introduce mycelium to the straw, evenly distributing it mixing with your hands. Make sure your hands are clean and moist before you touch the mycelium.

6. Perforate the sides and bottom of the plastic bag with holes every few inches or, if using a bucket, every 4 inches; perforate the top of either the bag or bucket every 4 inches to allow for some passage of air. (The holes are for drainage, intake of fresh air, and exhaust of carbon dioxide.)

7. Incubate for 1 to 3 weeks, depending upon the species. Once the straw is colonized, you can either fruit mushrooms from the myceliated straw or use the straw as spawn for inoculating more substrates.

8. Initiate fruiting (see chapter 14 for fruiting processes of individual species). Typically, this involves exposing the myceliated straw to the ideal temperature, watering, and exposure to indirect natural or fluorescent light.

An alternative to pure hydrogen peroxide is One Step, sodium carbonate complexed with hydrogen peroxide or iodaphor, a commercial disinfectant used by the brewing and winemaking industries. We have had success diluting 120 ml iodaphor in 10 gallons of water to create a dilution of approximately 50 ppm. After soaking in this solution overnight, the straw is twice flushed with fresh cold water. After the straw is drained, spores and/or spawn are introduced.

Caution: Concentrated disinfectants, including hydrogen peroxide, can be dangerous to handle. Follow the manufacturer's recommendations and take extra precautions to prevent accidental injury! Dispose of them in an environmentally responsible manner.

Heat Pasteurization

The method with the longest history of use, and the one preferred by the majority of cultivators, is to use heat to pasteurize straw before it is inoculated. Pasteurization is a process in which the material in question is heated to approximately 160°F and maintained at that temperature for a particular period of time, which depends upon a few factors. The mass and chop size of the straw affects density (and thus the length of time required for pasteurization), and heat penetration is further affected by the methods used. The 3 methods of heat pasteurization are submerging the straw in hot water, injecting the straw with steam, and using dry heat to elevate the substrate temperature. Once treated and cooled, the straw is then inoculated with grain spawn at a 10 to 20 percent inoculation rate. In most cases, if you inoculate the straw with enough grain spawn directly after cooling, colonization is complete within a week and microbial competitors don't have enough time to mount an effective offense.

Pasteurization by Submersion in Hot Water

The first method is to pasteurize dry straw by placing it in a food-grade metal drum (or other suitable container) that has been partially filled with water, and heating it from underneath to 160°F. (Many cultivators use propane-fueled burners or wood-fired ovens, but rarely electricity. Some growers in tropical countries use solar ovens but clouds that block the sun or changing sun angles can make it difficult to keep the temperature regulated for long enough periods.) After being submerged for an hour, the hot, wet straw can be removed and spread upon a cleaned or plastic-covered table for rapid cooling (see figures 170 to 174). Usually, a 55-gallon drum filled with straw will, after inoculation and drainage, yield 100 to 150 pounds of moist straw. Once the straw has cooled to below 100°F, spread spawn evenly over the surface and mix in thoroughly by hand.

Whichever method of pasteurization you choose, there is a limited of window of opportunity for successful mycelial grow-out—typically about a week to 10 days before competitor molds emerge. If you overpasteurize, cooking the straw at too high a temperature (more than 190°F), black and pink molds can appear in just a few days.

Steam-Injection Pasteurization

The second method of pasteurization is to inject steam into the straw. However, steam will not moisten dry straw sufficiently to allow mycelium to grow. Using the steam method, most cultivators first spread the straw onto a cement surface or plastic tarp and wet it down with a garden hose, periodically turning the straw to expose dry zones. Once the straw has absorbed sufficient water, typically 75 to 80 percent saturated, then the straw is exposed to live steam. Some cultivators put the straw in a suitably sized container or custom-built insulated box and inject steam from underneath, under pressure. Various stem-injection equipment can be used. Oftentimes a small hole must be made in the chamber to allow for steam injection and condensation drainage. Long-shafted or remote reading thermometers are inserted to measure core temperatures, usually the coldest regions and last to heat up. (Typically the coldest location is in the lower center region.) When the thermometer reads 160°F, the steam is slowed to maintain a temperature of between 160 and 170°F for 1 to 2 hours. At this point, the steam is stopped and the straw is allowed to cool in its container.

As cultivators increase the mass of substrate they process, keeping the air around the cooling straw clean becomes increasingly important. Some cultivators install fan-powered micron (HEPA) filters so that the container remains pressurized with spore-free air as its contents cool. Another method is to make the container practically airtight, with only 1 port of air entry, for instance the port through which the steam was injected. In front of this airport, a down-spraying water mist rinses the incoming air free of contaminant spores during cooling. Other alternatives include using cotton, gauze, or similar fabrics to help reduce contaminant spore load. Your experience will guide you. Once cooled to 90 to 100°F, inoculate the straw with mycelium as described with the hydrogen peroxide method.

Dry Heat Pasteurization

Growers who cultivate mycelium in trays particularly prefer this third method. Wooden, metal, or plastic trays are filled about 8 to 12 inches deep with very moist straw (or compost; about 80 percent moisture), placed in an insulated, water-resistant room, and stacked upon each other with several inches of space between them. (In order for this method to work, the substrate must have sufficient moisture, usually 5 to 10 percent more than that required in other methods. Dry heat pasteurization will make use of the substrate's natural moisture to help steam the room and its trays.) Once the trays are in place, the room is injected with dry heat, elevating the substrate temperature to 160°F. A propane-fired "jet" burner, inserted through a hole in a wall or door to draw air from the outside, is a convenient source of dry heat. Don't use kerosene-fueled units. The intake of oxygen to the burners should be drawing air from outside the steam room. Also, the propane tanks fueling the jet burners should be outside and downhill from the flame jets heating the pasteurization chamber. Otherwise the room can become dangerously deoxygenated, encouraging anaerobes to grow, or if there is a propane leak, the jet burner could ignite the fumes, causing a disaster.

Depending upon the mass of the substrate and the room's size and insulation, temperatures will slowly climb over several hours. Thermal layering will occur; you can expect the air temperature near the ceiling to be 20 to 30°F higher than the temperature near the floor. Some cultivators use circulating fans to push down the hot air so that the air and temperatures in the room are evenly mixed, although fans are notorious for their short lives under these conditions. Monitoring air temperature will give you an idea of heat input, but the temperature of the substrate, especially the tray on the bottom layer (the coolest), should be your guide. Remote thermometers make it easy to check the substrate's temperature.

How much heat is needed? Here is a baseline to help guide you: a room measuring 24 by 30 by 12 feet (8,640 cubic feet) that is filled to about a tenth of its volume with moist substrate will require the input of 300,000 BTUs per hour from propane burners over several hours. Trays are filled to a maximum depth of 12 inches, and in this case 750 to 850 square feet of trays are used. Once the coolest tray has reached 140°F, the heat should be reduced to a trickle, or even turned off. In most cases, thermal momentum will continue to heat the substrate to the targeted 160°F plateau. The goal is for the substrate to hold this temperature for 12 hours. (Most rooms with drywall will fall apart under these conditions. The surfaces must be waterproof and tolerant of heat up to 180 to 200°F.) Such a room can grow up to 5,000 pounds of fresh mushrooms, with 2,500 pounds being an achievable first run every 8 weeks, depending upon the species and other factors.

This type of dry heat system using trays can be used for growing oyster, button, Brazilian blazei, paddy straw, and many other mushrooms.

Additional Comments Regarding Heat Pasteurization

The key to successful pasteurization is to maintain the balance between the temperatures that neutralize competitors and those that preserve microbial allies. This balance is a tightrope for the cultivator, who must carefully monitor the process throughout. Many thermophilic organisms survive pasteurization. When the heated straw drops back down to room temperature, most of these thermophilic microbes go dormant. Nevertheless, they occupy an important niche, forestalling competition. Long ago, cultivators found that when the straw is cooked at 200°F for several hours, near or at sterilization temperatures, it is prone to contamination from several molds, especially black-pin molds in the genera *Aspergillus, Mucor,* and *Rhizopus*. See *The Mushroom Cultivator* (Stamets and Chilton 1983) for more information.

Once the straw cools, spawn is inoculated at a 10 to 20 percent rate and thoroughly mixed through the pasteurized substrate. Afterward, do not disturb the mycelium during its incubation. Mycelium needs a quiet place, protection from moisture loss, and access to oxygen. Keep out of sunlight until the time comes for initiating fruiting. The mycelium does better at a stable temperature, or within a limited range of temperatures. For many species, incubating at 65 to 75°F is ideal (see chapter 14 for optimum temperatures for some mushroom species). Wide fluctuations in temperatures hinder the mycelium and cause condensation to stream, carrying contaminants from one location to another.

Mushroom cultivation is an art. Pasteurization is a selective treatment, and straw is a selective substrate. Combining all 3 is an effective strategy for navigating around obstacles that limit success. Once the mycelium takes hold and springs back, mycelial momentum normally eclipses competition.

More capital-intensive methods used by small- to large-scale commercial growers are described in detail in my earlier book, *Growing Gourmet and Medicinal Mushrooms* (2000a).

⋏ **FIGURE 170**

Using a food-grade steel drum and crab-pot burner for hot-water or steam pasteurization of straw.

⋏ **FIGURE 171**

A specially designed stainless steel pasteurization chamber. Steam is injected through a port on the side. Once pasteurization is complete, the door is lowered and doubles as a ramp for unloading the straw.

 **FIGURE 172**

A cleaned table is an ideal surface for inoculating straw.

▲ **FIGURE 173**

Using a pitchfork, straw is tossed onto the clean table.

▲ **FIGURE 174**

Once the straw is cool, pure grain spawn is laid on top and then evenly mixed through. Approximately 8 pounds of grain spawn can make 20 to 40 bags (kits) weighing 6 to 8 pounds each. Each kit produces 1 to 2 pounds of fresh mushrooms 2 to 8 weeks after spawning.

 ◀ **FIGURE 176**

Once filled and closed using a twist tie, the bags are punctured using bladeless arrowheads mounted on a board.

▲ **FIGURE 175**

Damein Pack packs inoculated straw into bags.

Dusty Yao holds the phoenix oyster mushroom *(Pleurotus pulmonarius)* fruiting from pasteurized straw 2 weeks after inoculation with grain spawn.

Leached Cow Manure: A Readymade, Easy-to-Use Compost

Many years ago, I was introduced to leached cow manure as a readymade substrate for growing dung-loving mushrooms (Stamets 1978). Leached cow manure is obtained from manure collected by dairies from their milking barns. The manure is cleaned from the cement slabs in the barns and removed to a separator—a machine that separates the roughage from the liquid effluent; the latter is redistributed to the pastures as fertilizer. The roughage—a rich but fairly loose dark material similar in texture to peat moss—is typically piled up as waste material. Large dairies generate tons of this material per week but have little or no market for it, and so they will sell it at a very low price: just $5 to $10 for a typical truckload of about a thousand pounds.

The advantages of using leached cow manure include the following:

- It is a readymade compost ideal for many coprophilic mushrooms.
- Using this compost eliminates the need for expensive and often dangerous heavy machinery normally

⋏ FIGURE 178

La Dena Stamets holds the golden oyster mushroom *(Pleurotus citrinopileatus)* fruiting from pasteurized straw.

⋏ FIGURE 179

David Sumerlin examines oyster mushrooms *(Pleurotus ostreatus)* fruiting after immersion in a mycofiltration chamber. (See figure 72.)

needed for composting systems. The expense of this equipment limits entry into the market by smaller growers.

- The compost can be collected hot and self-heats to near pasteurization temperatures naturally. It can be used outdoors or indoors; if it's used indoors, transferring the compost into an insulated growing room for final pasteurization will require little additional heat. This third advantage is very important, since the natural warming by hungry thermophilic microbes offsets the cost of heating the substrate.
- The texture and composition allows for easy handling in filling and spawning.

The biggest disadvantage of using leached cow manure is that the yields are not as impressive as those now commonly touted by the button mushroom industry. Depending upon the species, mushroom yields from unsupplemented leached cow manure are 1 to 2 pounds per square foot, whereas button mushroom compost boasts 5 to 7 pounds per square foot. However, adding supplements like Spawn Mate can boost yields, and if you take into account the amount of start-up capital required of the home or small-scale commercial grower, leached cow manure seems economically attractive. Its ease of use makes it a preferred material of many who have tried it. In some ways, leached cow manure has been the best-kept secret in the mushroom cultivation industry.

The greatest limitation is availability. Is there a dairy operation near you that uses separators to disperse their leachate onto the grazing fields? Such dairy operations are usually detectable by their odor—you can smell them from afar. You can contact your local dairy association for a list of members in your area.

When you find a dairy that has a supply of leached cow manure, you'll notice that the manure is stored in high piles. This causes thermogenesis, or self-heating, often to internal temperatures of more than 140°F. This temperature is ideal for the growth of actinomyces, a group of thermophilic organisms that bind nitrogen over time, later to be exploited by the mycelium at room temperature. Remove the first 1 to 2 feet of the outer layer of the leached cow manure

and transfer the deeper, steaming-hot compost into an awaiting truck. Cover with a tarp and drive directly to your growing facility.

After the compost is trucked to the growing facility, it needs to be "finished off." What this means is that the compost needs to mature, outgassing ammonia and carbon dioxide. The compost can be laid out on a tarp in a mound (no more than 2 1/2 feet high), and covered. Over a few days, it will naturally heat up. After 5 to 7 days, the mound should be in-turned and restructured by flipping it with a round-headed shovel. The leached manure compost will mature a few days to a week thereafter, depending upon outside temperatures and how fresh (stinky) it is.

An alternative method pioneered and preferred by my good friend David Sumerlin (2004) is to load the compost into open beds, 4 feet by 4 feet and 8 inches deep, lined with plastic and overlaid upon a mesh of chain-link fence, all in a waterproofed room. The compost is loosened as it is added to the beds. Once the compost is mounded in each bed, the waterproof room is sealed and dry heat is injected using the methods described on page 164. David likes to bring the air temperature to 160 to 180°F, which is followed by a rise of bed temperature to 140 to 160°F for 24 hours. After the substrate temperature has been approximately 150°F for at least 8 hours, the heat is turned off (or during very cold weather slowly regulated) so the compost "coasts" in its temperature profile. Over 12 to 24 hours, the compost stays in the 140 to 150°F range, outgassing nitrogen. Then the compost is lowered to 118 to 130°F, an ideal temperature for actinomyces to bloom, for up to 5 days or until the ammonia smell dissipates. As the actinomyces proliferate, less ammonia gas is generated because nitrogen becomes incorporated into the cellular composition of these microbes. By the seventh day, the temperature of the compost has gradually cooled, and when it is at less than 100°F, spawning can begin.

Grain spawn is the preferred form of mycelium used for inoculation of leached cow manure. (To purchase grain spawn, go to www.fungi.com or consult the

The Brazilian blazei *(Agaricus brasiliensis* or *Agaricus blazei* ss. Heinemann)* fruiting from leached cow manure indoors, using the methods described here.

▲ FIGURE 181

The regal *Psilocybe cubensis* fruiting on leached cow manure outdoors. This mushroom is legal to cultivate in some countries but not in others.

Resources section in this book.) The rate of inoculation depends upon several variables: cost of spawn, strain vigor, and type of grain. As a baseline, $1/4$ to 1 pound of spawn is used per square foot of leached cow manure. If you can make your own grain spawn, then the cost of using 1 pound per square foot will be relatively low. (If you wish to make your own, see my book *Growing Gourmet and Medicinal Mushrooms* [2000a]). When using a high spawn rate, you can expect colonization to take place within a week, plus the grain itself becomes a supplement without making the substrate more susceptible to contamination, since it's already encapsulated by mycelium.

Once the spawn run is completed, usually in 1 to 3 weeks, the compost can be covered with a layer of moist peat moss $1/2$ to 2 inches thick. The mushroom mycelium is triggered into primordia formation—the first showing of baby mushrooms—just as the mycelium shows on the surface. This is typically a result of watering, lowering temperature, introducing fresh air to displace carbon dioxide, and exposure to light. For more information on these methods, see *The Mushroom Cultivator* (Stamets and Chilton 1983).

Methods making use of leached cow manure can be further refined to boost yields. Compared to the arduous tasks employed by the button mushroom industry—building compost ricks and turning the compost—using leached cow manure is far easier and succeeds in overcoming a major economic hurdle that has limited entry into the market by small-scale growers.

Using Spent Mushroom Compost: Amendments for Soil Enhancement

The word *compost* is so general, it's hard to define. One dictionary describes compost as "a mixture of decaying organic matter, as from leaves and manure, used to improve soil structure and provide nutrients" (www.dictionary.com). What this definition leaves out is the fact that hundreds of billions of organisms unite in compost's myriad forms. Mushroom compost, an excellent soil amendment for gardeners that is commonly acquired from mushroom farms, does not contain mycorrhizal fungi. Often mixed into other soils or used as a topdressing for plants, mushroom compost, and compost in general, is not usually used as soil itself for any number of reasons. Foremost among these is that it's too nutritious (too "hot") or not suitably balanced for the growth of the desired plants. Nevertheless, mushrooms and mushroom farms generate streams of compost that can form the basis for good soil.

The largest producers of mushroom compost are button mushroom *(Agaricus bisporus)* farms. To grow their mushrooms, these farms use straw as a structural base, to which manure (primarily from chickens, less frequently from horses, and more rarely from cows), calcium, cottonseed hulls, and other supplements are added. Since the button mushroom is a secondary decomposer—unlike the shiitake and oyster mushrooms, which are primary decomposers—the substrate must be composted, a process of microbial transformation, before it is suitable for growing them. After composting and "phasing"—a term used to describe the heat conditioning of the compost to around 120 to 130°F, the compost is cooled in preparation for inoculation with spawn. When the button mushroom crops diminish, the spent compost becomes a secondary by-product and is often sold off as a soil amendment product.

If the mushroom farms are certified organic—like our gourmet mushroom farm—the spent compost is assuredly free of pesticides. However, most non-organic mushroom farms use pesticides and these toxins survive in their composts. Because of increasingly restrictive pesticide regulations, some pesticides and fumigants, such as methyl bromide, Dursban, chlordane, parathion, and many others have been banned in the United States and in many other countries. But the chemical pesticides still in use also contaminate groundwater and can be toxic to nontargeted organisms. Similarly, many of the "inert" ingredients used in fertilizers and soil amendments come from toxic waste recycling centers that redistribute heavy metals and other inorganic ingredients. When considering using compost from a mushroom farm, ask to be informed regarding its pesticide and heavy metal content. You should be careful not to unwittingly import hidden toxic wastes into your soil while striving to improve soil heath.

Button mushroom composts are very different from the structurally coarse compost many gardeners make—replete with sticks, stems, food scraps, and so on. Gardeners tend to top-dress or mulch with compost that is in an intermittent stage of decomposition. In contrast, button mushroom compost is much farther along and qualitatively distinct. And the mushroom compost made by gourmet mushroom growers is different still from these others. Our compost is made from recycling aged shiitake-, oyster-, maitake-, and reishi-colonized sawdust "blocks" (what we call our mushroom kits and production units). See *Growing Gourmet and Medicinal Mushrooms* (Stamets 2000a).

Since most the mushrooms we focus on growing are primary saprophytes, we only use manure for specialty species. Our farm's main production systems are based on sawdust. Once the sawdust blocks flush with gourmet mushrooms, we throw them into a pile and let nature take over. (See figure 182.) Red worms gobble up the spent substrate, which is rich in mycelium. Birds come to feed on the worms. Constellations of other organisms cascade, working in concert with the mycelium, to transform the waste into rich humus. In 4 to 6 months, the compost is ready, resulting in a nutritious soil.

▲ FIGURE 182

Piles of mushroom compost at Fungi Perfecti made from spent sawdust fruitings. Once the mushrooms have fruited, the spent sawdust substrates (we call them "blocks") are thrown into a pile and turned every other month. The speed of this conversion of sawdust to dirt is remarkable—taking only 4 to 6 months—and illustrates the valuable role mycelium plays in building soil.

Any gardener with a compost pile has noted how much the pile shrinks over time. This is biological reductionism in the extreme: the composts lose mass as they outgas, releasing carbon dioxide and, to varying degrees, ammonia and exude water. One researcher, Zadrazil (1980), calculated that more than 50 percent of the carbon from straw is liberated as gaseous carbon dioxide while oyster mushrooms grow upon it.

Rapid composting, although it may seem alarming to those concerned about global warming, actually reduces carbon dioxide in the long term by providing a nutritional soil for oxygen-generating plants. The quick return of nutrients back into the food chain fortifies plant communities that consume increasingly more carbon dioxide as they proliferate. The dance between carbon dioxide producers, the fungi, and carbon dioxide consumers, the plants, is the foundation of the continuous flow of nutrients in evolving environments. Our modern lifestyles disrupt this balance by creating more debris fields than can be returned to the food chain, causing deficits in the nutrient cycles and destabilizing the harmony of healthy ecosystems.

Whether primary or secondary saprophytes produce soil, and whether the soil is already rich with microbes, without mycorrhizae the plants are weaker and more susceptible to opportunistic disease. Gardeners using mushroom compost need mycorrhizae to help their plants. Mycorrhizae can be introduced at any of several stages in the plants' life cycle, although earlier is better. For techniques on using inoculating with mycorrhizae and gardening with gourmet mushrooms, see chapters 9 and 12.

Now that we have explored growing mushrooms on straw and leached cow manure, let's progress to growing mushrooms on wood. The majority of mushrooms in nature used wood as a niche, and we can grow more saprophytic gourmet and medicinal mushrooms on wood than we can on straw and manure.

BY-PRODUCTS OF STRAW SUBSTRATE DUE TO CONVERSION BY *PLEUROTUS OSTREATUS*

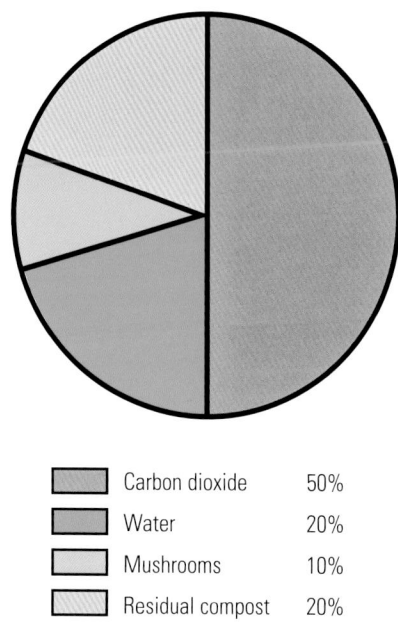

Carbon dioxide	50%
Water	20%
Mushrooms	10%
Residual compost	20%

▲ **FIGURE 183**

Pie chart showing proportion of carbon released relative to total mass. Carbon is either released as carbon dioxide, incorporated into mushroom tissue, or retained in the "spent" substrate prior to digestion by other microbes.

CHAPTER 11

Cultivating Mushrooms on Logs and Stumps

Traditionally, dowel ("plug") spawn has been used to inoculate logs and stumps, although shiitake cultivators sometimes also use sawdust. The advantage of dowel spawn is its staying power—the dowels survive longer than the sawdust, which is quickly consumed by the mycelium and munched by insects. The dowel's longer survival time gives the mycelium a better chance of infecting the log. However, automated sawdust inoculators are far faster than hand-delivered plug inoculation methods. Sawdust spawn is less expensive by weight than plug spawn, but sawdust spawn is less dense and is prone to falling out of the inoculation sites unless packed tightly. One's choice is a matter of experience and scale. For homeowners, gardeners, and those whose labor expense is not a determining factor, I recommend using dowel spawn for inoculating logs and stumps. In large commercial operations, the benefits of automation usually tip the scales in favor of sawdust spawn. In growing mushrooms on stumps, plug or sawdust spawn can be used, separately or in combination with rope spawn and spored oils (see pages 183–185).

Growers who use sawdust spawn to inoculate logs and stumps place a cap on the inoculation hole after inserting the sawdust in order to keep the sawdust from falling out. Some shiitake growers in Korea, for instance, sell equipment that automatically inserts sawdust spawn into the inoculation hole, in some cases capping it with a plug of Styrofoam. However, this method seems contrary to the philosophies of recycling, sustainability, and reducing pollution, tenets usually supported by those growing mushrooms naturally. Most log growers of shiitake—whether they inoculate the logs with sawdust or dowel spawn—cover the holes with a dab of melted wax. Cheese wax, bee's wax, and, less preferably, paraffin are used.

Another method of inoculation, developed by some innovative Japanese growers, is "wafer spawn"—thin cellulose wafers grown through with mycelium that are inserted into cuts in the log or stump. The cutting

▲ **FIGURE 184**

Plug spawn for inoculating logs. Mycelium grows in the grooves of the spiral dowel, helping it recover from the shock of inoculation.

equipment and wafer dimensions are sized so the task of insertion can be completed quickly, in some cases in a single pass. Although this innovation may be successful, I prefer dowel plug spawn because it's vigorous and easy to handle, and sawdust spawn for its economy.

Which Trees Are the Best for Cultivation?

The ideal trees for stump or log cultivation are those that the mushroom species grows upon naturally. The simplest way to find a mushroom's natural host is to consult a reputable field guide's description of the preferred habitat of that species. Many saprophytes, however, are opportunistic—if given the opportunity, they can adapt to types of wood not ordinarily found in their natural domain. I'll use shiitakes as an example to illustrate the selection and growing process of mushrooms on trees. Shiitakes grow wildly on a wide number of hardwoods in Asia, particularly oaks. Pioneers in shiitake cultivation long ago discovered that other woods would also support this mushroom—with a little encouragement. Fortunately, we have several decades of experience comparing shiitake strains and yields.

A good study, conducted by Joe Deden and the Minnesota Forest Resource Center, expanded the cast of characters of trees that will support shiitake to include woods like elm, maple, ironwood, and honey locust. Several themes emerged from this and subsequent studies. Although oak is the natural host for shiitake, alder, eucalyptus, and other rapidly decomposing hardwoods are readily colonized by this species. Generally, mushrooms are produced sooner on these than on the denser hardwoods that are slower to decompose. Shiitake mushrooms form on alder (*Alnus rubra*) in 6 to 12 months, whereas oak can take 1 to 2 years and the denser ironwood can take 2 to 3 years before first flushing. A significant advantage of denser woods like ironwood is a longer life span for fruiting, often lasting 5 to 10 years. Another major factor affecting life span is the bark layer. Many mushrooms inoculated onto rapidly decomposing hardwoods like alder, birch, and aspen stop fruiting once

the bark has peeled off. However, other mushrooms, such as *Hypholoma capnoides* and *Hypholoma sublateritium*, continue to produce well after the bark layer has fallen off. In fact, I have seen *Hypholoma sublateritium* fruit 8 years after inoculation on fallen alder logs when the logs had decomposed significantly, into loose, long, pulpy fibers.

On pages 174 to 176 is a list of tree species that are suitable for growing many of the saprophytic mushrooms listed in this book. This list is not exhaustive; many other tree species may also be suitable. (If you have successes with tree species not listed, please write me about your experiences.) As always, I encourage experimentation, but I recommend conducting mini trials before investing much time and money.

A couple of trees not listed are cedars and redwoods. The only mushroom species I know of that has the enzymatic strength to saprophytize these difficult-to-decompose woods is, again, the heroic clustered woodlover (*Hypholoma capnoides*). The cedar and redwood logs I have seen supporting this mushroom were aged at least 20 years. I suspect that the mycelium struggles to decompose these woods for a long time before a mushroom forms. For this reason, I would suggest using this tree-fungus pairing for experimental or environmental purposes and not for production.

In many cases, storms or catastrophic events provide "windfalls" of wood for mushroom growers, who can assist with efforts to clean up storm debris. In many urban and suburban areas, homeowners commonly pay people to remove their fallen trees! Alternatively, I know of several arborists who have added mycelium to their repertoire of services and products and thus set themselves apart from competitors, increasing their markets. These mycologically astute small business owners offer a variety of plug and outdoor spawn so homeowners can select their preferred fungal friends for recycling wood debris. Burning storm debris is increasingly unpopular and, in some municipalities, illegal. As an alternative to burning, which causes pollution, sudden emissions of carbon dioxide, and denutrification of soil, growing fungi on

List of Suitable Tree Species for the Cultivation of Gourmet and Medicinal Mushrooms

Scientific Name	Common Name	Scientific Name	Common Name
Abies spp.	Fir	*Carpinus japonica*	Japanese hornbeam
*Abies alba***	White fir	*Carpinus laxiflora*	Loose flower hornbeam
Abies amabilis	Pacific silver fir	*Carpinus orientalis*	Oriental hornbeam
Abies magnifica	California red fir	*Carpinus tschonoskii*	Korean hornbeam
Abies procera	Noble fir	*Carpinus turczaninowii*	
Acacia spp.		*Carya* spp.	Hickories
Acacia mangium	Forest mangrove	*Carya aquatica*	Water hickory
Acer spp.	Maples	*Carya cordiformis*	Bitternut hickory
Acer macrophyllum	Bigleaf maple	*Carya glabra*	Pignut hickory
Acer negundo	Box elder	*Carya illinoensis*	Pecan
Acer rubrum	Red maple	*Carya laciniosa*	Shellbark hickory
Acer saccharum	Sugar maple	*Carya ovata*	Shagbark hickory
Ailanthus altissima		*Carya texana*	Black hickory
Alniphyllum fortunei	Alders	*Carya tomentosa*	Mockernut hickory
Alnus spp.	White alder	*Castanea* spp.	Chestnuts
Alnus alba	European alder	*Castanea crenata*	Japanese chestnut
Alnus glutinosa	Gray alder	*Castanea henryi*	Chinese chestnut
Alnus incana	Japanese alder	*Castanea mollissima*	Blume Chinese chestnut
Alnus japonica	Red alder	*Castanea sativa*	Spanish chestnut
Alnus rubra	Hazel alder	*Castanea sequinii*	
Alnus serrulata		*Castanopsis* spp.	Chinquapins
Alnus tinctoria	Tree of heaven	*Castanopsis accuminatissima*	Berangan
Altingia chinensis		*Castanopsis argentea*	Sarangan
Arbutus spp.	Madrones	*Castanopsis caerlesii*	
Arbutus menziesii	Pacific madrone	*Castanopsis chinensis*	Chinese chinquapin
Betula spp.	Birches	*Castanopsis chrysophylla*	Golden chinquapin
Betula alleghaniensis	Yellow birch	*Castanopsis cuspidata*	Shii tree
Betula dahurica		*Castanopsis fabri*	White beam
Betula lenta	Sweet birch	*Castanopsis fargesii*	
Betula nigra	River birch	*Castanopsis fissa*	Chestnut oak
Betula papyrifera	Paper birch	*Castanopsis fordii*	
Betula pendula	European birch	*Castanopsis hickelii*	
Betula pubescens	Hairy birch	*Castanopsis hystrix*	Katus
Carpinus spp.	Hornbeams	*Castanopsis indica*	
Carpinus betulius	European hornbeam	*Castanopsis lamontii*	
Carpinus caroliniana	American hornbeam	*Castanopsis sclerophylla*	
Carpinus fargesii	Asian hornbeam	*Castanopsis tibetana*	Tibetian chinquapin

Scientific Name	Common Name	Scientific Name	Common Name
Cinnamomum camphora	Camphor laurel	*Larix lyallii*	Subalpine larch
Cornus spp.	Dogwoods	*Larix occidentalis*	Western larch
Cornus capitata	Flowering dogwood	*Liquidambar* spp.	Sweetgums
Cornus florida	Flowering dogwood	*Liquidambar formosana*	Formosa sweetgum
Cornus nuttallii	Pacific dogwood	*Liquidambar styraciflua*	
Corylus spp.	Filberts	*Liriodendron tulipifera*	Tulip poplar
Corylus americana	American hazelnut	*Lithocarpus* spp.	Tanoaks
Corylus avellana		*Lithocarpus auriculatus*	
Corylus heterophylla	Siberian hazelnut	*Lithocarpus calophylla*	
Corylus maxima		*Lithocarpus densiflorus*	Tanbark oak
Distylium myricoides		*Lithocarpus glaber*	Japanese oak
Distylium racemosum	Isu tree	*Lithocarpus lanceafolia*	
Elaeocarpus chinensis		*Lithocarpus lindleyanus*	
Elaeocarpus japonicus		*Lithocarpus polystachyus*	
Elaeocarpus lancaefolius	Bhadrase	*Lithocarpus spicatus*	
Engelhardtia chrysolepis		*Mallotus lianus*	
Eriobotrya deflexa	Bronze loquat	*Malus* spp.	Apples
Eucalyptus spp.	Eucalyptus	*Morus alba*	White mulberry
Eucalyptus globulus	Tasmanian blue gum	*Morus rubra*	Red mulberry
Eucalyptus grandis	Grand eucalyptus	*Nyssa sylvatica*	Tupelo
Eucalyptus saligna	Blue gum	*Ostyra* spp.	Ironwood (hop hornbeam)
Eucalyptus urophylla		*Ostyra carpinifolia*	European hop hornbeam
Eurya loquiana		*Ostrya virginiana*	American hop hornbeam
Fagus spp.	Beeches	*Pasania* spp.	Shii tree
Fagus crenata	Japanese beech	*Peltophorum africanum*	African wattle tree
Fagus grandifolia	American beech	*Pinus contorta*	Lodgepole pine
Fraxinus spp.	Ashes	*Pinus lambertiana*	Sugar pine
Fraxinus americana	White ash	*Pinus ponderosa*	Ponderosa pine
Fraxinus latifolia	Oregon ash	*Platycarya strobilacea*	Asian walnut
Fraxinus nigra	Black ash	*Populus* spp.	Cottonwoods and poplars
Fraxinus pennsylvanica	Green ash	*Populus balsamifera*	Balsam poplar
Fraxinus velutina	Velvet ash	*Populus deltoides*	Eastern cottonwood
Garcinia multiflora	Saptree	*Populus fremontii*	Fremont cottonwood
Hevea brasiliensis	Rubber tree	*Populus grandidentata*	Bigtooth aspen
Juglans spp.	Walnut	*Populus heterophylla*	Swamp cottonwood
Juglans nigra	Black walnut	*Populus nigra*	Black poplar
Lagerstroemia subcostata		*Populus tremuloides*	Quaking aspen
Larix spp.	Larches	*Populus trichocarpa*	Black cottonwood
Larix laricina	Tamarack	*Prosopis* spp.	Mesquite

List of Suitable Tree Species for the Cultivation of Gourmet and Medicinal Mushrooms, *continued*

Scientific Name	Common Name	Scientific Name	Common Name
Prosopis juliflora	Honey mesquite	*Quercus palustris*	Pin oak
Prosopis pubescens	Screw-pod mesquite	*Quercus phellos*	Willow oak
Prunus cerasifera	Cherry plum	*Quercus prinus*	Swamp oak
Prunus domestica	European plum	*Quercus rubra*	Northern red oak
Pseudotsuga menziesii	Douglas fir	*Quercus semiserrata*	
Pyrus spp.	Pear	*Quercus serrata*	Nara oak
Quercus spp.	Oaks	*Quercus spinosa*	
Quercus acuta	Japanese evergreen oak	*Quercus variabilis*	Gulcham namu
Quercus acutissima	Sawtooth oak	*Quercus virginiana*	Live oak
Quercus agrifolia	Californa live oak	*Rhus* spp.	Sumac
Quercus alba	White oak	*Rhus glabra*	Smooth sumac
Quercus aliena	Oriental white oak	*Rhus succedanea*	Arkol sumac
Quercus bella		*Robinia* spp.	Locust
Quercus berberidifolia	Scrub oak	*Robinia neomexicana*	New Mexico black locust
Quercus brandisiana		*Robinia pseudoacacia*	Black locust
Quercus chrysolepis	Canyon live oak	*Salix* spp.	Willows
Quercus crispula	Japanese white oak	*Salix amygdaloides*	Peachleaf willow
Quercus dentata	Emperor oak	*Salix exigua*	Sandbar or coyote willow
Quercua dumosa	California scrub oak	*Salix fragilis*	Crack willow
Quercus emoryi	Emory oak	*Salix geyerana*	Geyer willow
Quercus fabri		*Salix lasiandra*	Pacific willow
Quercus falcata	Southern red oak	*Salix lasiolepis*	Arrow willow
Quercus gambelii	Gambel oak	*Salix nigra*	Black willow
Quercus garryana	Oregon white oak	*Salix scouleriana*	Scouler willow
Quercus glauca	Japanese blue oak	*Sapium discolor*	
Quercus grosseserrata	Mongolian oak	*Sloanea sinensis*	
Quercus kelloggii	California black oak	*Taxus* spp.	Yews
Quercus kerii		*Taxus brevifolia*	Pacific yew
Quercus kingiana		*Tilia* spp.	Lindens
Quercus laurifolia	Laurel oak	*Tsuga canadensis*	Eastern hemlock
Quercus lobata	California white oak	*Tsuga heterophylla*	Western hemlock
Quercus lyrata	Overcup oak	*Ulmus* spp.	Elms
Quercus michauxii	Swamp chestnut oak	*Ulmus americana*	American elm
Quercus mongolica		*Ulmus campestris*	English elm
Quercus muehlenbergii	Chinquapin oak	*Ulmus glabra*	Mountain elm
Quercus myrsinae		*Ulmus laevis*	Smooth elm
Quercus nigra	Water oak	*Ulmus parvifolia*	Chinese elm
Quercus nuttallii			

wood allows for the entrainment of nutrients back into the ecosystem, improvement in soil structure, and the gradual release of carbon dioxide over many years.

When to Cut Logs, and Which Logs Are Best?

Cutting your logs in the late winter or early spring when the sap begins to run helps to ensure high sugar content. The precise time depends upon your location. However, disregarding conventional wisdom, I have inoculated alder cut at all times of the year without producing noticeable differences in yield. Your results may differ, depending upon the trees and location.

Choose logs that are healthy. When choosing logs to inoculate, avoid woods with any sign of decay, wounds, evidence of blight, or preexisting mushroom growth. You can tell which logs are already infected by looking at their freshly cut ends. The ends of healthy logs show growth rings without mottling. The ends of infected logs show a marbling, since the mycelium runs parallel to the cell walls. If a side branch or an infection is seen near the cut zone, you can sometimes salvage the log by cutting away the infected zones. I recommend that you cut a full meter away from the last detectable area of discoloration, and that you inoculate the log at double the usual rate.

The length of the log is up to you. Most log growers cut logs approximately 1 meter long, stacking them off the ground until the time of use. (When logs make ground contact, many other organisms jump from the soil, causing infections.) With some species, such as reishi, a preferred method is to make "pony" logs, 2 feet or so in length that, after inoculation, are placed into black pots, which are then filled with gravel or sand (see figures 192 and 193).

How Are Logs Inoculated?

The standard method for inoculating logs is to drill holes and insert plug spawn or sawdust spawn into the holes. Molten wax is dabbed on the mycelium-filled hole to seal it and prevent beetles and other bugs from eating the mycelium. The hole size is usually around $5/16$ inch (8 mm). A drill is used to bore holes that are deeper than the length of the plugs. If you are inoculating only a few logs, the least expensive method is to use a drill or modified grinder holding a $5/16$-inch or 8 mm drill bit. (Modified grinders are far faster than drills.) In the Far East, cultivators use specialized equipment designed to automate the process. Although inoculating logs is labor-intensive, the experience is socially bonding and fun, especially for families and children.

Logs of many different diameters can be used. Often, they average 4 to 8 inches in diameter. A sufficient quantity of plug spawn for this size log would be 30 to 50 plugs. Logs greater than 10 inches in diameter are notoriously slow to fruit and require much more spawn. The plugs are spaced 4 to 6 inches apart in long rows along the log's length (see figure 185). The rows are staggered to form a checkerboard pattern of inoculation points, ensuring full and even coverage.

As the mycelium grows from the plugs, island colonies form and enlarge over time. Large-diameter logs have proportionately more mass than smaller ones, and hence the mycelium has greater real estate to conquer; naturally, fruitings on these logs begin much later. (I occasionally inoculate logs of very large diameter just to see what will happen. In some cases, I'm still waiting!) Most 4- to 8-inch-diameter logs fruit in 1 year from the time of inoculation, with some producing in only 6 months while others take up to 2 years.

Generally speaking, mushrooms growing on logs are damaged by direct sunlight and are best placed into a forestlike setting or under a porous canopy (see figure 187). Since humidity is high nearer to the ground and radically diminishes with elevation, many mushrooms benefit from being placed close to the ground or directly in contact with it rather than being stacked. On the other hand, ground-dwelling logs rot faster and are more subject to insect damage, and some species, such as shiitake, do not fair as well as when they are elevated. Consequently, placement should depend on the preferences of the species.

▲ **FIGURE 185**

Logs are positioned on a table, and drills equipped with $5/16$-inch drill bits are used to make holes. By making the holes $1^1/_2$ times deeper than the plug is long, the grower forms a "cave" deep within the log, allowing the mycelium to fill the chamber.

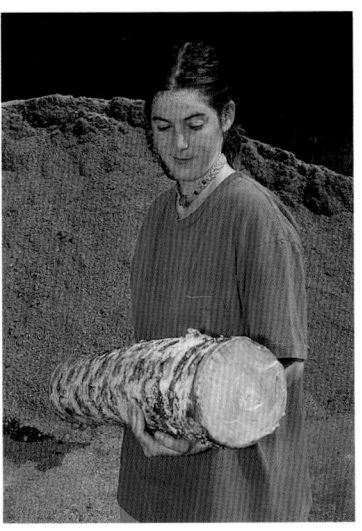

▲ **FIGURE 186**

Like some cultivators, Virginia Fraser likes to dip the ends of the logs to prevent invasion from competitor fungi. The holes are dabbed with molten cheese wax. A gourmet stainless steel baster is a good tool for efficiently placing a measured dose of wax in each hole.

▲ **FIGURE 187**

Once inoculated, the logs are laid to rest on a pallet or stacked in shady areas. For drier locations, I recommend covering them with a shade cloth or tarp to buffer against wild fluctuations in humidity.

◀ **FIGURE 188**

After 6 to 12 months, mycelium can be seen at the ends, showing discolorations, a sure sign that the logs are colonized end to end with mycelium. Note that the mycelia flow inward from the outer surface, where the spawn dowels were inserted.

◄ **FIGURE 189**

Damien Pack coaxed a shiitake flush from this alder log by immersing it overnight in water 2 weeks before.

▲ **FIGURE 190**

Steve Cividanes holds an alder log fruiting with reishi (*Ganoderma resinaceum*).

Fruitings occur 2 or 3 times a year, depending upon strain, wood, temperatures, and other conditions. About 1/2 to 2 pounds of mushrooms may be cropped from each log every year. By the fifth year, alder logs slow in their fruitings, and by the tenth year oak logs usually expire and fall apart. Maximum production is usually seen in the first third of the log's life span. In the end, the log, now made up of pulped wood, is a mere fraction of its original weight and disintegrates into long white fibers upon handling.

When the logs are grown through, the ends will show bands of discoloration caused by the emerging mycelium. Once the mycelium surfaces at the ends, the log is ready to fruit. (Several students of mine forgo fruiting altogether and incubate logs specifically to create "tiger" or "spaulded" maple, for instance, especially favored by cabinetmakers for its unique zonations and sheen. At one company I know of, the workers use reishi on maple to create a beautiful shimmering pattern.)

When one strategy fails to result in mycelium, using a combination strategy might overcome resistance. Having 4 options for inoculation is an advantage when dealing with the unknown. The combination method is not much more labor-intensive but is clearly beneficial because of the synergistic effect of complementary inoculations. For instance, when

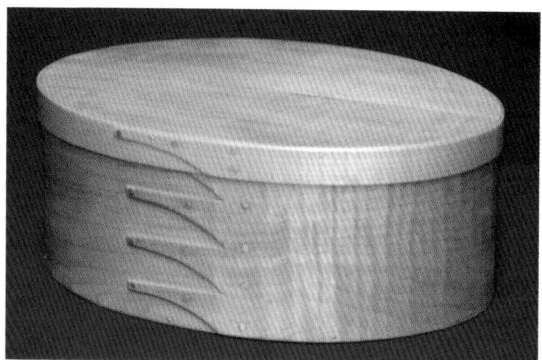

▲ **FIGURE 191**

"Spaulded" (or "tiger") maple shaker box, made of wood from a log inoculated with reishi (*Ganoderma lucidum*). When using the beefsteak fungus (*Fistulina hepatica*), the wood is stained red, which increases the wood's value, especially oak, substantially. Many other species can be used for colorizing wood. The logs are usually milled 2 to 4 years after inoculation.

▲ **FIGURE 192**

Once these pony logs are inoculated with plug spawn and the holes waxed over, Virginia Fraser places them into black plastic pots, which are then filled with sand. They can be conveniently stored underneath the benches of a standard greenhouse, benefiting from the added moisture and the stabilizing influence of contact with the ground.

▲ **FIGURE 193**

About 6 months later, the mushrooms emerge, in this case reishi *(Ganoderma lucidum).*

▲ **FIGURE 194**

David Brigham used a chain saw to scar these logs with wounds into which spawn will be placed.

cutting logs for shiitake inoculations, having spores of the shiitake strain in the bar oil of the chain saw ensures that the faces of the cut logs will have spores of this mushroom resident, helping prevent competition from other fungi. These logs are then inoculated with plug or sawdust spawn of the same strain. This combination of methods using matching species accelerates colonization and overcomes resistance barriers.

▲ FIGURE 195

Sawdust spawn of nameko *(Pholiota nameko)* covers the wounds and spans the gaps between the logs.

▲ FIGURE 196

Fresh wood chips are placed atop the raft of spawned logs.

Growing Mushrooms on Stumps

Stumps and their root systems can be massive, often weighing hundreds of pounds. Once stumps are inoculated, colonization can occur for years before mushrooms form. Once fruiting begins, mushrooms can sprout for prolonged periods, sometimes decades, before the stump totally decomposes. Growing mushrooms in wood chips or on logs is far faster. But this apparent disadvantage of using stumps to grow mushrooms also foretells of its advantage: mushroom fruiting can persist on a stump for many years longer than on wood chips and logs. I have seen a stump produce woodlovers, for instance, every October for more than 10 years. Stumps that are interspersed amongst overshadowing stands of trees have the best chance of success.

Cultivation issues are much the same for stumps as for logs. Like logs, if the stumps show fungal activity (signified by discolored zones of growth) and/or sport mushrooms (most likely during the rainy season), they are not good candidates; the wood should be bright and homogeneous in color. In addition, stumps of recently felled trees are by far preferred and are best inoculated within a few days of their creation. The reason for this is that wild mushroom spores constantly rain down, especially during mushroom season; if they fall onto exposed wounds, wild fungi may colonize the faces of cut stumps and logs first.

Stumps with fine fissure cracks (called *checks*) running through them are more quickly colonized with mushroom mycelium than those without, especially if mycelium is placed directly into these cracks. The method for inoculating stumps with plug spawn

◀ FIGURE 197

Once covered, the bed is left alone for a year. First fruitings can take up to 2 years, but this mycelial raft can produce mushrooms for a decade before exhausted. This method works well for species of *Hypholoma*, *Psilocybe*, *Stropharia*, *Agrocybe*, *Pleurotus* (oysters), *Ganoderma* (reishis), and others. I do not recommend this method for shiitake.

▲ FIGURE 199

The best place to inoculate a stump is the inside of the peripheral edge, into the sapwood.

▲ FIGURE 198

Inoculating stumps with plug spawn.

▲ FIGURE 201

Oysters *(Pleurotus ostreatus)* and honey mushrooms *(Armillaria mellea* species) fruiting from the same stump. Such events suggest that oyster mushrooms, which are saprophytes, can be good competitors against honey mushrooms, which have a dual nature, first parasitic, killing trees, and then saprophytic, growing upon their dead tissue.

▲ FIGURE 200

The clustered woodlover *(Hypholoma capnoides)* has fruited from this same stump for more than 10 years. Once a gourmet mushroom is established on a stump, you have years of delicious fungi to enjoy!

is similar to that for logs, except the cut face of the stump is a huge wound allowing the entry of competitors and should be inoculated with no more than 4 inches of spacing across the face. In addition, using rope spawn and spored oils increases the probability of complete colonization. Such combinations of methods are described next.

Quadruple Inoculation Method (QIM) of Stumps

Combined inoculation methods do a better job of colonizing stumps than do single methods used alone. For mycorestoration projects in particular, I recommend using a quadruple inoculation strategy. Although any of the described inoculation methods can be triumphant, using a quartet of methods is more likely to result in success. As mentioned, plug spawn may be more attractive for home growers than professionals. Here are the methods that I recommend for mycorestoration warriors on the front lines:

- Spored oils: Use while cutting the trees, creating logs and stumps.
- Rope spawn: Girdle and wrap the stumps with rope spawn.
- Plug spawn: Drill and inoculate stumps and logs with plug spawn.
- Sawdust spawn: Use sawdust spawn in a fungal "sandwich."

You will need the following tools:

- remote drill or modified grinder with backup batteries or a ready source of electricity, such as a small generator
- $^5/_{16}$-inch (8 mm) drill bits
- chain saw and safety gear
- hammer and 2- to 4-inch flat-head nails

In the first method, you'll use chain-saw oil infused with mushroom spores ("spored oil") when cutting trees and logs so that the open cuts are exposed to the selected fungus immediately upon contact (see figure 130).

The second method is similar to that of inoculating logs using plug spawn (using spawn of the same lineage

▲ **FIGURE 202**

Oyster mushrooms *(Pleurotus ostreatus)* fruiting from coil of hemp rope. Rope spawn is an effective way to inoculate stumps.

▲ **FIGURE 203**

Cauliflower mushrooms *(Sparassis crispa)* fruiting from hemp rope. This species, as well as woodlovers *(Hypholoma capnoides)*, an edible mushroom; *Hypholoma fasciculare*, a poisonous mushroom; and turkey tail *(Trametes versicolor)*, a medicinal mushroom, fight *Armillaria* root rot.

as the spores in the chain-saw oil). The difference is that inoculations are concentrated just inside the peripheral edge of the stump (see figure 199). This sugar-rich zone provides food for the mycelium, which then streams in all directions in between the bark and cambium.

The third method is to use a chain saw to girdle the stump 4 to 12 inches above the ground, and then insert rope spawn into the groove.* Painting the exposed rope with hot wax helps to protect the mycelium from bugs, keeps the moisture in, and encourages the mycelium move into the wood. You can make your own by soaking the rope in peroxide using the technique described on page 162 or using prolonged hot-water pasteurization (described on page 163), and inoculating it with grain or sawdust spawn. Simply sprinkle grain or sawdust spawn over the rope and place it into a plastic bag. Let the inoculated rope sit at room temperature for a month, or until fully colonized with mycelium.

For the grand finale, the fourth method, you can cut off the top 12 inches of the stump, remove this cylindrical section, spread spawn (preferably sawdust spawn) evenly across the surface, and return the cut section to the spawn-covered face, making a spawn "sandwich."** Covering the outside edge of the cut with wax or fabric can prevent bug invasion. If bugs gain access to the implanted mycelium, especially mealy bugs or beetles, they can take up residence. However, if your goal is to attract blight-causing insects with the intent of stopping an insect migration, then using a mycopesticidal fungus may be a good strategy. For instance, blighting insects could be drawn to the mycelium-sandwiched stump and become infected with an entomopathogenic fungus, stopping the further migration of forest-devastating insect plagues. A ring of such stumps could provide a line of defense (see figure 204 and chapter 11).

By introducing mycelia via several vectors, you maximize the likelihood that the mycelium will take hold. Mycelium at each point of inoculation grows outward, eventually connecting and enveloping the stump beneath the bark layer in a contiguous mat of mycelium from which mushrooms will sprout for years to come.

▲ **FIGURE 204**

Two weeks after this log had been inoculated with sawdust spawn. Mealy bugs had already climbed up 2 feet to take up residence with the oyster mushroom mycelium, consuming it.

The QIM is the approach I prefer to use to overcome acceptance barriers. The labor per stump, depending upon size, takes 5 to 20 minutes, a small investment considering the long-term rewards. Members of a skilled crew can work in tandem, speeding up the process. When the goal is to create a leading edge of mycelium encircling many acres, using a plurality of inoculation methods tips the balance in favor of the intended mycelium. Once established, the mycelium may remain resident for decades.

The forms of spawn used in the spored oil, plug, rope and sawdust all pivot around a predominant phenotype (a single individual) while the spores in the chain-saw oil provide for alternative phenotypes originating from the same mother mushroom. Having these additional phenotypes improves the adaptability of the species entering into the stump's complex microbial

---

* Using a chain saw and falling trees are both highly dangerous activities. Do not attempt unless you are skilled at doing both. Take all appropriate precautions!

** We use spent oyster sawdust or straw after it has produced mushrooms indoors. For many, a simple method is to use the leftover mycelium from a mail-order mushroom kit after it has produced mushrooms.

⚊ **FIGURE 205**

The author girdles a stump.

⚊ **FIGURE 206**

Rope spawn is tapped into the groove using a hammer.

universe. Which phenotype dominates is influenced by environmental factors. In the process called *anastomosis* or "same-self fusion," the mycelium seeks itself and fuses with compatible substrains, and spores and mycelium marry into a single organism.

The multiple inoculation method creates a matrix of inoculation points, preforming a netlike checkerboard pattern. From each inoculation point, the mycelium grows to meet other satellite colonies and become a communal web, benefiting from the shared nutrition. The many points of inoculation all synchronize for rapid colonization.

The QIM is ideal for mycoforesters, particularly those interested in corralling a habitat with a mycological species buffer. For instance, if a root-rot plague is sweeping a forest, by inoculating stumps on the peripheral edge of the nearby woodlands a mycoforester can establish a zone of resistance, quashing or limiting the disease by creating a mycelial fence. Similarly, veins of particular mushroom species can be threaded through the forest; interweaving these threads creates a fabric of species. Using such techniques, mycoforesters can project "smart" species mosaics of mushrooms to move silently through landscapes, governing the health of fungal, insect, plant, and animal communities.

⚊ **FIGURE 207**

Using ⁵/₁₆-inch-diameter rope spawn, slightly swelled with water, makes insertion easy, since the chain saw cut is also ⁵/₁₆-inch. If inoculating a live tree, two parallel girdles stuffed with rope spawn are recommended. (Note: Girdling a tree will kill it.)

Initiating the Flush: Encouraging Mushrooms to Form on Logs and Stumps

When growing mushrooms outdoors, mimicking rainy natural weather conditions helps promote fruiting. About 6 months to 2 years after inoculation, most logs myceliated with shiitake, oyster, reishi, and other species will flush mushrooms soon after an overnight submersion in cold water. This timely soaking of your logs is a technique called an *initiation strategy*. The best time is when temperatures and rain patterns outdoors are within the natural fruiting range of the chosen species. Your role is to increase the frequency of rain showers. (See chapter 14 for the preferred temperature ranges for many species.) Afterward, keep the logs loosely tarped or, if ideal weather conditions prevail—wet and in the right temperature range—uncover the logs, watering them 2 or 3 times a day. Replace the tarp with a 70 percent shade cloth after watering. Partially buried logs should be watered 2 or 3 times a week, depending on weather conditions. (Use a drip line to conserve water; or locate your logs below your house's downspouts, but not if your roof tiles are made of asphalt or contain petroleum or other toxic products). Generally, I don't water stumps, allowing nature to induce fruitings during the rainy seasons.

With some luck, mushrooms will begin forming on logs within 2 weeks of initiation. Mushrooms will often first form near where the holes were originally drilled, and they will also cause the bark to blister and crack as they emerge. Stumps fruit when they want to, but providing shade and exposure to rain will help promote mushroom formation. Watering must be budgeted according to the needs of the developing mushrooms. The mushrooms' princely forms will tell you if you are under- or over-watering. Most cultivators quickly understand how much moisture needs to be applied by carefully observing the flush develop. Watching a population of mushrooms emerging from adolescence to adulthood and beyond gives you a spread of individuals to guide you. If 10 of 10 mushrooms that form grow to full maturity without blemishes or drying, then the grower has succeeded in coaxing a flush. If only 2 of 10 grow healthily, then the grower adversely affected the flush's development.

Mushrooms are best harvested before they heavily sporulate—before populations of microscopic basidia emerge and bear spores (see chapter 2). Most mushrooms are at this stage when they are adolescents or young adults; for example, gilled mushrooms have generally reached this stage when they are convex in shape and not yet flattened. You can easily see spore dust collect around the mushrooms when they mature. You will gain a good understanding of the beginnings and endings of the mushroom's cycle (which spans a week or more) from simple observation.

When harvesting mushrooms, usually it is best to trim them as close as possible to the substrate; leftover stems often attract insects and molds. Don't throw away your stem butts without first thinking of using them as spawn for further inoculations! (See chapter 9). Sometimes keeping the stem butts attached to the plucked mushrooms has advantages. Growers have found that selling mushrooms with stem butts attached, oftentimes with a clump of substrate, prolongs shelf life compared to trimmed mushrooms. Knowing this gives cultivators more options in managing their crops for personal consumption versus selling them to market. Stem butts from some store-bought mushrooms can be used as spawn for outdoor inoculations. Viability declines with age.

After you have harvested your first flush, your logs and stumps will usually lapse into a period of dormancy. If they do not fruit again immediately after the first flush, let them sit for 2 to 3 weeks before reinitiating them by heavily soaking them in water. Stumps often fruit continuously during the rainy season in fall. In regions where water is scarce, collecting rainwater from roofs and channeling it to your mushroom patches is a convenient way to provide them with necessary moisture. Installing understory and overstory shade plants also helps to increase and preserve humidity.

CHAPTER 12

Gardening with Gourmet and Medicinal Mushrooms

Mushrooms enhance edible landscapes with a bounty and diversity of flavors, complementing the feast from plants. Further, fungi in mycological landscapes can be designed to fruit according to the seasons. As with plants, the timing of "flowering" can be sequenced so that mushrooms are available throughout much of the year. This method works especially well when the mycological landscape is interspersed with low-lying leafy shrubs and trees. Here are some general fruiting times of mushroom species in northern temperate climates of North America, Europe, and Asia. Please note that the fruiting times often reach into the next season or can start late in the prior one. And as global warming progresses, seasons will shift.

All gardeners are mushroom growers already—most just don't know it. Companion cultivation with saprophytic and mycorrhizal mushrooms is becoming increasingly popular as gardeners learn about the benefits of mycological plantings. Using fungi in the garden increases yields, reduces the need for fertilizers, and builds soil structure for the long term. Mycelium also loosens the soil as mass is reduced, enhancing aggregation while creating micro spaces that absorb and fill with water. The carbon dioxide outgassed by mycelium, heavier than air, saturates the soil and fuels developing plants. Much of the carbon from the carbon dioxide is incorporated into plant tissue. This closed circuit empowers both fungi and plants in an ongoing but

Fruiting Seasons

Spring

Morel (*Morchella angusticeps* and *Morchella esculenta*)

Spring oysters (*Pleurotus ostreatus* and allies)

Summer

Reishis and allies (*Ganoderma species*, including *G. lucidum, G. curtisii, G. oregonense, G. resinaceum, G. tsugae*, and *G. applanatum*)

Garden giant (*Stropharia rugoso annulata*)

Elm oyster (*Hypsizygus ulmarius*)

Shiitake (*Lentinula edodes*)

Late Summer to Early Fall

Button and meadow (*Agaricus bitorquis, A. campestris, A. subrufescens*, and *A. brasiliensis*)

King oyster (*Pleurotus eryngii*)

Maitake (*Grifola frondosa*)

Parasol (*Macrolepiota procera* and *Chlorophyllum rachodes*)

Rocky mountain enoki (*Flammulina populicola*)

Cubies (*Psilocybe cubensis*)

Late Fall to Early Winter

Oyster (*Pleurotus ostreatus* and *Pleurotus pulmonarius*)

Shaggy mane (*Coprinus comatus*)

Sacred psilocybes (*Psilocybe azurescens, Psilocybe cyanescens*, and *Psilocybe cyanofibrillosa*)

Enokitake (*Flammulina velutipes*)

Blewitt (*Lepista nuda*)

flexible dance of nutrients. As sugars flow between plant and fungi, other biochemical interchanges occur at such complex levels that scientists have little understanding of them. While all of this goes on, we reap soil as the reward for fostering mycelium.

Habitats having mycelium can support richer biodiversity, and they blossom as mycosystems. The nutrient-return cycles of habitats lacking mycelium are essentially stalled and must be artificially supported using external input (such as fertilizers) to offset their natural deficits. Such artificially supported habitats are prone to disease and are ultimately unsustainable. We must reverse the course of conventional farming practices so that soil is created, not depleted. Mushroom mycelium fulfills this need, providing a balance between input and output of nutrients between the kingdoms.

The following sections first address the design and placement of mushroom patches, then discuss some experiments we have conducted pairing mushrooms and plants, and finally consider the use of mycorrhizae. Once you understand each method, you can customize your own combination of methods in order to create the best strategy for your circumstances.

Designing and Installing a Mushroom Patch

This section will show you how to create a personal mushroom patch outdoors. Know that the land around you is filled with interfusing, interconnected mycelial networks, which are vibrant islandlike lenses in the soil. Lenses of mycelium are called *mother patches* when they become mycoengines for the exponential expansion of mycelial mass. We can amplify these native mycelial lenses by feeding them plant debris, or construct a new mycological landscape of our own design using wood chips, for instance. In effect, we are going to create cellular membranes that travel underground, surfacing, like whales, when they seek air for fruiting. When the mycelium surfaces, we can expand the patch or fruit it, weather permitting.

Once you understand these techniques, you can incorporate them into your surrounding landscape or into your garden. You could even replicate this model on a larger scale to put into practice mycoforestry, mycoremediation, and mycofiltration practices.

Siting with Respect to Lighting

Site your mushroom patch with particular attention paid to exposure to the sun. Many mushrooms actually benefit from indirect sunlight, especially in the northern latitudes. In the northern hemisphere, site locations on eastern and northern slopes are generally preferred to those with southern and western exposure. Canopies of shade-providing plants can mitigate the effects of light exposure, allowing for the cultivation of mushrooms irrespective of slope aspect.

Pacific Northwest mushroom hunters have long noted that mushrooms grow most prolifically not in the darkest depths of woodlands but in environments of both shade and dappled sunlight. Light stimulation is absolutely necessary for the healthy fruiting of most saprophytic mushrooms. I have seen mushroom beds thrive when they get 1 to 2 hours of direct sunlight while adjacent to them the same type of mycelium in total shade barely fruits. When sunlight moves across a bed of wood chips, moisture moves up to the surface, warms, and flows out in the form of short-lived evaporation. This wicking effect aids fruiting provided condensation occurs soon after the beams of sunlight have passed.

Studies examining sensitivity to light have established that different species respond optimally to different wavelengths; see my book *Growing Gourmet and Medicinal Mushrooms* (2000a). Most grassland and woodland mushrooms prefer dappled sunlight but require moisture protection at the interface where primordia form. For terrestrial mushrooms, the primordia form near the soil-air interface, while log species create primordia underneath the bark layer.

A few mushrooms are heliotropes—sun lovers— and do well in exposed areas, often in association with grasses. Good examples are the garden giant, the shaggy mane, and many of the *Psilocybe* mushrooms. Although these mushrooms will form on barren wood chips, often late in the season, they benefit from the microclimates of grasses, helping them appear weeks or even months earlier. The grass, with its long vertical shoots, provides shade right above the ground, and the stems act as conduits for collecting condensation, sending beads of moisture to the soil-interface zone where primordia form. I have hypothesized—from years of observation—that primordia originate within these dewdrops that collect at the base of grass stems. The constant wicking of water away through evaporation, combined with the replenishment of moisture from dew or raindrops streaming down the grass shoots, provides a nurturing environment for the birthing of baby mushrooms.

Other grassland mushrooms that are heliotropes include giant puffballs (*Calvatia gigantea*), meadow mushrooms (*Agaricus bernardii* and *Agaricus campestris*), shaggy manes (*Coprinus comatus*), and many of the *Psilocybe* species. Bear in mind that these recommendations for locating your mushroom patch are not strict. You can grow heliotropes in the shade and log-friendly mushrooms in sunny areas, but your yields will be a fraction of what they could have been. To succeed, you must become skilled at managing moisture at the microscopic level. Let the mycelium guide you to the best site for growing.

Choosing the Site

Places where grasslands and woodlands meet allow for easy access and tend to be ideal sites for many species; many saprophytes are edge runners. A subsurface flow of moisture, even if it is only occasional, goes far in fueling the mushroom mycelium. A swale found in the upland of a ravine leading into a narrowing watershed is a good example. Some of the best places for growing many mushrooms are at habitat interfaces where light, shade, water, humidity, and plants are in constant transition. By introducing plants of various heights, you can design a shaded habitat with humidity levels increasing nearer to the ground. As environmental architects, mycologists can design habitats patterned after the diversity found in the mushrooms' native habitats.

Be forewarned that if your mushroom patch is far afield from your daily walking route you may miss the fruitings. This has happened to me more than times than I wish to remember. Mushrooms can first be seen as primordia on Monday, mature by Friday, and be past their prime by Sunday. Primordia are really cellular explosions in not-so-slow-motion; they can be born, burst, and break down in a week's time.

Creating a Mushroom Patch in Your Backyard

Each cultivator develops preferences and methods based on his or her successes and failures. I have learned more than a few lessons, which means I failed many times. The simplest method that works best for me is to use a bed of wood chips in a moist location that is exposed to dappled sunlight.

The area must be prepared in order to give your mycelium a foothold. Extraneous debris is removed first. (In effect, you are surgically removing native fungi that will compete with your implanted mycelium.) Once the mycelium establishes a foothold, it naturalizes and increases in its appetite for more food as the mat differentiates, enlarges, and migrates. More animal-like than plantlike, these digestive cellular networks often achieve great masses.

Establishing the largest possible mycelial mat before the fruiting season is the shortest path to fruiting mushrooms outdoors. Unlike the process of growing mushrooms on sterilized or pasteurized substrates, spawning by equally distributing fragments of mycelium evenly throughout a substrate—called *through spawning*—is not as effective, in many cases, for outdoor cultivation. For outdoor plantings, placing a continuous layer—or sheet—of spawn or naturalized mycelium quickly creates a common plateau that will give rise to fruitings far sooner than the through spawning methods used by indoor commercial cultivators. Most outdoor beds of wood chips are inoculated with sawdust, dowel, or stem butt spawn. Simply disperse spawn over the surface of wood chips rather than mixing it through. When the mycelium has formed a sheet, competitors will be less likely to penetrate through. Once a mycelial membrane has been established, its capacity for growth—especially horizontal growth—is greatly enhanced.

I have seen mycelial mats several inches deep emanate outward from a small path of about 4 feet in diameter to 30 feet in 4 months' time. The resulting fruitings, in this case of the garden giant (*Stropharia rugoso annulata*), were mammoth. As the patch matured over several years, the central regions died back, to be replaced by secondary saprophytes, whereas the outer peripheral zones continued to produce garden giants as the mycelial wave moved outward. This behavior is reminiscent of mushrooms that form fairy rings, and of the growth mycologists see in sterile cultures in media-filled petri dishes.

My preferred method for projecting mycelium over an expansive landscape is to create 2 parallel horizontal waves of mycelium combined with fragmented mycelium interspersed between the layers. (Usually, spawn fragments naturally fall through into the depths of the substrate.) This is sometimes called the "lasagna" method, but I like the phrase "mycelial wave" because it calls to mind the coordinated action of mycelial cellular energy whose life force has become synergized with millions of its once-fragmented cellular colonies. The mycelial planes and the fragments seek each other and connect, and the organism is strengthened as it integrates. Once a mycelial wave begins, the cellular momentum crosses over inhospitable barriers. The strength of projection is a direct function of the ability of the mycelial mat to sequester and channel nutrients to the tips of the emerging mycelium, which form the leading edge of the mycelial wave. As the tips form, cell walls and nuclei pair up, and each hypha enlarges, penetrating forward, exuding water-absorbing extracellular sugars, and encouraging the mycelium to surge. The high art of a mushroom cultivator is to surf these mycelial waves by creating debris fields they can strive for. In effect, you can steer mycelium through an ecosystem by incorporating its favored debris into your landscape design.

MAKING A MUSHROOM BED

1. Clear the site of debris down to the mineral earth, which in many cases is clay. Lay down wood chips to a depth of 2 inches, spreading evenly. Obviously, choosing fresh wood chips without fungal contamination is better than using wood chips already run through with other fungi.

2. Before applying spawn, moisten the wood chips for a few minutes. For every 100 square feet, I recommend using 5 to 10 pounds of spawn, broadcast evenly across the surface of the 2-inch-deep bed of wood chips. You can use commercially made sawdust or chip spawn, dowel spawn, or homemade bunker spawn. Avoid grain spawn, since bugs may steal your mycelium for their own purposes. (However, using grain spawn in burlap sacks to create bunker spawn is a preferred method for preventing insect invasion, since the fabric acts as a barrier to insects.)

3. Once the first layer of spawn has been placed, lay down more wood chips to a depth of 2 to 4 inches, moisten the chips, and then add another plane of spawn. In much the same fashion as before, place more wood chips on top. Each layer should be separated by approximately 2 to 3 inches of wood chips, but in most cases the bed should not be more than 6 inches deep. Saturate the bed with water, if possible.

4. Torn-up pieces of cardboard can be placed over the top to seal the bed and to protect it from long-term moisture loss. Additionally, you can top-dress the bed with loose straw as added protection; however, straw harbors some aggressive saprophytic mushrooms belonging to the genera *Coprinus*, *Conocybe*, *Bolbitius*, and *Psathyrella*, which can dominate your planted mycelium. Straw used for mulch should be bright and coarse, not soft, and appear free of fungal growth. By the time the straw softens and begins to rot, the implanted mycelium should have colonized the substrate sufficiently to limit competition.

5. For a fruiting bed, sprinkle a small handful of grass seed (*Ammophila*, *Phalaris*, timothy, bent grass, blue grass, fescue, and others) over the bed—about 10 to 20 percent of what is recommended for a "healthy" lawn. The germinating grass stimulates mycelial growth by providing nutrients and shade, increasing condensation, and channeling moisture. (If you are making a spawn bed, do not seed with grass.)

As the seasons cool, the days shorten, rains begin, and the grasses die back, fueling the mushroom mycelium, which reaches upward to exploit the newly available dead plant tissue. This ascension and decline of grasses goes far in encouraging mycelial growth toward the surface of the wood chips. Due to the nature of the habitat, the grasses often look sparse and weak after an initial burst of good growth, but as the wood chips rot, the grasses recover and regreen. These climax and decline cycles stimulate mycelial growth and mushroom formation.

Leave the bed undisturbed for at least 6 months. Often, the initial leap-off is very slow, since the mycelium needs time to acclimate. When colonization is complete, initiate fruiting by soaking the mushroom bed with water. I have used a rainbird-style sprinkler set on a timer for one hour in the morning and early evening. Since timing is critical, many of my

⋏ **FIGURE 208**

Removing debris from future mushroom patch and laying down cardboard.

⋏ **FIGURE 209**

Distributing wood chips across the surface.

⋏ **FIGURE 210**

A layer of spawn is dispersed and 1 to 2 inches of more chips are added on top, sandwiching the underlying spawn.

◄ **FIGURE 211**

Top-dressing with a shallow layer of fresh straw (1 to 2 inches).

friends have their beds in place between March 1 and April 1 in anticipation of fall rains, which stimulate fruitings in September and October.

Figures 208 to 213 depict the construction of a mushroom bed of the garden giant *(Stropharia rugoso annulata)*.

Mushrooms producing rhizomorphs, especially *Hypholoma, Psilocybe,* and *Stropharia* species, are particularly suited for this technique. Once the mushrooms of your choice are growing, not only do you have spores for creating satellite colonies far and wide, but also the stem butts can be planted like flower bulbs using the methods I described on pages 146 to 153.

▲ **FIGURE 212**

A year later, David Sumerlin looks for mycelial growth and finds some surfacing. Note that the straw has mostly decomposed.

Invasive species, once they sporulate, may compete with your implanted mycelium. If an undesirable saprophyte attempts to take over your mushroom bed, you can let the habitat mature and the invasive mycelium will die back in a few years, or you can scoop up the top 4 inches of soil, remove it to another location, and replant with the species you desire. In the extreme, you can start a bonfire on the site, rake out the coals, and once cooled, begin anew. This task is not difficult when your patch is modest in size.

Companion Cultivation of Mushrooms with Garden Vegetables

Garden vegetables benefit from the activities of both saprophytic and mycorrhizal mushrooms. First we'll explore the use of saprophytic mushrooms. After discussing our experiments and experiences with these, I'll move on to using mycorrhizae to aid the garden.

Using Saprophytic Mushrooms to Aid the Growth of Garden Vegetables

For years I have heard stories about the garden giant mushroom *(Stropharia rugoso annulata)* being used for intercropping in cornfields in eastern Europe, primarily Hungary, where inoculated straw is mixed with corn stubble and ploughed into the field every few years when farmland goes fallow. The mycelium goes to work and nourishes the soil. When the fields are replanted, the farmers are delighted to find garden giants coming up between the rows of corn. Once resident, the garden giants tend to perpetuate themselves. In our garden we confirmed that corn and garden giants grow well together, with the mushrooms often selecting to fruit near the emerging cornstalks.

◀ **FIGURE 213**

David scoops up an island colony of mycelium, a mycelial lens that can be transported and expanded by a factor of 10 in a single year. This mycelium—originally from pure culture sawdust spawn—is now naturalized, competing well with outdoor microbes, and running fast.

In the spring of 1999, Christiane Pischl came from the University of Innsbruck, Austria, to live at our farm and study companion cultivation of mushrooms and vegetables for her master's thesis. We purchased some local soil that, upon analysis, proved to be low in essential nutrients, and so we used this soil for the experiment. We paired the mycelia of saprophytic mushroom species with assorted vegetables to see which union would prove to be most beneficial to both fungus and plant. Once the plants were in the soil, we allowed them to recover for a few days before beginning the treatments.

We chose 9 mushroom species: pioppino *(Agrocybe aegerita)*, Brazilian blazei *(Agaricus brasiliensis)*, almond Agaricus *(Agaricus subrufescens)*, maitake *(Grifola frondosa)*, elm oyster *(Hypsizygus ulmarius)*, blewit *(Lepista nuda)*, the king oyster *(Pleurotus eryngii)*, oyster *(Pleurotus ostreatus)*, and garden giant *(Stropharia rugoso annulata)*. We made sterilized sawdust spawn of each species. We selected 4 vegetables: Brussels sprouts, broccoli, peppers, and beans. In total, 36 beds, 2 by 4 feet each, were constructed and planted. In this experiment, some vegetables cohabited beds containing only 1 species of mushroom mycelium.

Upon 27 of the 36 beds we placed a layer of untreated alder *(Alnus rubra)* sawdust, a sprinkling of spawn (approximately 2 pounds), another layer of sawdust, additional spawn, more sawdust, and a topping of loose straw. The beds were wetted and the young plants were partially uncovered for exposure to sunlight.

In a month, elm oysters began to fruit, selecting sites near the emerging plants. The plants appeared healthy and undeterred by the close contact with the mushrooms. After several months, other mushroom species appeared. After 4 months, harvests of both plants and mushrooms were completed. Christiane reported that 2 mushroom species had failed to produce visible mycelium and had presumably died: maitake *(Grifola frondosa)* and king oyster *(Pleurotus eryngii)*. She reported that 4 mushroom species produced fruitbodies, listed here with their average dry weights: almond Agaricus (26 g), pioppino (49 g), oyster (114.5 g), and elm oyster (170 g). Typically a

▲ **FIGURE 214**

Baby garden giants *(Stropharia rugoso annulata)* huddled next to rapidly growing corn.

▲ **FIGURE 215**

Christiane Pischl plants starts of several vegetable species in undernourished soil.

▲ **FIGURE 216**

Adding pure sawdust spawn.

▲ **FIGURE 217**

Laying down a layer of sawdust over the test beds.

▲ **FIGURE 218**

Top-dressing with loose straw.

mushroom's dry weight is one-tenth of its fresh weight. This means that the elm oyster produced 1.7 kg, or 3³/4 pounds fresh weight. The garden giant and Brazilian blazei showed mycelial growth, with the former showing dense rhizomorphic growth near plant roots. (Typically the garden giant fruits 1 to 2 years subsequent to planting, so its lack of appearance the first year is normal.) In subsequent years, the beds flourished with garden giants, which were planted with corn (see figure 214).

The most remarkable results occurred in connection with the matching of the elm oyster with Brussels sprouts and broccoli. In total weight, the beds with the elm oyster mushroom yielded 4 to 6 times as much vegetables as those without. In contrast, oyster *(Pleurotus ostreatus)* had a diminishing effect on plant growth and yield, although nearly a kilogram of fresh oyster mushrooms were harvested. The chart on page 196 summarizes the net fresh weight and dry weight values.

Christiane Pischl's work is some of the first that I am familiar with that shows that saprophytic mushroom mycelium existing in the woody top layer above the root zones influences plant development. Caesar-TonThat and other researchers (2000) also showed that mushroom-forming fungi aid farm soils, although

▲ **FIGURE 219**

Once the beds are covered, straw is cleared from the plant stems to help trap water and aerate.

▲ **FIGURE 220**

Three weeks later, the elm oyster *(Hypsizygus ulmarius)* fruits adjacent to the emerging plants, which appear healthy from this fungal pairing.

▲ **FIGURE 221**

About a pound of fresh elm oyster mushrooms *(Hypsizygus ulmarius)* were picked from this bed a month after the bed was created.

▲ **FIGURE 222**

Christiane Pischl hard at work maintaining the companion mushroom garden experiment at Fungi Perfecti, Kamilche Point, Washington.

▲ **FIGURE 223**

With Brussels sprouts and other plants, we noticed significant increases in output, root wad development, and stem length when grown with mycelium.

➤ **FIGURE 224**

The plants having contact with mycelium of the elm oyster *(Hypsizgus ulmarius)* produced much better than the controls.

Companion Planting Yield Results

Total yields comparing production from unmulched, mulched, mulched with mycelium of the elm oyster mushroom *(Hypsizygus ulmarius)*, and mulched with mycelium of oyster mushrooms *(Pleurotus ostreatus)*. d.w. = dry weight.

Bed Treatment	Yield Plants (d.w.)	Yield Mushrooms (d.w.)	Total (d.w.)	Total Fresh (g)
Control unmulched	100.5	0	100.5	904.5
Control mulched	73.7	0	73.7	663.3
Pleurotus ostreatus	42.6	114.5	157.1	1,413.9
Hypsizygus ulmarius	248.6	170.1	418.7	3,768.3

The garden giant *(Stropharia rugoso annulata)* also enhanced crop production within 25 percent of the elm oyster *(Hypsizygus ulmarius)* but did not fruit the first year.

the species have not yet been identified. Pischl's work also shows that species affect neighboring plants differently, and, in the case of the elm oyster, that some vegetables clearly benefit from nearby mycelial growth and mushroom formation.

This discovery may represent a new class of saprophyte-plant mutualism, and this possibility logically leads to much broader questions about species pairings. We are continuing to study combinations to find additional beneficial matches. Christiane and I both believe that matching plants with complementary saprophytic mushroom species increased flow of nutrients as the fungi digested the sawdust. As we learn more, new companion cultivation strategies can be put into practice, helping growers in impoverished lands and landscapes and reducing the need for fertilizers.

Another group of fungi, the mycorrhizae, are increasingly used to lessen dependency upon fertilizers. For supporting horticulture, I recommend using saprophytic, endophytic, and mycorrhizal fungi in concert.

Using Mycorrhizal Mushrooms to Aid the Growth of Garden Vegetables

Vegetable seeds seek mycorrhizal fungi immediately upon germination. (Mycorrhizal fungi are abundant in nature, but many are difficult to collect and only

some, like those of the genera *Glomus* and *Rhizopogon* are suitable for most trees and plants.) This pairing of seed sprouts and fungi helps the seeds gather nutrients and prevent parasitization. The fungi benefit from the plant's waste products, sugars, hormones, and dead tissues that flake off. When seeds are dusted with a mix of mycorrhizal spores (primarily *Glomus aggregatum*, *Glomus intraradices*, *Glomus mosseae*, *Pisolithus tinctorius*, and assorted *Rhizopogon* species), the spores and seeds germinate simultaneously, thriving within the protecting and nurturing mantle of mycelium.

To use mycorrhizal spores, simply dust the seeds with spores prior to germination using the recommended dosages—a 1-ounce commercial mixture is enough for 1 pound of seeds (or about 10,000 spores per gram of seeds). The simplest way is to mix the spores and seeds in small container or sack. Once the seeds are dusted, plant them as you would normally. Seeds with mycorrhizal spores often germinate quickly and grow more luxuriantly than those without. This head start against competitors gives the plants an obvious advantage.

In figures 226 to 227, the plants were grown from pasteurized soil with and without mycorrhizae.

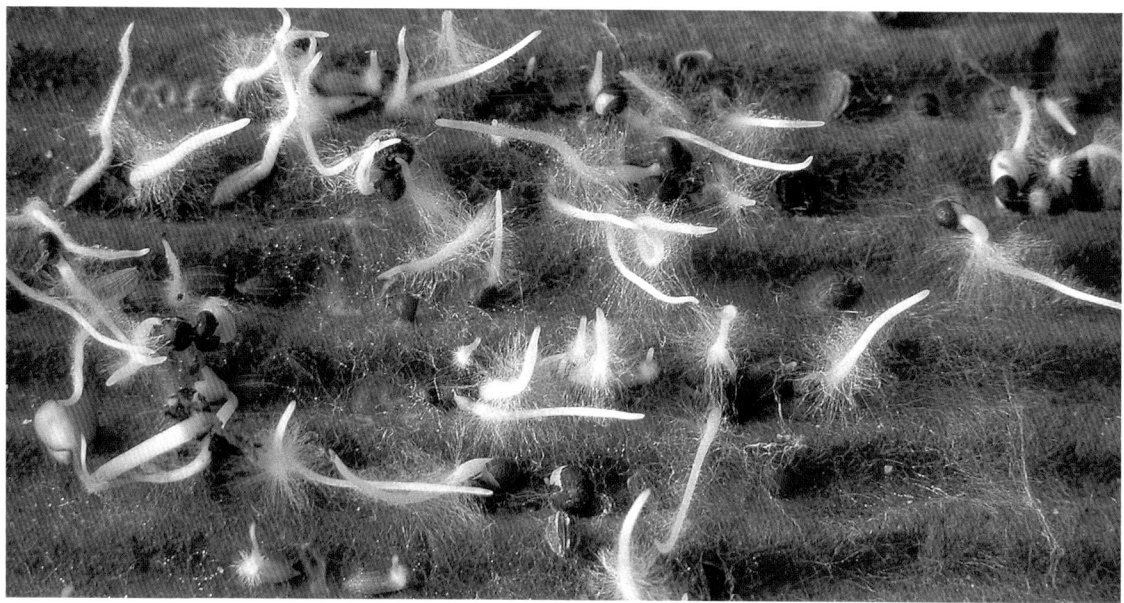

▲ **FIGURE 225**

Seeds of 8 garden vegetable species germinate 5 to 6 days after being soaked in water and are embedded within a common mycelial network that nurtures their growth.

▲ **FIGURE 226**

Comparison of California poppies with and without mycorrhizal spores.

▲ **FIGURE 227**

Comparison of potatoes with and without mycorrhizal spores.

Arugula roots encased with saprophytic mushroom mycelium, which helped this plant absorb nutrients.

USING MUSHROOM COMPOST OR NURSERY-BAGGED SOILS

Soil purchased from your local nursery is usually mycorrhizally anemic. Similarly, young starts of vegetables are usually potted in pasteurized soils, devoid of mycorrhizae. Compost from mushroom farms also lacks mycorrhizae. This means that mycorrhizal spores must be introduced into pasteurized soils or mushroom compost if plants are to benefit from their activity. When tablets enriched with mycorrhizal spores are added to the root zone, the introduced fungi quickly associate with the plant. In addition, if young starts are placed adjacent to already naturalized plants, the neighboring mycorrhizae jump to the newly planted starts. However, if the adjacent plants are diseased, planting unmycorrhized starts nearby puts them in jeopardy of infection. To be safe, add mycorrhizal spores every time you purchase seedlings, especially in new gardens. Most manufacturers of mycorrhizae sell spores in tablet, powdered, or liquid form, and each differs in its recommended uses. For sources of mycorrhizae see the Resources section.

USING NATIVE MYCORRHIZAE

Most gardeners who have developed beautiful gardens over the years have soils that are already resplendent with mycorrhizal fungi. When purchasing soil amendments, gardeners should consider their microflora as well as their nitrogen-phosphorus-potassium composition. The best garden soil is a microbial matrix, with organic debris fused with mycelium—plants spring from such soils.

Several companies sell mycorrhizal spore mixes customized for different groups of plants. However, keep in mind that spores quickly lose viability over time. For the first 3 years, they have the highest rate of germination. Some of the spores I've tried are commonly 90 percent fertile the first year, but germination gradually declines approximately 50 percent per year thereafter. Aged spores, or those kept under adverse storage conditions, germinate less. Decreased spore vitality can be partially compensated for by increasing the number of spores used per plant or container, although each circumstance is unique. When in doubt, use more spores than recommended.

Mycorrhizologists harvest spores from wild or greenhouse-cultivated mycorrhizal puffball- or truffle-shaped mushrooms from fungi like *Glomus*, *Pisolithus*, and *Rhizopogon*. Mature mushrooms are selected, pulverized into a powder, and then shaken through varied-size mesh screens until the fine spores fall through the bottommost screen (with the smallest openings). Centrifugation further separates the chaff from the spores, which are heavier.

If you live in northwestern North America, for example, using mycorrhizal fungi from this region is far better than risking the use of fungi collected from afar—for instance, New Zealand. The nuances of climate, host plants, and complex microflora are factors determining success and failure. In the future, foresters may see the usefulness of gathering spore stock from the forest before they cut down trees so that native fungi can be used to help in the replanting.

I hope more mycologists will spearhead spore-collecting forays to help foresters and habitat restorers do their jobs better. Although *Glomus* and *Rhizopogon* are so ubiquitous, paying attention to ecotypes of the strains used is important, since strains differ from region to region. That these species are easy to acquire in the wild and are acceptable to such a wide range of plant hosts makes them attractive for use in mycogardening and mycoforestry. As our knowledge expands, more species and combinations will emerge. Perhaps elaborate suites of species will one day be taken from healthy environments to restore nearby devastated lands.

The Potential Use of Endophytic Fungi to Aid Agriculture

Most plants have growing within them, between their cell walls, fine filaments of a unique group of fungi. Marshland grasses show enormous diversity of hundreds of these species, particularly in the mid-tidal regions. These fungi baffled mycologists for many years. Eventually, this new group of fungi was given the name *endophytes* and was born into the lexicon of mycology. As more plants were explored, endophytes were found in the vast majority of plants surveyed.

Because they grow below the epidermal surfaces of the plants, they are usually masked from view. Plants "infected" with endophytes sometimes show discoloration in their leaves, but they are not sick. Endophytic fungi help plants defend against insect, fungal, and bacterial parasites. They promote root and shoot growth, increase foliage, and increase tolerance to environmental extremes. Recently, mycologists have begun using them in trials to enhance agricultural crop production.

Endophytic fungi belong to many genera, including *Colletotrichum*, *Curvularia*, *Piriformospora*, *Pezicula*, *Neotyphodium*, *Pseudorobillarda*, *Pyricularia*, *Nodulisporium*, *Pestalotiopsis*, *Phomopsis*, and the liverwort-loving *Xylaria*. There are so many endophytes, with so many unique secondary metabolites, that a new hunt has begun: to search them for novel medicinal compounds. More than 50 percent of the biologically active secondary metabolites isolated from endophytes are new to science.

The U.S. patent application by Henson and others (2004) describes the use of a pipette to disperse 10,000 to 30,000 spores to the zone between the crown and the first leaf in 3-leafed seedlings of hot springs panic grass *(Dichanthelium lanuginosum)*—in effect an aerial application of a spore-mass slurry. Their patent describes similar applications to corn, wheat, watermelon, and mustard seedlings, whereby heat tolerance and drought resistance was conferred. Clearly, this is an effective method for introducing endophytic fungi to a broad range of plant hosts. (For general discussion of endophytes, see page 31.)

Potentiating the Activities of All Three: Saprophytic, Mycorrhizal, and Endophytic Fungi

Gardeners and farmers are biological orchestra conductors. This book adds more musicians to the stage by showing the importance of integrating saprophytic, mycorrhizal, and endophytic fungi. Soils are generated and replenished by the saprophytic fungi.

Upon germination, seeds seek mycorrhizal partners. Endophytic fungi rain down from the sky to associate with accepting plant hosts. This is the way of the natural world. The opportunity we have now is to synergize these 3 groups of fungal populations in order to support plant communities, which in turn provide habitats and fuel the carbon cycle. We are on the verge of a mycological revolution in the use of complex, complementary fungal systems to help horticulture. I encourage readers to become users of these techniques and perfect them. This book is only a gateway to greater knowledge. It is our job to expand the body of knowledge for future generations. I will keep you updated on this subject in future editions.

CHAPTER 13

Nutritional Properties of Mushrooms

Many species of mushrooms are highly nutritious to animals, humans included. The Food and Drug Administration (FDA) has officially designated mushrooms as "healthy foods." Andrew Weil, MD, author of *Natural Health, Natural Medicine* (1998) strongly encourages the ingestion of medicinal mushrooms to help prevent or treat disease. The late diet celebrity Robert Atkins, MD, promoted mushrooms as healthy foods for weight management in more than 100 of his selected recipes in his books and literature. The National Institutes of Health (NIH) are funding research into the medicinal properties of mushrooms. Doctors worldwide are recognizing that mushrooms are a medicinal food rich in nutrition.

Nutritionists often misunderstand mushrooms as a food. Although most fresh mushrooms are 90 percent water, they can vary in their individual moisture content, so it's best to look at them in terms of dry weight. Mushrooms are rich in protein, very low in simple carbohydrates, rich in high molecular weight complex carbohydrates (polysaccharides), high in antioxidants, and very low in fat. They lack cholesterol, vitamin A, and vitamin C. They are a good source of some B vitamins—riboflavin (B2), niacin (B3), and pantothenic acid (B5)—as well as ergosterols (which upon exposure to ultraviolet light convert to provitamin D2). They're high in dietary fiber, with edible varieties ranging from 20 percent fiber (by dry weight) for *Agaricus* species (such as button mushrooms) up to

50 percent for *Pleurotus* species (such as the phoenix oyster). Mushrooms are good sources of essential minerals—especially selenium, copper, and potassium—elements important for immune function and for producing antioxidants to reduce free radicals (some mushrooms hyperaccumulate selenium; see page 110). Mushrooms also contain numerous medicinal compounds such as triterpenoids, glycoproteins, natural antibiotics, enzymes, and enzyme inhibitors that fortify health.

Protein content of mushrooms ranges from 3 percent for the tough, inedible agarikon (*Fomitopsis officinalis*) to 33, 34, and 35 percent for shiitake, nameko, and portobello, respectively. Mushrooms are rich in complex carbohydrates—high molecular weight polysaccharides. Our analyses show that beta-glucans range from 8.9 percent in the almond portobello (*Agaricus brasiliensis*) to 14.5 percent in the maitake (*Grifola frondosa*) to 41 percent in the reishi (*Ganoderma lucidum*). Fat content ranges from 0.3 to 4 percent, with polyunsaturated fats making up 10 to 30 percent of the total dry weight.

A 20 g (dry) or approximately 200 g (wet) serving of fresh maitake provides approximately 75 calories and 5 g of protein. This serving has 0.8 g of fat, made up of about 70 percent linoleic acid, an essential omega-6 fatty acid, and up to 15 percent ergosterol. Such a serving provides the following percentages of your reference daily intakes (RDIs): 17 percent selenium, at least 30 percent vitamin D, 8 percent pantothenic acid, 87 percent niacin, 4 percent thiamine, and 464 percent potassium. This 2-handful serving provides 10 percent of the protein needed by a 140-pound person or 8 percent needed by a 180-pound person. (See www.fda.gov/fdac/special/foodlabel/rdichrt.html.)

The FDA states that if 20 percent of your daily nutritional needs are met by consuming a single serv-ing of certain food, then that food is rated "excellent"; that food is "good" if a single serving supplies 10 per-cent of your needs. Given the FDA's definition of "healthy foods," mushrooms rank "good" to "excel-lent" in several categories of essential nutrients. (It is not unusual for my wife and me to each consume a half pound of freshly cooked mushrooms in one meal—combinations of shiitake, maitake, and oysters,

The Nutritional Properties of Mushrooms	Calories	Protein, g/100g	Fat, g/100g	Polyunsaturated Fat, g/100g	Total Unsaturated Fat, g/100g	Saturated Fat, g/100g	Carbohydrates, g/100g	Complex Carbohydrates, g/100g	Sugars, g/100g	Dietary Fiber, g/100g	Cholesterol, mg/100g	Vitamin A, IU/100g	Thiamine (B$_1$), mg/100g	Riboflavin (B$_2$), mg/100g	Niacin (B$_3$), mg/100g
Agaricus bisporus Button mushroom	340	33.48	2.39	0.41	0.44	0.26	46.17	24.27	21.90	19.10	0	0	0.23	3.49	38.50
Agaricus bisporus Portobello	355	34.44	3.10	1.43	1.46	0.30	47.38	24.68	22.70	20.90	0	0	0.27	4.13	69.20
Agaricus brasiliensis Brazilian Blazei	362	35.19	3.39	1.51	1.72	0.37	47.70	26.50	21.20	21.00	0	0	0.26	2.40	58.50
Flammulina populicola Enokitake	346	26.59	3.06	1.08	1.22	0.23	52.95	30.55	22.40	25.80	0	0	0.35	1.69	60.60
Ganoderma lucidum Reishi	376	15.05	3.48	0.50	1.20	0.27	71.00	69.30	1.70	66.80	0	0	0.06	1.59	12.40
Ganoderma oregonense Oregon polypore	367	13.27	2.52	0.21	0.48	0.01	72.79	72.09	0.70	72.00	0	0	0.20	1.49	20.90
Grifola frondosa Maitake	377	25.51	3.83	1.12	2.08	0.34	60.17	41.37	18.80	28.50	0	0	0.25	2.61	64.80
Hericium erinaceus Lion's mane	375	20.46	5.06	0.83	1.85	0.76	61.80	40.90	20.90	39.20	0	0	0.16	2.26	11.80
Lentinula edodes Shiitake	356	32.93	3.73	1.30	1.36	0.22	47.60	31.80	15.80	28.90	0	0	0.25	2.30	20.40
Pholiota nameko Nameko	364	33.65	3.91	1.01	1.29	0.17	48.36	29.26	19.10	28.10	0	0	0.28	3.06	106.00
Pleurotus djamor Pink oyster	356	30.20	2.86	0.91	0.97	0.16	52.76	29.66	23.10	43.80	0	0	0.26	2.45	65.80
Pleurotus ostreatus Pearl oyster	360	27.25	2.75	1.16	1.32	0.20	56.53	38.43	18.10	33.40	0	0	0.16	2.04	54.30
Pleurotus ostreatus var. *columbinus* Blue oyster	355	24.64	2.89	1.05	1.18	0.16	57.61	35.31	22.30	34.10	0	0	0.16	2.14	48.30
Pleurotus pulmonarius Phoenix oyster	355	19.23	2.70	0.53	0.62	0.11	63.40	51.60	11.80	48.60	0	0	0.10	1.68	23.80
Pleurotus tuber-regium King tuber	329	14.97	0.31	0.04	0.05	0.02	66.68	66.68	0.00	65.50	0	0	0.07	0.65	7.30
Trametes versicolor Turkey tail	369	10.97	1.51	0.27	0.32	0.06	77.96	76.06	1.90	71.30	0	0	0.07	1.06	9.30

for instance—which comes to more than 20 g (dry) per person per meal. Since mushrooms are so versatile—they can be baked, broiled, grilled, or sautéed, and they go well in soups, stir-fries, patés, and teas—it is easy to consume many different kinds of mushrooms prepared many different ways on a regular basis. (See Resources for recommended mushroom cookbooks on page 308.)

According to the USDA, 84 g of fresh, or about 8 g of dried, button mushrooms constitutes a single serving. To simplify the math for the hungry mycophile, I rounded the daily serving to 100 g fresh. The following tables were created from the analyses of 100 g samples of *dried* mushrooms. Each nutrient is listed as a percentage of total mass. To examine the nutrients in a single daily dietary serving, simply divide each percentage by 10 to see how much nutrition you get from eating a serving of the listed species.

Medicinal Properties of Mushrooms and Their Enzymes

Mushrooms producing enzymes and enzyme inhibitors are useful to medical practitioners and nutritionists composing menus customized for their patients. Suites of enzymes are secreted by the mycelium as extracellular sweat (see figure 58). These enzymes—laccases, cellulases, lignin peroxidases, manganese superoxide dismutases—are well-known for their power in decomposing plant fibers in nature. Mushrooms also produce enzyme inhibitors. Shiuan Chen and colleagues (1997) tested many foods for aromatase inhibitors and found several mushrooms with especially high concentration of this substance, which interrupts the conversion of androgens to estrogens, significant for postmenopausal women at risk for breast cancer. Similarly, some mushrooms inhibit 5-alpha-reductase, an enzyme that converts testosterone to dihydrotestosterone, which stimulates growth of the prostate. Increases in dihydrotestosterone are associated with the growth of prostate cancer. Dr. Chen tested a small sample, about a dozen species of our mushrooms, at the Beckman Research Institute at the City of Hope National Medical Center; wider samples of mushrooms are likely to yield more medicinally valuable species. Figures 229 and 230 are summaries from Dr. Chen's study of the mushrooms and their hormone inhibiting properties.

Western medical practitioners are starting to recommend mushrooms as preventive or adjunct therapies

Pantothenic Acid (B_5), mg/100g	Vitamin C, mg/100g	Vitamin D, IU/100g	Calcium, mg/100g	Copper, mg/100g	Iron, mg/100g	Potassium, mg/100g	Selenium, mg/100g	Sodium, mg/100g
21.70	0	26	9	20.80	4.8	4800	0.066	3
12.70	0	235	23	4.33	2.1	4500	0.415	52
14.20	0	737	36	4.28	1.9	5200	0.35	43
10.90	0	113	14	0.61	8.3	3100	0.054	19
2.70	0	66	37	1.30	13.0	760	0.014	6
2.10	0	32	18	1.10	4.3	850	0.039	2
4.40	0	460	31	1.88	7.6	2300	0.056	14
7.40	0	57	8	1.66	6.0	2700	0.091	4
11.60	0	110	23	1.23	5.5	2700	0.076	18
17.50	0	38	18	1.60	16.0	2500	0.103	4
33.20	0	136	5	1.61	11.0	4600	0.175	13
12.30	0	116	20	1.69	9.1	2700	0.035	48
13.70	0	214	3	1.19	5.2	4400	0.083	31
8.80	0	178	9	1.03	6.5	2600	0.09	16
3.20	0	65	12	0.13	3.5	500	0.092	2
1.70	0	62	34	0.65	8.7	570	0.007	6

for fortifying health and dealing with several medical conditions. Mushrooms are appropriate in diets for treating obesity, adult-onset (non-insulin-dependent) type 2 diabetes, and immune disorders. Mushrooms are also some of the best sources of ergosterols, which are thought to inhibit angiogenesis, the proliferation of blood vessels supporting tumors. A better understanding of the biochemical pathway for creating and converting ergosterols may lead to other factors that also limit carcinogenesis. Mushrooms contain several

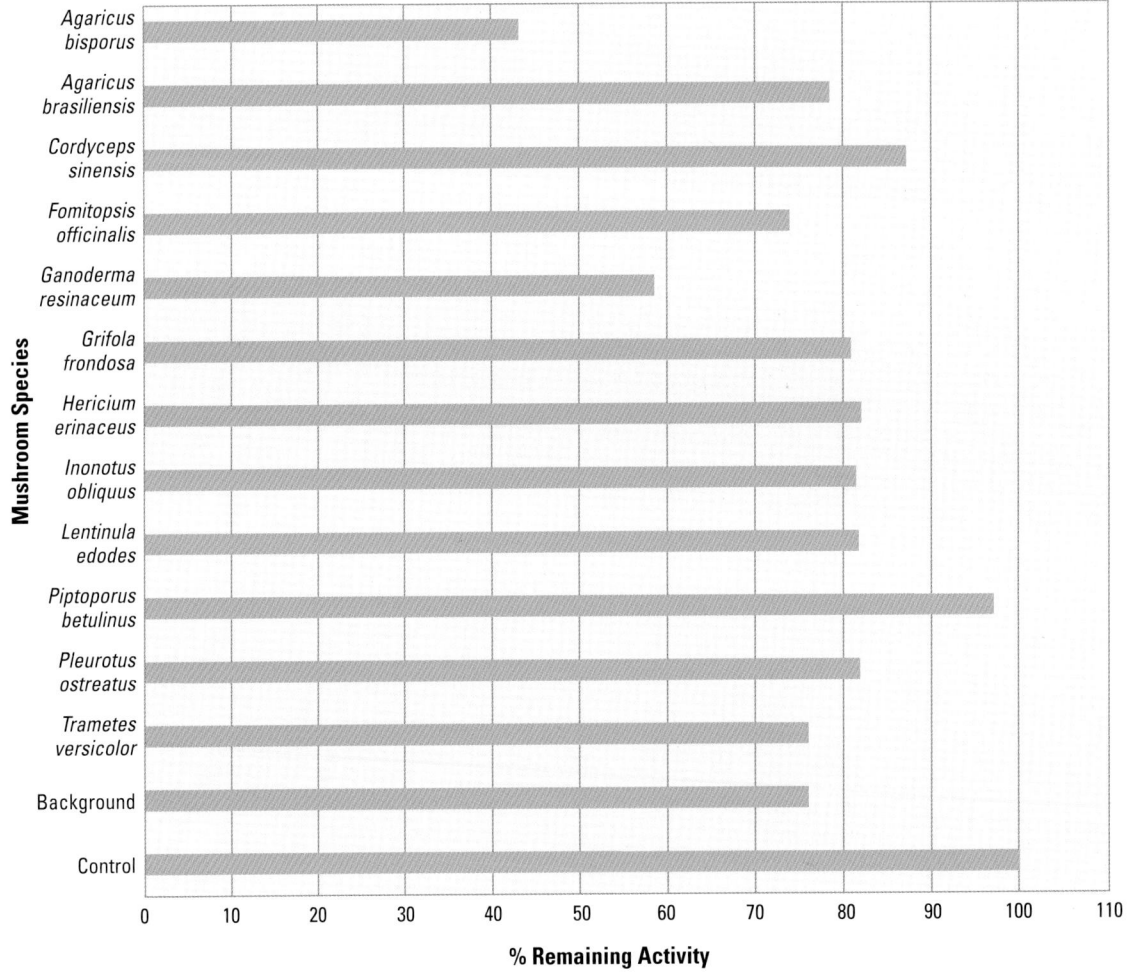

Aromatase Inhibition in Medicinal Mushrooms

▲ **FIGURE 229**

Aromatase inhibition by mushroom species. The lesser the value, the greater the inhibition of aromatase, significant for limiting the growth of breast cancer.

families of medicinally active constituents. For more details on the medicinal properties of mushrooms, see the descriptions of each species in chapter 14, *MycoMedicinals: An Informational Treatise on the Medicinal Properties of Mushrooms* (Stamets and Yao 2002), and Christopher Hobbs's *Medicinal Mushrooms* (1995). *The International Journal of Medicinal Mushrooms*, a peer-reviewed quarterly journal published by Begell House, reports on state-of-the-art research in this field.

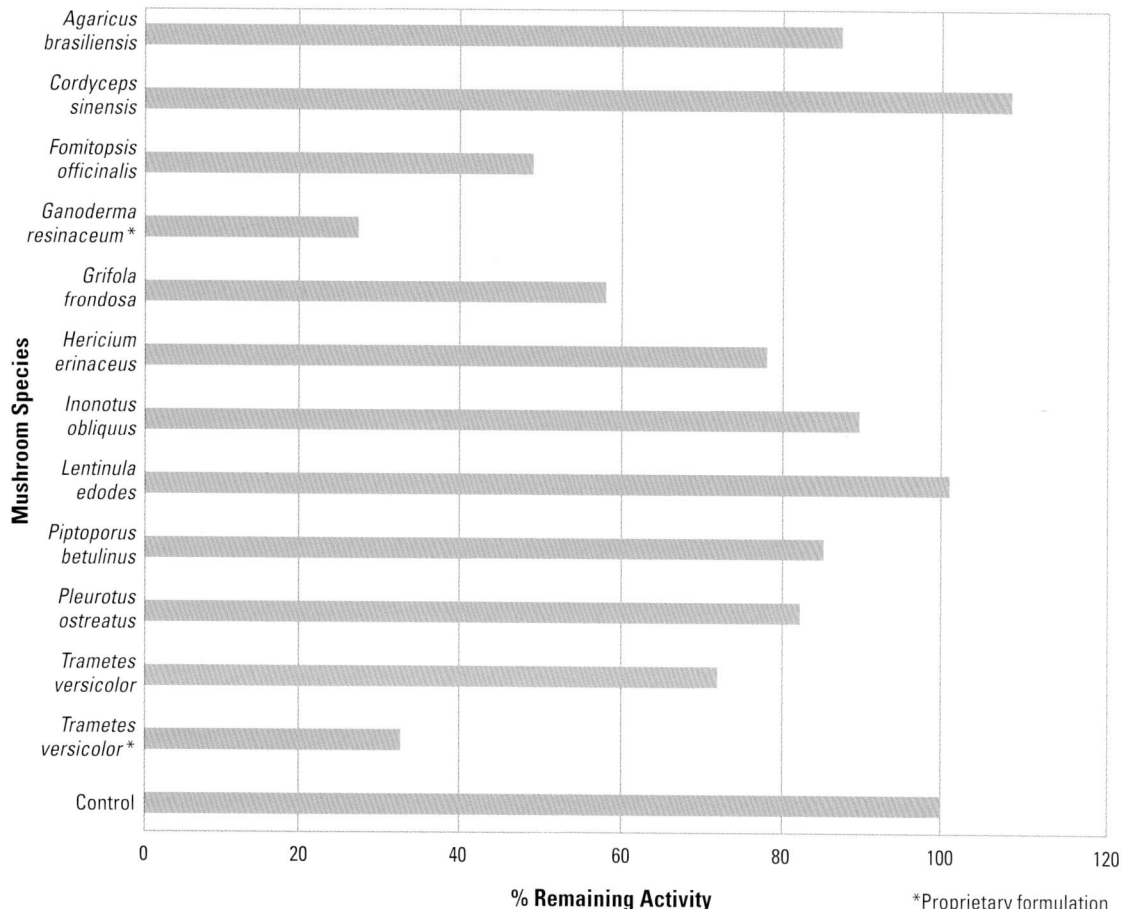

5-Alpha-Reductase Assay of Medicinal Mushrooms

▲ **FIGURE 230**

5-alpha-reductase inhibition by mushroom species. Mushrooms having lower values show greater inhibition of 5-alpha-reductase, significant for limiting the growth of prostate cancer.

The Influence of Habitat on Mushrooms' Nutritional Content

Even within a single mushroom species, nutrient and mineral levels can vary greatly, influenced by habitat and the growing medium. For instance, specimens of a particular strain of oyster mushrooms grown on sawdust have 32 g of protein per 100 g dry weight and 89 mg of niacin per 100 g dry weight. The same oyster mushrooms grown on straw consist of 27 g protein per 100 g and 54 mg of niacin per 100 g (see pages 202 and 203). Note that some nutrients and active medicinal compounds degrade with time. (Some of our samples in the chart on pages 202 and 203 were nearly a year old, stored at room temperature, at low or no light.) For updates on nutritional information regarding mushrooms, consult www.fungi.com or the listings compiled by the United States Department of Agriculture at www.nal.usda.gov/fnic/cgi-bin/nut_search.pl.

Some mushrooms concentrate minerals more than others, depending on where they are cultivated. Dr. Beelman (2003), of the Nutrition Research Advisory Panel of the American Mushroom Institute, found that the region in which button mushrooms are cultivated causes selenium content to vary. Crops originating from Texas and Oklahoma have significantly higher concentrations of selenium than samples from Florida and Pennsylvania. (As noted in chapter 7, some selenium is good for your immune system, but too much can be toxic.) Many elements in habitats influence the nutritional profiles of mushrooms, a fact that nutritionists should consider.

The Influence of Light Exposure on Vitamin D Content of Mushrooms

Most gourmet and medicinal mushrooms, unlike button mushrooms, require light. Light exposure influences vitamin content in mushrooms, particularly the conversion of ergosterol to provitamin D_2, a strong antioxidant. In the human body, UV light transforms calciferol, but not ergosterol, into vitamin D. In mushrooms, UV light transforms ergosterol into vitamin

D_2. Hence, you can supplement your vitamin D levels by ingesting light-exposed mushrooms if you do not get enough sun exposure to manufacture your own vitamin D. We conducted a series of experiments growing mushrooms indoors and measuring their vitamin D levels as affected by sunlight as they dried. We grew all of the mushrooms listed at Fungi Perfecti.

Mushrooms, upon harvesting, were either dried indoors, in darkness, or, as indicated, exposed to sunlight from 10 A.M. to 4 P.M. between June and September 2004 at Fungi Perfecti Research Laboratories, coordinates N. 47.14970 and W. 123.03905. Once exposed to sunlight, fruiting bodies were harvested and dried indoors by commercial dyers or outside under summer sun. The products were then subjected to standardized high-pressure liquid chromatography analysis in conformity with the *Official Methods of Analysis of AOAC International*, 17th edition (Horowitz 2000).

Freshly picked indoor-grown shiitake mushrooms, when dried indoors, had only 110 IU vitamin D, but when placed outdoors to dry in the sun produced an astonishing 21,400 IU of vitamin D per 100 g (IU stands for international unit; 1 IU of vitamin D is equal to 40 mcg of vitamin D). In comparison, mushrooms, from the same strain, when grown outdoors in sunlight and dried in the dark, produced only 1,620 IU, a 13-fold difference. We also found surprising differences depending on how the mushrooms were oriented to the sun. When the spore-bearing surfaces (gills or pores) of fresh mushrooms faced up toward the sun, much more vitamin D was produced than when these fertile surfaces were face down. When shiitake were dried with gills facing the sun, the vitamin D soared to the highest levels found in our tests, 46,000 IU, compared to 10,900 IU with gills down. Most surprising to me was that maitake, after being grown indoors in filtered light and dried indoors in the dark, produced 31,900 IU vitamin D upon sun exposure for approximately 6 hours, up from an ambient level of just 460 IU. Currently, the FDA recommends a daily dose of 400 IU of vitamin D; some physicians recommend as much as 800 IU. (Dr. Andrew Weil recommends

Influence of Sunlight on Vitamin D Content in Mushrooms

Species	Form	Substrate	Growth and Drying Conditions	Vitamin D Content (IU/100g)
Shiitake *(Lentinula edodes)*	Fruitbodies	Sawdust	Grown in dark, dried in dryer	134
Shiitake *(Lentinula edodes)*	Fruitbodies	Supplemented sawdust	Grown in dark, dried in dryer	15
Shiitake *(Lentinula edodes)*	Fruitbodies	Supplemented sawdust	Normal growth conditions, filtered light, dried in dryer	110
Shiitake *(Lentinula edodes)*	Fruitbodies	Supplemented sawdust	Normal growth conditions, filtered light, sun-dried	21,400
Shiitake *(Lentinula edodes)*	Fruitbodies	Supplemented sawdust	Sun grown (fruiting from composted kit), dried inside	1,620
Shiitake *(Lentinula edodes)*	Fruitbodies	Supplemented sawdust	Normal growth conditions, filtered light, sun-dried, gills down	10,900
Shiitake *(Lentinula edodes)*	Fruitbodies	Supplemented sawdust	Normal growth conditions, filtered light, sun-dried, gills up	46,000
Shiitake *(Lentinula edodes)*	Stem butts	Supplemented sawdust	Normal growth conditions, filtered light, dried in dryer, ground into powder, no sun exposure	137
Shiitake *(Lentinula edodes)*	Stem butts	Supplemented sawdust	Normal growth conditions, filtered light, dried in dryer, ground into powder, 6–8 hours sun exposure	939
Shiitake *(Lentinula edodes)*	Mycelium	Rice	Grown inside, freeze-dried, no sun exposure	<20
Shiitake *(Lentinula edodes)*	Mycelium	Rice	Grown inside, freeze-dried, 6–8 hours sun exposure	<20
Reishi *(Ganoderma lucidum)*	Fruitbodies	Supplemented sawdust	Normal growth conditions, filtered light, dried in dryer, no sun exposure	66
Reishi *(Ganoderma lucidum)*	Fruitbodies	Supplemented sawdust	Normal growth conditions, filtered light, dried in dryer, 6–8 hours sun exposure	2760
Maitake *(Grifola frondosa)*	Fruitbodies	Supplemented sawdust	Normal growth conditions, filtered light, dried in dryer, no sun exposure	460
Maitake *(Grifola frondosa)*	Fruitbodies	Supplemented sawdust	Normal growth conditions, filtered light, dried in dryer, 6–8 hours sun exposure	31,900

1,000 IU and personally consumes 1,200 IU per day.) If you consumed only 1 g dried of sun-dried maitake, about 10 g fresh or $1/3$ ounce, you would be getting close to the FDA's recommended levels. Here is an example of a food becoming a supplement. Another way of looking at this is that 500 grams of fresh maitake (more than 1 pound), or 50 g dried, would provide enough vitamin D for 50 people! What this means is that in populations where vitamin D is seriously deficient, sun-exposed dried mushrooms can help address a serious health issue.

Ingesting excessively high levels of vitamin D may be too much of a good thing. Experts vary in their opinions on this matter, but there is general agreement that vitamin D intake should not exceed 2,000 IU per day. Skin rashes, as sometimes have been reported from consuming shiitake may, in fact, be caused or exacerbated by excessive intake of the mushroom's vitamin D. Keep in mind, however, that most mushrooms are grown in the shade, and many are dried in food dehydrators, so much of their embedded ergosterol remains unconverted to vitamin D.

Mau and fellow researchers (1998) showed that outdoor-grown shiitake mushrooms contained 5 to 7 times more vitamin D than the indoor-grown variety, and that shiitakes grown in latitudes closer to the equator naturally had more vitamin D than those grown in northern regions. In this study, artificial UV exposure from a germicidal lamp for only 1 minute tripled concentrations, with a corresponding decrease in ergosterols. However, exposure to UV for 2 hours decreased vitamin D by 12 percent, because the radiation began breaking down the vitamin. In our experiments, we found that sun exposure between day 2 and day 3 caused a drop in vitamin D, apparently due to UV degradation from overexposure. Our data also confirmed Perera's (2003) work that determined that vitamin D production is most concentrated in the gills of shiitake mushrooms, with half as much in the caps, and a third as much in the stems. Efficiency in converting ergosterol to vitamin D was optimized when mushrooms were at 70 percent moisture. To increase vitamin D in mushrooms, artificial ultraviolet light could be used in mushroom driers. What this collective research shows is that vitamin D production from ergosterols in mushrooms is controlled by light (UV) exposure, and light levels can be manipulated to affect vitamin D concentrations.

Earlier in human history, we got our vitamin D from chemical processes in our skin triggered by exposure to sunlight. As our ancestors migrated from areas near the equator to regions with shorter days and colder areas where we wore more UV-blocking clothing, our bodies produced less vitamin D. One well-known disease from vitamin D deficiency is rickets, which afflicts mostly children; another effect is decreased bone density. Deficiency of vitamin D may be a cofactor in the growth of breast, prostate, and colon cancers and some immune disorders. Mushrooms, particularly maitake, shiitake, reishi, and turkey tails, have been the subject of research papers addressing these specific cancers. In 2005, a Harvard University team led by Dr. David Christiani reported that from a survey of 456 lung cancer patients, of which about only 10 percent had radiation treatment or chemotherapy, those with higher levels of vitamin D from sunlight or food supplements had a 3-fold better chance of surviving lung cancer surgery. I suspect that mushrooms are nature's best land-based food source for vitamin D. Beyond vitamin D, mushrooms are packed with other immune-enhancing agents. For any medicinal mushroom research study on immunity, knowing the influence of vitamin D in the mushroom sampled would be critical for drawing conclusions on the sources of the immunopotentiating effects. (For an overview of medicinal mushrooms and their properties, see Stamets and Yao 2002.)

Sunlight drives vitamin D synthesis in fungi, fish, plankton, reptiles, and mammals. There are 2 distinct pathways for synthesizing vitamin D. In fungi, short-wave ultraviolet light (UVB) converts ergosterols to provitamin D, called *vitamin D2* or *ergocalciferol*. In parts of the world where vitamin D deficiencies are common, the occasional eating of fresh shiitake exposed to sunlight may boost vitamin D levels and mitigate deficiency-related diseases. Another form of vitamin D is vitamin D3, or cholecalciferol, which is manufactured from 7-dehydrocholesterol in the human skin during sun exposure. In summertime, a young Caucasian person's body can make 10,000 IU with just 30 minutes of sun exposure.

Your body protects itself from excess vitamin D through the activities of the parathyroid gland and its hormones. If too much vitamin D is present, then a reverse pathway denatures vitamin D. In the process, calcium and phosphorus, whose intake had been enabled by vitamin D, are now eliminated from the body. In the event of a severe vitamin D overdose, damaged bone density and blood chemistry could result as calcium and phosphorus levels plummet. This bidirectionality of vitamin D regulation points to the importance of knowing how mushrooms are grown and processed, how much to consume, and how much vitamin D you need. As is often said, the difference between a medicine and a poison is dose.

Culinary Uses of Mushrooms

Culinary artists are experimenting with new mushrooms. From the less known Canadian tuckahoe *(Polyporus tuberaster)*, appealing for its large size, palatable mild flavor, and long shelf life, to the more popular and stronger flavored morels, each mushroom has a unique taste, a flavor and texture personality.

Recently, we sampled a delightful new use of porcinis *(Boletus edulis)*—baking them into spongy mushroomy bread. I have also tried mushroom-infused beers and wines, mushroom cookies, mushroom chocolates, and even mushroom ice cream: Jerry Greenfield (of Ben and Jerry's ice cream fame) and I concocted a chocolate-*Cordyceps* ice cream, which I thought was quite good. Distributors of nutraceuticals (health supplements) continually seek mushrooms for new products. The mushroom industry as a whole, once dominated by buttons and portobellos, is now flush with diverse choices of fungal products, and inventive recipes seem to be proliferating like, well, mushrooms.

In general, mushrooms should be cooked before consumption because their tough cell walls, when eaten raw, are largely indigestible. Uncooked mushrooms pass through the digestive tract largely intact, imparting little if any nutritional benefit. When cooked, however, they are highly digestible and are excellent sources of nutrition, since heat softens their tissues and allows your digestive enzymes to more easily reach through the chitinlike cell walls.

Why Grow Mushrooms Organically?

If there ever was a case for growing food organically, mushrooms make it. Mushrooms can benefit the immune system (Stamets and Yao 2002; Wasser and Weis 1999), but ingesting those carrying industrial toxins sabotages their health-enhancing effects. Depending upon ambient levels, mushrooms may contain some heavy metals and certain pesticides, as do many other foods, from spinach to salmon. But when mushrooms are grown organically and processed according to good manufacturing standards, they are safe to eat and will bolster your health—if the substrate is free of contaminants. Of course, practicing organic methods does not guarantee toxin-free produce. Organic methods lessen risk, but all habitats are imperiled by the inevitable upwardly creeping concentrations of toxins in our overall environment. In China, the Almond Portobello (*Agaricus brasiliensis*, a species from Brazil formerly known as *Agaricus blazei* ss. Heinemann) are sometimes cultivated using organic methods, but turn out to nevertheless contain cadmium at levels far above the standards set by the European and American health agencies. (Stijve et al. 2003). In particular, air pollution rains down heavy metals into the food chain, and consequently mushrooms can accumulate toxic levels. (See chapter 7.) The bottom line: If your habitat is contaminated, so are your mushrooms, whether you use organic methods or not.

And now, let's move on to growing mycelium and letting mycelium lead the way!

CHAPTER 14

Magnificent Mushrooms: The Cast of Species

The following cast of characters is just the beginning of a catalogue that may eventually encompass dozens of species. What these species have in common is that they can be naturalized outdoors in your backyard and most have proven beneficial medicinal and ecological properties. Mycological populations are constantly in a state of flux, and the understanding of how they move within ecological systems is rapidly developing as an art and a science. The most useful information often comes from field observations—particularly from skilled amateurs—whose close encounters of the mycological kind may result in the development of new methods. For us, the bemushroomed, this field of research is exciting, rewarding, and often downright fun. Our studies sometimes lead us to discover and investigate the bizarre, and many of us suspect that mushrooms possess a peculiar sense of humor.

When choosing a mushroom species, first match species with their preferred habitats; this will help guide your choices for each location. Once you add spawn in an area and establish a mother patch, you can expand and guide the mycelial lens by feeding it

Since this book will forever be a work in progress, I welcome emails and photographs telling me about your mushroom fruiting experiences of the bizarre, the intentional, and the unintentional. I am particularly interested in companion cultivation strategies and the population dynamics of saprophytes as they influence biological communities.

debris. Natural weather cycles trigger fruitings, but a timely application of water in the weeks prior to the rainy season will usually set the stage for bountiful crops. Note that these recommendations are for outdoor inoculations. When working indoors, cultivators have more options; see *Growing Gourmet and Medicinal Mushrooms* (Stamets 2000a).

Agaricus brasiliensis Wasser, Didukh, de Amaozonas, and Stamets

Common Names: Brazilian blazei, royal sun Agaricus, himematsutake, kawariharatake, songrong, cogumelo de Deus (mushroom of God), cogumelo de sol (mushroom of the sun), Murrill's Agaricus (ABM), or almond portobello.

Taxonomic Synonyms and Considerations: The name *A. brasiliensis* is a new moniker for a mushroom known to most of the mycological world as *A. blazei*. In 1945 the American mycologist W. A. Murrill found a new species in the lawn of a Mr. R. W. Blaze in Gainesville, Florida, whereupon he gathered a collection of these mushrooms and published it as *Agaricus blazei*. Decades later, specimens forwarded to Pennsylvania State University from Brazil via Japan were identified by P. Heinemann as best conforming to Murrill's species concept *Agaricus blazei*. Recently, Wasser and other researchers (2002) distinguished this species from *A. blazei* and published the new taxon *Agaricus*

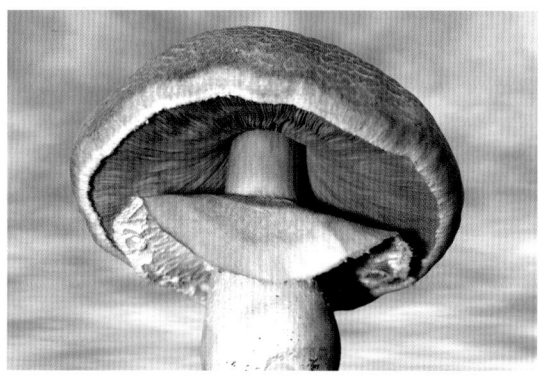

▲ **FIGURE 231**

Formerly called *A. blazei* by many, the regal *A. brasiliensis* is a popular medicinal mushroom with a strong almond flavor.

brasiliensis. This new name describes the mushroom "misidentified" by Heinemann. (Heinemann did not really misidentify it; he placed it into the nearest species concept published at that time.) Now we believe that *Agaricus brasiliensis* is synonymous with *Agaricus blazei* ss. Heinemann but not *Agaricus blazei* ss. Murrill.

The mushroom that it most closely resembles is the slender but tasty *A. subrufescens*, its taxonomic cousin, which differs slightly in the shape of the spores. The spores of the similar *A. augustus* are much larger, 7.5–10 µm, compared to the smaller spores, 4–5 µm, seen in *A. brasiliensis.* Freshly picked *A. brasiliensis* usually bruises bright yellowish when cut, while *A. subrufescens* bruises only along the outer cuticle, if at all. Once *A. brasiliensis* has been harvested, the yellowing reaction diminishes and is replaced by a browning reaction. The difference in the staining reaction between these two species may or may not be taxonomically significant. From my experience, the staining reaction in *A. brasiliensis* is not a dependable taxonomic character. However, I have noticed another difference of undetermined taxonomic significance. The freeze-dried, powdered mycelium of *A. brasiliensis* is light tan in color upon sterilization, while *A. subrufescens* is a darker, cinnamon brown. DNA tests will further illuminate the lineages of these taxa.

Introduction: Japanese mycologists pioneered the cultivation of this species from specimens collected in Brazil and are credited for bringing this species to the forefront. *A. brasiliensis* has the general appearance of an oversized portobello mushroom, with a beguiling almond fragrance and flavor, but it bruises yellow, sometimes in a bright hue. This mushroom is comparable in many ways to the prince, *A. augustus*, but has, in my opinion, far better texture. This mushroom is a natural for mitigating the outflows from horse paddocks, stables, cattle feedlots, or other sources of manurial leavings, especially those outflows mixed with straw or sawdust. A good species for the garden, this mushroom prefers moist, warm climates, with fruitings coinciding with the summer months in temperate climates. In subtropical zones, this mushroom appears during the warm, wet seasons.

Description: A classic large *Agaricus* species, this mushroom is grander in culture than most wild forms. Cap 7–25 cm broad, convex at first, soon hemispheric, then broadly convex, eventually flattening; often cap margin smooth, white, and splitting only in age. Cap surface covered with brownish fibrillose patches. Partial veil membranous, cottony, with patches of the veil typically tearing to form a median membranous ring, but sometimes with remnants attached to the margin at maturity. Gills pallid at first, soon gray, and then chocolate brown when mature. Stem cylindrical, solid, tall, whitish, smooth; flesh thick, quickly staining yellowish when bruised. Growing singly or in clusters, arising from clusters of dense rhizomorphs leading to and often attached to the stem bases. Imparting strong scent of almonds, especially during cooking. Spore deposit dark chocolate brown, 5.7–7 by 3.8–4.6 µm.

Distribution: First collected in Brazil in the Sal Hose do Rio Preto district, northwest of Sao Paulo city, this mushroom is common in fields and mountainous regions. Probably more widely distributed than the literature presently indicates.

Natural Habitat: Growing in soils with plentiful lignin-rich debris, in mixed woods, well-composted soils, and along forest edges. This mushroom is a complex sapro-phyte and prefers rich and actively composting soils.

Type of Rot: Mottled white rot.

Fragrance Signature of Mycelium: Musty grain with almond overtones.

Natural Method of Cultivation: Mounds of leached cow manure or similarly supportive compost can be inoc-ulated and then planted with cover crops to promote growth. Good companions are kale, zucchini, squash, potatoes, melons, and other leafy green vegetables.

Season and Temperature Range for Mushroom For-mation: Fruiting when temperatures hover between 70–95°F; classically a summer mushroom in the northern latitudes.

Harvest Hints: This mushroom is best harvested while the partial veil is intact but stretched, usually at the stage when the mushrooms are hemispherically shaped, before they expand to convex or plane.

Nutritional Profile: Protein content can range from 33 to 48 percent (making this species one of the most protein-rich of all cultivated mushrooms; content fig-ures based on dry weight). Our analysis of this mush-room (see chart on pages 202 and 203), when dried down to 9.66 percent moisture, shows the following per 100 g: protein 33.48 percent; fat 2.39 percent; polyunsaturated fat 0.41 percent; total unsaturated fat 0.44 percent; saturated fat 0.26 percent; carbohydrates 46.17 percent; complex carbohydrates 24.27 percent; dietary fiber 19.10 percent; cholesterol 0 percent; vita-min A 0 IU; thiamine .23 mg; pantothentic acid (B5) 21.70 mg; vitamin C 0 mg; vitamin D2 26 IU; calcium 9 mg; copper 20.8 mg; iron 4.8 mg; potassium 4,800 mg; niacin 38.50 mg; riboflavin 3.49 mg; selenium 0.06 mg; sodium 3 mg; ash 8.30 g. An unsubstantiated report on the Internet states that this mushroom has more than 50 times the beta-glucans of the common button mushroom, which I find doubtful.

➤ **FIGURE 232**

A. brasiliensis fruits from a tray filled with pas-teurized leached cow manure. This majestic species produces mushrooms on a wide array of substrates— from supple-mented sawdust to a variety of composts.

▲ **FIGURE 233**

A. brasiliensis emerges from a cluster of rhizomorphic mycelium.

Medicinal Properties: This potent medicinal mushroom produces beta-1,6-glucans and beta-1,3-glucans, polysaccharides currently being studied for immunopotentiation. The literature reports beta-glucan levels of up to 14 percent. (An analysis of our *A. brasiliensis* showed this mushroom contains 9 percent beta-glucans and approximately 27 percent total polysaccharides.) Its unique beta-glucans promote natural killer cells and are selectively cytotoxic to tumor cells. This mushroom has been the subject of numerous analyses isolating constituents, both tumoricidal and immunomodulatory, for the treatment of cancers (Fujimiya et al. 1998; Ito et al. 1997; Itoh et al. 1994). The cultured mycelium also produces antitumor compounds (Mizuno et al. 1999). That this mushroom produces compounds specifically increasing apoptosis in cancerous cells but not in healthy cells, and also triggers an immune response, is notable. A yellowish metabolite exuded by the mycelium apparently has bactericidal properties. For current information on studies of the medicinal properties of this mushroom, refer to *MycoMedicinals* (Stamets and Yao 2002), a booklet summarizing medicinal studies on mushrooms, which is periodically updated.

▲ **FIGURE 234**

David Sumerlin prepares to dry freshly picked *A. brasiliensis*.

Flavor, Preparation, and Cooking: The almond flavor of the fresh mushrooms is so powerful that it easily overpowers the flavors of other foods. I prefer to baste the mushrooms in olive oil and cook them separately with rosemary and onions, either sautéed or grilled. Thinly sliced, they are also good in soups. To the newly initiated, this mushroom gets high marks for its culinary value. We growers can easily tire of its all-too-potent almond flavor. Hence, it is best treated as a delicacy and not a daily consumable. Since the dried mushrooms can have a terrible taste, marketing these mushrooms fresh best suits the culinary market. Dried mushrooms are usually powdered and presented in capsule, tablet, or extract form for medicinal purposes.

Mycorestoration Potential: Good candidate for recycling of manure wastes, especially where cows, chickens, and horses concentrate. This species absorbs cadmium, copper, lead, and mercury (Stijve et al. 2003). Also a consumer of *Escherichia coli* and other coliforms, *A. brasiliensis* works well in warmer climates for outdoor cultivation, and mushroom beds can double as mycofiltration buffers around point-source pollution centers, such as dairies, chicken farms, and slaughterhouses.

Comments: As popular as this mushroom has become, it is also at the center of controversy. Fueled by competition amongst Brazilian producers, and between growers in China and Japan, the fervor with which it is marketed approaches that of a fad. Ikekawa (2003) questions the veracity of Takashi Mizuno's seminal work on the medicinal properties of this mushroom (Mizuno 1995). However, subsequent research has identified medically significant anticancer constituents and immune enhancement. Worth noting is that in one test, Stijve (2003) determined that *A. brasiliensis* mushrooms grown in China had significantly more cadmium than those grown in the United States.

Agrocybe aegerita (Brigantini) Singer

Common Names: Pioppino, black poplar mushroom, the swordbelt Agrocybe, south poplar mushroom, yanagimatsutake (Japanese), zhuzhuang-tiantougu (Chinese).

Taxonomic Synonyms and Considerations: A variable mushroom that may well be split into several distinct taxa after more research is done, *A. aegerita* was once called a *Pholiota (Pholiota aegerita)*. Other synonyms are *Pholiota cylindracea* Gillet (Singer 1986), and *A. cylindracea* (DC. ex Fr.) Maire, a name still preferred by Asian mycologists. *A. molesta* (Lasch) Singer and *A. praecox* (Pers. ex Fr.) Fayod are notable species related to *A. aegerita* and can be cultivated using the same methods described here.

Introduction: This species is an excellent candidate for stump recycling and log cultivation and is one of our favorite edible mushrooms.

Description: A substantial mushroom, with cap often up to 12 inches in diameter. Cap convex to hemispheric, expanding to plane at maturity; smooth, yellowish gray to grayish brown to tan to dingy brown, darker toward the center. Gills gray at first, becoming chocolate brown with spore maturity. Stem white, adorned with a well-developed membranous ring. Spores brown, measuring 9–11 by 5–7 μm. The spores usually color the upper surface of the membranous ring brown.

Distribution: Growing in Mexico and the southern United States (Mississippi, Louisiana, and Georgia), this mushroom is also common across southern Europe and in similar climatic zones of the Far East.

Natural Habitat: Growing saprophytically, often in clusters, on stumps in the southeastern United States and southern Europe. Preferring hardwoods, especially cottonwoods, willows, poplars, maples, box elders, and in China tea-oil trees.

Type of Rot: Not clear; thought to be a white rot.

Fragrance Signature of Mycelium: Mealy, farinaceous, but not pleasant.

Natural Method of Cultivation: Easily cultivated on the stumps of the above-mentioned trees, or on

▲ **FIGURE 235**

A star cluster of *A. aegerita* primordia.

▲ **FIGURE 236**

A. aegerita fruits from a mushroom kit, also known as a sawdust production block.

half-buried logs or upon logs laid on the ground. A quick method of inoculation is to scarify logs with a chain saw and then pack sawdust spawn into the wounds between the adjoining logs.

Season and Temperature Range for Mushroom Formation: Primarily fall. Temperature: 60–70°F.

Harvest Hints: A more fragile mushroom than it initially appears, this mushroom should be encouraged to grow in clusters. If mushrooms are harvested before the veils break, shelf life is prolonged.

Nutritional Profile: Our analysis of a 100 g serving shows the following: calories: 347; protein: 29.90 g; fat: 3.24 g; polyunsaturated fat: 1.20 g; total unsaturated fat: 1.34 g; saturated fat: 0.19 g; carbohydrates: 49.50 g; complex carbohydrates: 46.08 g; sugars: 3.40 g; dietary fiber: 32.00 g; cholesterol: 0 mg; vitamin A: 0 IU; thiamine (B_1): 0.17 mg; pantothenic acid (B_5): 9.10 mg; vitamin C: 0 mg; vitamin D: 233 IU; calcium: 9 mg; copper: 3.79 mg; iron: 3.7 mg; potassium: 3,200 mg; niacin: 57.00 mg; riboflavin: 2.35 mg; selenium: 0.125 mg; sodium: 27 mg; moisture: 9.71 g; ash: 7.65 g.

Medicinal Properties: Yoshida and other researchers (1996) reported that the administration of 3 high molecular weight polysaccharides (carboxymyethylated (1,3)-alpha-D-glucans) from this mushroom induced the ratio of macrophages more than 50 percent resulting in tumor regressions in mice, reversing peritoneal cancer. Another notable report concerns the water fractionation and isolation of 2 polysaccharides from this mushroom with remarkable hypoglycemic activity in diabetic mice (Kiho et al. 1994). Kim and fellow researchers (1997) isolated 2 new indole derivatives that acted as free radical scavengers. Zhang and others (2003) isolated an alcohol-soluble fatty acid fraction and ergosterols and mannitols with enzyme-inhibiting antioxidant properties against COX-1 and -2. A lectin isolated from this mushroom showed inhibition of the tobacco mosaic virus, a property shared by *Fomes fomentarius* (Sun et al. 2003); while Zhao and others (2003) found similar antitumor lectins that induced apoptosis in multiple cancer cell lines in vitro.

Flavor, Preparation, and Cooking: Absolutely delicious! Finely chopped and stir-fried, cooked in a white sauce and poured over fish or chicken, or baked in a stuffing, this species has a nutty flavor and a wonderful, crunchy texture. This mushroom is versatile in many recipes known to the mycoculinary arts. Those who have eaten this mushroom, including me, become hooked (or "pioppino'd").

◀ **FIGURE 237**

Buried logs, arranged as a "log raft," fruiting with *A. aegerita*. The sawdust spawn—from an expired mushroom kit—was introduced the year before by placing it between the logs and against wounds created by a chain saw. (See page 181.) The logs produced to 3 fruitings per year for more than 5 years. I suspect many *Agrocybe* species can be grown this way.

Mycorestoration Potential: An excellent species for mycoforestry, especially for stump recycling (in the humid southeastern United States in particular) and for outdoor log-raft cultivation. Inoculated logs can frame the perimeters of garden beds or landscaped areas.

Comments: Of the many species we grow, this mushroom has become one of the favorites of our employees and my family. The willow-populated swamps of Louisiana seem an ideal setting for the deliberate cultivation of *A. aegerita*. Regions of Chile, Japan, and the Far East, as well as southern Europe, also have weather patterns and ecosystems conducive to growth.

Chlorophyllum rachodes (Vitt.) Vellinga

Common Name: Shaggy parasol.

Taxonomic Synonyms and Considerations: Recently transferred to the genus *Chlorophyllum* based on DNA characters (Vellinga 2003; Vellinga et al. 2003), this mushroom was previously most commonly listed in field guides as *Lepiota rachodes* (Vitt.) Quelet, later to become known as *Macrolepiota rachodes* (Vitt.) Singer. A variety of this mushroom, *Lepiota rachodes* var. *hortensis*, from western North America, has greenish tinges in the gills and between the gills and the stipe, and can cause gastrointestinal distress in some people (although I have never had any problems with it). This variety may in fact be a separate species from what we now know as the edible *C. rachodes*. Generally, in my experiences I have found that *C. rachodes* is a stouter, shorter-stemmed mushroom, darker in color and with more abundant scales than the longer-stemmed, lighter-colored, less ornate but more majestic *Macrolepiota procera*. Now sharing the genus with the poisonous *C. molybdites*, a mushroom that casts a green spore print, the shaggy parasol, *C. rachodes*, produces a whitish spore print. To avoid accidental poisoning from *C. molybdites*, it is absolutely essential to make spore prints for identification.

Introduction: Common in the Pacific Northwest of North America, this has long been one of my favorite mushrooms. I often see it on roadsides, as its large, white form is discernible from great distances. Not as large as *M. procera*, this mushroom has one notable advantage: it digests grass clippings from lawn mowing. I first found this mushroom, which is also friendly to thatch ants (*Formica* species), growing out of anthills amongst the Mima glacial mounds near Tenino, Washington. Sargara (1992) reported a similar co-occurrence in England. Recalling this experience, I found that inoculating ant mounds with spawn of this mushroom established fruiting colonies, as with *M. procera*. The mycelium of *C. rachodes* is generally not as densely matted as with *M. procera*, and cultures in the laboratory tend to die out, necessitating the gathering of new strains every few years. Strains growing under trees are not as easy to capture in culture as those growing from composting debris piles.

Description: Cap 5–15 cm in diameter, convex, expanding to broadly convex and plane in age. Brown to reddish brown, lightening with maturity. Surface breaks up in maturity. Surface coarsely scaly with shaggy, fibrous scales. Flesh white, bruising orangish yellow when cut or handled. Gills white, closely

▲ **FIGURE 238**

C. rachodes, also known as *Lepiota rachodes*, is great for decomposing grass clippings and fir needles.

FIGURE 239

After mowing our yard, La Dena Stamets gets ready to dump the grass clippings.

FIGURE 240

Sawdust spawn of *C. rachodes* is broadcast over the surface of the moist, rank grass.

FIGURE 241

La Dena places grass clippings on top of the spawn, creating a "mycelial sandwich." This patch is located in the shade beneath a fir tree along the edge of our yard.

FIGURE 242

Half a year later, a delighted La Dena Stamets rejoices as the first *C. rachodes* mushrooms appear. "I did it Dad!" she exclaims. The mushrooms have continued to fruit, without additional spawning, for more than 4 years at this same location. Feeding the mushroom patch regular lunches of cut grass is essential for keeping it alive.

arranged, broad, becoming pallid to dingy brown in age. Stem 10–20 cm long by 1–3 cm thick, stout relative to the breadth of the cap, often thickly bulbous and curved at the base. Partial veil thick, membranous, falling with age to form a well-developed, movable sheathlike ring on the upper regions of the stem. Spores white in deposit, 6–10 by 5–7 µm.

Distribution: Widespread throughout the temperate northern regions of the world. Found in northern Europe and England and throughout much of northern North America.

Natural Habitat: Growing along roads, in lawns, along borders to woods, and in anthills.

Type of Rot: White.

Fragrance Signature of Mycelium: Forest musty.

Natural Method of Cultivation: Having found this mushroom where grass clippings have been dumped into compost piles, and knowing its preference for living under fir trees on the edges of yards, I cloned a mushroom and inoculated grass clippings in a very laissez-faire fashion. A bag of spawn was spread over grass clippings that had been raked out to about 2 inches in depth. After being covered with 2 to 4 inches of fresh grass, that zone was then left alone. Additional grass clipping deposits were placed on the edges of the inoculated grass pile. I found that if the grass is piled more than 6 inches deep, an anaerobic environment predominates in the core. I like to create shallow mounds with attenuating edges so the mycelium can seek the best depth for mushroom formation.

Alternatively, as with *M. procera*, anthills can be inoculated. The mycelium can be placed directly upon the top of the anthill in one pile. The ants will busily move the mycelial sawdust, incorporating it into their mound, and by doing so spread the mycelium. Over time, the mycelium will grow throughout the nest, most often producing mushrooms in autumn, when the nest is abandoned.

Season and Temperature Range for Mushroom Formation: Mostly fall, September to November, although we have had a few fruit in early July. Temperature: 50–70°F.

Harvest Hints: Best harvested when the mushrooms are tightly convex, before spore maturity and before insect larvae develop.

Nutritional Profile: Not yet known to this author.

Medicinal Properties: Not yet known to this author. I suspect that the mycelium produces antibiotics bolstering the defense of ant colonies against parasitic diseases, much in the same manner as described by Currie and others (2003). (See page 30.) Suay and other researchers (2000) report that this species has moderate activity in limiting the growth of *Staphylococcus aureus*.

Flavor, Preparation, and Cooking: *C. rachodes* is a much more moist mushroom than its cousin *M. procera*. Some varieties of this mushroom have a strong nutty flavor, while others do not. The variety I cultivate is delicious, but I have had other forms that are less pleasing to the palate. Cook in a fashion similar to that described for *M. procera* on page 267.

Mycorestoration Potential: Thus far, the greatest strength of this species for mycorestoration is recycling grass and garden debris and aiding ant communities.

Comments: Here, the strain makes all the difference. Find a good strain from the wild and propagate by creating your own mycelium from stem butts or spores. Or buy spawn of a proven strain from a spawn laboratory. This mushroom is a gardener's ally, running on edges adjacent to decomposing debris. I have long been curious about the ability of this species to thrive despite vast complexes of competitors occupying the same niche. Its tendencies beg more questions than I can answer. A curious and delicious friend for your property, this species is fun to grow, tending to reoccur within the same 2-week period every year—in our case the second to fourth weeks of September—provided you replenish the patch with additional clippings that are appealing to this fungus.

See page 265 for information on a sister parasol mushroom, *M. procera*

Coprinus comatus (Muller: Fries) S. F. Gray

Common Names: Shaggy mane, inky cap, lawyer's wig, maotou-quisan (Chinese).

Taxonomic Synonyms and Considerations: *C. comatus* is considered a safe species for most amateurs, since it can be easily identified. A number of related species pivot around this species concept, whose differences are largely unimportant to those interested in the culinary applications. Ecotypes vary, however, and this is an important consideration for those implementing mycorestoration strategies.

Introduction: The shaggy mane has long been a favorite of mycophagists in North America and Europe. Easy to identify and often growing in massive quantities, this brilliantly white, shaggy mushroom is not likely to be confused with poisonous species. Its mild but excellent flavor makes it one of the most popular of edible mushrooms amongst hikers and mushroom hunters.

Having experimented with cultivation, I am pleasantly surprised at how well this species adapts to a wide variety of indoor and outdoor substrates. Although commercial cultivation is limited by its tendency to disintegrate into an inky mess, this mushroom is fantastic for those who can consume it within 2 days of picking.

Description: Cap 4–15 cm high by 3–5 cm thick, vertically oblong, dingy brown at first, soon white, and decorated with ascending scales. Gills crowded, white to pale, and long, broad, and slightly attached to or free from the stem. Stem 6–12 (15 cm) long by 1–2 cm thick, equal in diameter, hollow, bulbous at the base, and adorned with a movable, membranous collarlike ring that separates from the cap margin as the mushroom enlarges. As the mushroom matures, the gills blacken, or deliquesce, transforming into a black, spore-laden fluid that drips from the rapidly receding cap margin. The cap eventually totally deliquesces, leaving only the stem. Spores black, 11–15 by 6.0–8.5 µm.

Distribution: Growing in the late summer and fall throughout the temperate regions of the world.

◄ **FIGURE 243**

C. comatus are stately mushrooms; they're easy to identify and are prime edibles.

◄ **FIGURE 244**

Dusty Yao picks *C. comatus* on the ski slopes above Telluride, Colorado.

C. comatus, a delicious choice edible mushroom and an indicator of habitats in transition. Here, Scott Oliver picks one of hundreds that came up in his lawn for many years. When he stopped fertilizing, the mushrooms did not reoccur. Some strains of *C. comatus* commonly form in yards in response to nitrogen-rich fertilizers.

Natural Habitat: In lawns and meadows, around barnyards, in wood chips, along roadsides, and in hard-packed and enriched soils.

Type of Rot: White rot.

Fragrance Signature of Mycelium: Farinaceous and mildly sweet.

Natural Method of Cultivation: Inoculation of spawn directly into manure-enriched soils or 4- to 6-inch-deep beds of hardwood sawdust. Newly laid or fertilized lawns that are frequently watered are perfect habitats for shaggy manes. Cow or horse manure, mixed with straw or sawdust, is also ideal. Hardwood sawdust spawn should be used as inoculum for establishing outdoor patches.

Season and Temperature Range for Mushroom Formation: Spring and fall temperature: 40–60°F.

Harvest Hints: Since this mushroom deliquesces from the end of the gills upward to the stipe, mushrooms should be picked before the appearance of the slightest hint of the gills turning black. If picked when no basidia have matured, mushrooms can be kept in cold storage for 4 to 5 days. Any mushrooms that begin to deliquesce should be removed from the fresher fruitbodies, since the enzymes secreted by a single deliquescing mushroom will decompose adjacent mushrooms, regardless of age.

Nutritional Profile: 25–29 percent protein; 3 percent fat; 59 percent carbohydrates; 3–7 percent fiber; 1.18 percent ash (Crisan and Sands 1978; Samajpati 1979).

Medicinal Properties: The natural antibiotic coprinin, also known as coprinine, has been isolated from this species. This species has not yet been thoroughly studied for its medicinal properties.

Flavor, Preparation, and Cooking: Shaggy manes were the first mushrooms that seduced me into the art of mycophagy. Then and now I still prefer to eat this mushroom for breakfast: I like frying thinly cut slices (stems included) in a frying pan with onions and a light oil; once they're slightly browned, the mushroom slices are folded into an omelet. Or, fry the mushrooms in butter over medium heat, lightly salt, and serve on whole wheat toast. Since this mushroom has considerably more moisture than shiitake, for instance, it should be cooked until the water has evaporated before other ingredients are added to the frying pan.

Mycorestoration Potential: This species is surprising in its adaptability to a variety of habitats, its number of varieties and closely related allies, and its broad-spectrum antimicrobial properties. The shaggy mane is a "first responder" to habitats after a disturbance. It proliferates around construction of housing developments and highways, often when the ground has been scraped down to gravel. When ecosystems recover, this is one of the first mushrooms to co-occur with pioneering grasses and shrubs. However, shaggy manes can persist when fertilizers, high in nitrogen, are added to a lawn, for instance. Showing antimicrobial activity against *Aspergillus niger*, *Bacillus* species, *Candida albicans*, *Escherichia coli*, *Pseudomonas aeruginosa*, and *Staphylococcus aureus*, *C. comatus* is good habitat buffer, waste decomposer, and environmental indicator. A bioaccumulator of arsenic, mercury, and to a lesser degree cadmium, the shaggy mane often appears in polluted soils. Ecotypes can vary substantially in their selectivity and performance. Many ephemeral, smaller-statured species that quickly deliquesce or disintegrate in a couple of days pop up when straw is placed upon the ground.

Comments: Shaggy manes are great mushrooms to grow in compost piles and in your yard. Once an outdoor patch is established, these mushrooms fruit for many years. For mycological landscapers not concerned about territorial confinement of their mushroom patch, the shaggy mane is an excellent companion to many garden plants, especially those liking a hot, nitrogen-rich topdressing. I have seen fruitings of this mushroom numbering in the thousands on sports fields. This is also a road-warrior mushroom, often hugging highways in the gravelly soils along the edges. In the 1980s, on a mountain highway in Colorado, shaggy manes literally broke through newly laid asphalt, causing tens of thousands of dollars worth of damage and necessitating replacement of a stretch of the highway.

▲ **FIGURE 246**

Never underestimate the power of mushrooms! *C. comatus* busts through asphalt. They are well-known as road wreckers. Such Herculean events occur because mushrooms exert enormous upward forces when they fruit. Mycelium absorbs and pumps water into cells which are composed of a helical polysaccharide matrix, thus causing the mushrooms to push upward with pressures great enough to break asphalt, sometimes cement.

Historically, this mushroom has served a double purpose for collectors: first as a delicious edible, and second as ink used in medieval times for inscribing manuscripts. For more information, consult Van de Bogart (1976–79), Mueller et al. (1985), Stamets and Chilton (1983), and Stamets (2000a).

Flammulina velutipes (Curtis ex Fries) Singer

Common Names: Enokitake (Japanese for "the snow peak mushroom"; *enokidake* is an alternative spelling), yuki-motase (Japanese for "snow mushroom"), the winter mushroom, the velvet foot, furry foot, fuzzy foot, golden needle mushroom, the golden mushroom.

Taxonomic Synonyms and Considerations: Formerly known as *Collybia velutipes* (Fr.) Quel., this mushroom is at the center of a constellation of species that includes *F. populicola*, an aspen-loving Rocky Mountain and northern European species that fruits at warmer temperatures. Two other similar species, *F. fennae* and *F. elastica*, are distinguished from *F. velutipes* primarily due to spore size. (For more information, see *The Genus* Flammulina: *A Tennessee Tutorial* by Ronald H. Petersen, Karen W. Hughes, and Scott A. Redhead, viewable online at http://fp.bio.utk.edu/mycology/Flammulina/default.html#INDEX.)

◀ **FIGURE 247**

Wild *F. velutipes* bears little resemblance to its cultivated forms (see figures 249 and 250).

▲ FIGURE 248

I cloned this strain from a mushroom picked from a log just below tree line above Telluride, Colorado. Identified as *F. populicola*, a sister species to *F. velutipes*, this strain produces more than 2,000 mushrooms from a 5-pound mushroom kit made mostly of sawdust!

▲ FIGURE 249

Indoor-cultivated *F. velutipes* has become a large business. The uniformity of mushrooms grown in this fashion is remarkable.

Introduction: The Japanese led the charge in popularizing this mushroom. In the wild, *F. velutipes* is a short, fuzzy-footed mushroom. Enokitake metamorphoses into a different form while seeking light and oxygen cultured in chilled growing rooms; abnormally small caps and long stems form in response to elevated carbon dioxide levels and limited light exposure. This "unnatural" shape makes the harvesting of enokitake from bottles—cultivator's container of choice— economically feasible.

▲ FIGURE 250

Bottle culture of *F. velutipes* using heat-treated sawdust is the preferred commercial method primarily because of ease of harvesting and long shelf life. Enoki mushrooms change in form in response to environmental conditions to optimize spore release. The stems elongate as carbon dioxide outgasses from the mycelium's decomposition of the sawdust; the caps are small as a result of low light conditions.

Description: Cap 1–5 cm in diameter, convex to plane to upturned in age; smooth, viscid when wet; bright to dull yellowish to yellowish brown to orangish brown. Gills white to yellow, attached to the stem. Stem usually short, 1–3 inches, yellow to yellowish brown, darkening with age and covered with a dense coat of velvety, fine brown hairs near the base. In culture, the morphology of this mushroom is highly mutable, being extremely sensitive to carbon dioxide and light levels. Cultivated specimens usually have long, yellowish stems and small white to yellowish caps. When spores mature, the caps darken to brown. Spores white, 7–9 by 3–6 µm.

Distribution: Widespread throughout the temperate regions of the world.

Natural Habitat: Found on hardwoods and less often on conifers, commonly growing in the late fall through early winter. This mushroom can freeze, thaw, and continue to grow. One translation of the Japanese word *enoki* is "huckleberry," implying that this mushroom grows in association with forestlands populated with this plant. Cultivators have found that white fir wood, when sterilized, is a good medium for growing this mushroom.

Type of Rot: White rot.

Fragrance Signature of Mycelium: Musty, unpleasant.

Natural Method of Cultivation: Stump culture is possible, as evidenced by the penchant of this species for logs, stumps, and wood debris in the wild. Grows on a wide variety of hardwoods (oak, alder, poplar, cottonwood, aspen, willow, birch, beech, and so on) and some softwoods (Douglas fir and white fir). The pH range for fruiting is 5 to 6. Enokitake also grows on a wide variety of paper products. Outdoor natural culture should not be attempted by those who cannot distinguish enokitake mushrooms from small, wood-decomposing poisonous mushrooms such as the deadly members of the genera *Galerina* and *Conocybe*.

Season and Temperature Range for Mushroom Formation: Late fall through midwinter, thriving in temperatures of 40–60°F.

Harvest Hints: Mushrooms picked in the wild do not resemble the ones grown indoors. The wild forms have short stems, are covered with dark brown fuzz, and usually sport a cap as wide as the stems are long. Store-bought enokitakes have long stems and small caps, a form elicited by growing this species under low-light and high carbon dioxide conditions. Mushrooms in the wild should be harvested before the caps fully expand and white spores are released.

Nutritional Profile: Our analysis of a 100 g serving of *F. populicola*, a close relative of *F. velutipes*, shows the following: calories: 346; protein: 26.59 g; fat: 3.06 g; polyunsaturated fat: 1.08 g; total unsaturated fat: 1.22 g; saturated fat: 0.23 g; carbohydrates: 52.95 g; complex carbohydrates: 30.55 g; sugars: 22.40 g; dietary fiber: 25.80 g; cholesterol: 0 mg; vitamin A: 0 IU; thiamine (B1): 0.35 mg; pantothenic acid (B5): 10.90 mg; vitamin C: 0 mg; vitamin D: 113 IU; calcium: 14 mg; copper: 0.61 mg; iron: 8.3 mg; potassium: 3,100 mg; niacin: 60.60 mg; riboflavin: 1.69 mg; selenium: 0.054 mg; sodium: 19 mg; moisture: 10.64 g; ash: 6.76 g.

▲ **FIGURE 251**

Dusty Yao holds a tray of harvested *F. populicola*. Since these mushrooms were in a lighted greenhouse, their caps enlarged. With enokitake and many other mushrooms, light controls the diameter of the cap; carbon dioxide controls the length of the stem.

Medicinal Properties: Komatsu (1963) and Watanabe and others (1964) first reported that *F. velutipes* showed antitumor properties, leading to the discovery of the medicinally active protein-bound polysaccharide commonly referred to as FVP (*Flammulina velutipes* polysaccharide). The water-soluble polysaccharide flammulin was subsequently isolated in this species. Ikekawa and others (1969, 1985) reported anticancer activity from extracts of this mushroom after conducting an extensive epidemiological study in Japan, wherein a community of enokitake growers near the city of Nagano had unusually low cancer rates; frequent enokitake consumption by the workers and their families was thought to be the cause. Ikekawa later isolated another immunomodulating agent of comparatively low molecular weight, proflamin (Ikekawa 2001). Studies of the antitumor properties of this mushroom have also been published by Zeng and other researchers (1990), Qingtian and others (1991), and Zhang and others (1999).

Flavor, Preparation, and Cooking: This mushroom is enjoyed for its texture and its mild flavor, both of which enhance soups. Traditionally, enokitake is lightly cooked, served in soups or in stir-fries with vegetables and fish or chicken. The stems are often left long, thus posing some interesting problems in chewing and swallowing. I prefer enokitake finely chopped and then cooked over high heat for a short period of time. The addition of finely chopped enokitake to a cream sauce, stems and all, results in a *crème supérieur*.

Mycorestoration Potential: I like this species for the fact that the mycelium will run at low temperatures that inhibit most other fungi, making it a prime candidate for cold mycoremediation of toxins that become unstable at higher temperatures. Although I know of no reports of such a use to date, I encourage exploration of cold extremophiles in habitats prone to repeated freezing or for deactivating heat-sensitive explosives. This species also inhibits *Staphylococcus aureus*.

Comments: Enokitake mushrooms (*F. velutipes* and relatives) are known for their tolerance of cold weather and appetite for deciduous and coniferous woods. *F. velutipes* is a flagship species: one of the first mushrooms with antitumor agents to be brought to the attention of Western scientists. Its cold-weather tolerance, circumpolar distribution, ease of cultivation, edibility, history of use, and diversity of active medicinal constituents should continue to draw the attention of researchers for years to come.

Wild enokitake mushrooms bear some resemblance, especially to the untrained eye, to a few poisonous species. In particular, *Galerina* mushrooms (which have rusty brown spores), some of which are deadly, can look similar to the white-spored *Flammulina* mushrooms. Both can darken toward the stem and grow on wood. Anyone growing or collecting wild enokitake mushrooms should be certain of their identification and take special note of spore color before harvesting and ingesting.

Fomes fomentarius (L.:Fr.) J. Kickx

Common Names: Amadou, tinder conk, hoof fungus, hoof conk, touchwood conk, ice man polypore, tsuriganetabe (Japanese).

Taxonomic Synonyms and Considerations: *F. fomentarius* L.:Fr., was first described taxonomically in 1821. Its look-alike, *F. fasciatus*, is similar in form but tropical in its distribution.

Introduction: From China to Europe, humans have used this hoof-shaped mushroom for practical purposes for ages. Remnants of this mushroom have been found at Stone Age sites dating back to 11,600 B.C.E. It is the oldest-known manipulated natural (biological) product associated with Paleolithic humans. The first written record on *F. fomentarius* was authored by Hippocrates (460–377 B.C.E.) who mentioned its topical use for cauterizing wounds and for externally treating inflamed organs. The famous 5,000-plus-year-old ice man (nicknamed "Otzi") found on the slopes

▲ **FIGURE 253**

F. fomentarius fruiting on downed birch near Burlington, Vermont.

▲ **FIGURE 252**

F. fomentarius is a perennial polypore. This conk formed several months after we inoculated sterilized birch, and in a year's time, we observed no fewer than 6 growth rings when slicing the mushroom in half. Conventional wisdom is that only 1 growth ring forms per year, reflective of the seasons, similar to tree rings. We find that if temperatures are conducive, spore layers form every couple of months.

➤ **FIGURE 254**

Otzi's fire kit included this highly flammable mycelial wool made from boiling and separating the fibers within *F. fomentarius* conk. Also called the fire-starter mushroom, our ancestors' survival partially depended upon this species. This material was also useful as punk for black powder pistols and rifles, helping revolutionize warfare.

of the Alps in the fall of 1991 had *F. fomentarius* "wool" with him as well as whole fruitbodies (Capasso 1998). This wooly mass, actually dissociated mycelium (Stamets 2002b), feels like felt (and is sometimes called "German felt"); it's made by boiling the mushrooms, pounding them, and peeling them apart to uncover the fiberlike understructure inside the conk.

Amadou is one of the fire-starter mushrooms known as tinder conks. The internal wooly mass is highly flammable. A hole can be burrowed into the dried conk, and if embers from a fire are packed into it, fire can smolder for hours, possibly days, allowing fire to be transported. As our prehistoric ancestors migrated from Africa into European birch forests, their possession of this knowledge ensured their survival. The fire keepers of the clan, in a position of enormous importance for the clan's survival, knew how to find and prepare these mushrooms. Now, this nearly lost art has been rediscovered.

With invention of gunpowder by the Chinese and flint-spark guns by the Europeans (whose projectiles pierced body armor), demand for *F. fomentarius* soared, since it was the best source of punk, a preparation used to ignite gunpowder in primitive weapons. The fact that wood conk mushrooms helped in the development of warfare is another peculiar twist in the interactions of fungi and humans. Other mushrooms, including *Ganoderma applanatum*, *Inonotus obliquus*, *Phellinus igniarius*, and *Piptoporus betulinus*, can also be used to start fires.

Description: Cap 5–15 cm. Broad, zonate, growing and expanding downward with new tube layers forming underneath, compounding seasonally with the old tube layers composing the body of the mushroom as it amasses. Not vertically cracked, growth rings concentric, horizontal, hooflike in shape. Mushrooms can be dark grayish black in color, lightening with age, and are often grayish white at maturity. Spores white, 15–20 by 4.5–7 μm.

Distribution: Widespread throughout the boreal woodlands of the world—northern North America, Europe, and temperate forests—on dead and living trees.

Natural Habitat: A saprophyte, *F. fomentarius* grows on dying or dead birch, aspen, willow, and alder.

Type of Rot: White rot.

Fragrance Signature of Mycelium: Sweet fungal-forest scent; pleasing, rich, and reminiscent of many other wood rotters.

Natural Method of Cultivation: Plug or rope spawn inoculated into hardwood stumps, logs, or standing trees.

Season and Temperature Range for Mushroom Formation: Primarily a summer mushroom.

Harvest Hints: Outdoors this acts as a perennial mushroom, growing for 5 to 10 years before climaxing. However, when grown indoors, fruitbodies erect and form successive tube layers bimonthly. My experiences growing these mushrooms indoors on sterilized sawdust have taught me that the mushrooms mature after a half dozen or so tube layers have developed.

Nutritional Profile: Our analysis of a 100 g serving shows the following: calories: 376; protein: 15.05 g; fat: 3.48 g; polyunsaturated fat: 0.50 g; total unsaturated fat: 1.20 g; saturated fat: 0.27 g; carbohydrates 71.00 g; complex carbohydrates: 69.30 g; sugars: 1.70 g; dietary fiber: 66.80 g; cholesterol: 0 mg; vitamin A: 0 IU; thiamine (B1): 0.06 mg; pantothenic acid (B5): 2.70 mg; vitamin C: 0 mg; vitamin D: 66 IU; calcium: 37 mg; copper: 1.3 mg; iron: 13 mg; potassium: 760 mg; niacin: 12.40 mg; riboflavin: 1.59 mg; selenium: 0.014 mg; sodium: 6 mg; moisture: 8.36 g; ash: 2.11 g.

Medicinal Properties: Traditionally used as a styptic to stop bleeding and prevent infection, this mushroom, surprisingly, has yet to be thoroughly explored for its medicinal properties. A bitter diuretic tea can be made from boiling the conks for several hours in water.

The Okanogan-Colville Indians of British Columbia and Washington traditionally used this mushroom to treat arthritis (Turner et al. 1980). An antimicrobial, immune-boosting water extract or tea can be made by boiling the mushrooms in water. Traditionally, the conks can be dried, cooked, and pounded into a powder for use as a poultice to stifle infection or to alleviate pain from swollen joints.

Antitumor polysaccharides from the mycelium were isolated and tested by Ito and others (1976). Novel ergosterol peroxides have been isolated from this fungus (Rosecke and Konig 2000). Water extracts of *F. fomentarius* have strong antiviral properties (M. Aoki et al. 1993; Piraino and Brandt 1999) and effectively inhibit the reproduction of *Bacillus subtilis* (Suay et al. 2000) and possibly many other species of *Bacillus* and bacteria (Hilborn 1942). Vole et al. (1985) examined 40 species of fungi and found *F. fomentarius*

and *Trametes versicolor* to have the highest enzyme activity of converting D-glucose into dicarbonyl sugars, important considerations for fermentation manufacturing and for those employing fermentation methods for enzyme production, a market of increasing commercial interest.

Flavor, Preparation, and Cooking: Add about 1 pound of conks to 5 gallons of water, and bring to a rolling boil. Boil for at least 4 hours, or until the broth is reduced to 1 gallon. Drain, reserving the broth, and pound the softened conks with a wooden mallet, separating the fibers using a curved knife or other sharp edge. Dry the separated fibers on an absorbent cloth. Mat the fibers using methods similar to those employed for nonwoven fabrics, and hang to dry. The reserved boiled extract can be used for medicinal purposes. The dried fibers can be used to cauterize woods (or, traditionally, to extract evil spirits, perhaps what we know as microbes).

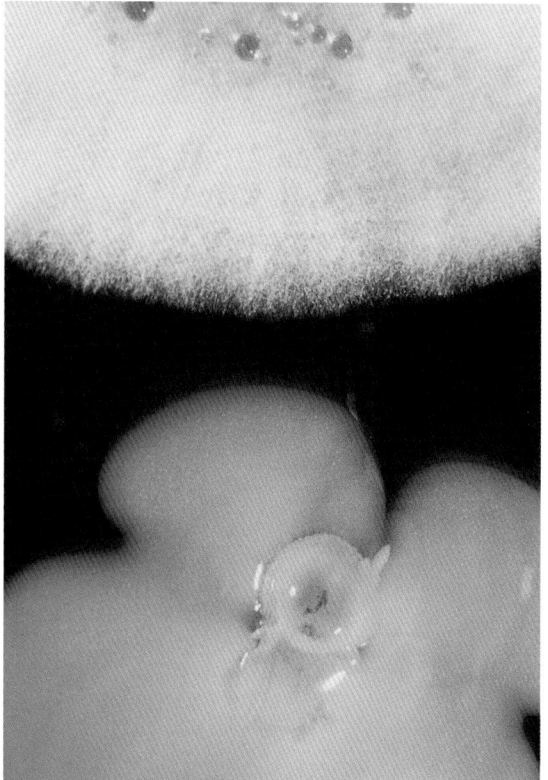

▲ **FIGURE 255**

Bacterial colonies of *Escherichia coli* and mycelium of *F. fomentarius* grow toward each other on sterilized agar media.

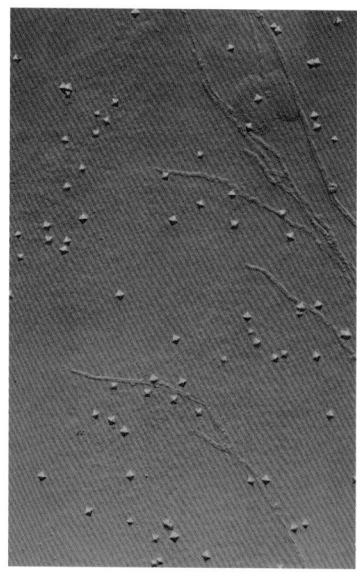

▲ **FIGURE 256**

Highly magnified octahedral "messenger crystals" secreted by *F. fomentarius* mycelium, advancing ahead of its leading edge prior to contact with coliform bacteria. I hypothesize that these crystalline entities disintegrate upon contact with microbial adversaries and leave a chemical scent trail that alerts the mother mycelium to the enemy in its path.

➤ **FIGURE 257**

F. fomentarius generates secondary, larger crystals subsequent to its mycelium encountering the chemical scent trails from the disintegration of the primary crystals featured in figure 256. *Escherichia coli,* a dreaded coliform bacterium (rodlike cells in photo) pathogenic to both mushrooms and humans, is attracted to these secondary crystals, stunned, and then consumed as food by the encroaching mycelium. I was a member of the Battelle mycofiltration and mycoremediation team that first made this discovery using my proprietary strain of *F. fomentarius.* A predator-prey relationship also exists between the garden giant *(Stropharia rugoso annulata)* and this bacterium. These mushroom species can help in the mycofiltration of coliform bacteria in the runoff from pollution sources. The birch polypore *(Piptoporus betulinus)* secretes similarly shaped crystals, but I don't know if they also stun *E. coli.*

⬣ **FIGURE 258**

The outer conks are wild *F. fomentarius;* the inner one is cultivated. The hat is made by boiling these mushrooms, pounding them, and pulling apart their internal fabric—in essence separating the hyphae and then relayering to form a mycelial felt. Originating in Romania and Transylvania, this method of making hats from wood conks has been passed down through the centuries. *F. fomentarius* can be cultivated to produce felt just as silk factories use silkworms. I envision the rebirth of this age-old industry.

Mycorestoration Potential: Helpful as an antimicrobial agent, this mushroom could be incorporated into woodland landscape buffers to mitigate the effects of coliforms and pseudomonads threatening downstream habitats that suffer from surface surges of bacteria when heavy rains and erosion occur. *F. fomentarius* may be an endophyte (Baum et al. 2003), infecting trees and thereby warding off devastating blights from competing fungi by potentiating the defense of the host forest, especially forests stressed from infiltration of pollutants.

Comments: A mushroom with many uses, *F. fomentarius* is a prominent species in the boreal forests of the world, being particularly fond of birch, maple, beech, and occasionally alder trees. As described, the mycelium has multiple purposes. That it has an appetite for *Escherichia coli* and perhaps other coliforms suggests to me that this species may have been used as an antimicrobial in ancient times, perhaps preventing stews and soups, for instance, from souring. In modern times, the antimicrobial properties of the mycelium may prove useful in a variety of mycofiltration and mycoremediation strategies.

◄ **FIGURE 259**

Steve Cividanes visits a massive artist conk deep in the old-growth forest on the Olympic Peninsula. This mushroom may produce enough spores to encircle the Earth (see page 134). A small fragment of this conk was cut from the leading edge and brought to my laboratory, where I cloned it and created a culture.

➤ **FIGURE 260**

This statuesque snag hosts *G. applanatum*. Three years later, the fungus had disintegrated this snag, leaving a feathery white pulp in its wake. The mycelium of this mushroom attracts many forestland insect species and is crucial in the recycling of nutrients in forestlands.

Ganoderma applanatum (Pers. ex Wallr.) Pat.

Common Names: Artist conk, giant shelf fungus, tree tongue, white mottled rot mushroom, kofukitake (Japanese).

Taxonomic Synonyms and Considerations: *Polyporus applanatus* (Pers.) Wallr.; *Fomes applanatus* (Pers.) Gill.

Introduction: A widespread and often mammoth species known and used for millennia, this mushroom has many attributes that were useful to our Paleolithic ancestors and their descendants. *G. applanatum* is a perennial polypore that can live for 40 to 50 years, perhaps longer. The mushroom produces a large, lateral shelf, flat in profile. Its spores fall from the pores on the underbelly of the conk; due to electrostatic and thermal differentials, many of the spores float upward to settle on the top of the cap, dusting the upper surface with a brown powder (see figure 261).

    This is one of the largest mushrooms in the world, holding a third-place position behind *Rigidoporus*

▲ **FIGURE 261**

G. applanatum grows annual concentric rings that harden and become part of its internal structure; the leading edge builds new cells as it grows outward. Note the brown spores collecting on the upper surfaces.

ulmarius (the largest) and *Bridgeoporus nobilissimus* (the second largest). This massive mushroom is prominent in old-growth forests and is effective as an antimicrobial agent and immune enhancer; for these reasons, it can be said that *G. applanatum* serves as a steward not only for the ecological health of woodlands but also for improving health of their human inhabitants.

Description: A planar wood conk, 5–90 cm broad and 3–20 cm thick; surface crustlike, smooth, usually covered with a layer of brown spore dust. The underlayer is pored and white, staining brown when touched. Flesh above pore layer is corky, .5–5 cm. thick. This conk lacks a stem, growing perennially for several decades but usually climaxing within 10 years. Mushrooms can grow to up to 1 meter in width. Spores chocolate brown, 6.5–9.5 by 5–7 μm.

Distribution: Widely distributed throughout the world wherever trees grow. This is one of the most common mushrooms in the world.

Natural Habitat: Growing on many hardwood tree species but also on conifers, especially old-growth Douglas firs in the Pacific Northwest.

Type of Rot: White, mottled.

Fragrance Signature of Mycelium: Musty, woodsy, pleasant, sweetly fungoid.

Natural Method of Cultivation: Plug or rope spawn, preferably inoculated into larger-diameter conifer trees or their stumps, especially hemlocks and Douglas firs. The mass of the tree or stump directly influences the size of the fruit body.

Season and Temperature Range for Mushroom Formation: Since this mushroom is a perennial, it can be found throughout the year. Growth spurts are usually confined to the warmer months of the year, usually summertime in the northern latitudes, when temperatures hover between 65 and 95°F

Harvest Hints: When the spore layer fails to produce during the summer season, this mushroom has completed its life cycle. Simply break off the mushroom from the tree. The resident mycelium may produce another conk in or near the location of the original specimen. Using a spray of water, rinse the spores from the top of the cap into a bucket. You can gather billions from a well-developed conk in summer, which are useful in mycorestoration strategies.

Nutritional Profile: Not yet known to this author. In the future, we plan to fully analyze this species.

Medicinal Properties: Sasaki and fellow researchers (1971), along with the pioneering mycologist Ikekawa (1969), first studied *G. applanatum* for its antitumor properties. Protiva and others (1980), Tokuyama and others (1991), and Chairul and Hayashi (1994) identified triterpenes and steroids from *G. applanatum*. Smania et al. (1999) found that a methanol extract of *G. applanatum* fruitbody, further fractionated with hexane and ethyl acetate, showed significant activity against a wide range of bacteria, including *Escherichia coli* and *Staphylococcus aureus*. Other studies have looked into constituents of these mushrooms that may have antimicrobial activity against *Escherichia coli* (Thomas et al. 1999; Suay et al. 2000). In addition, Ming and others (2002) identified a new aldehyde, one of many novel constituents recently discovered within this species.

▲ **FIGURE 262**

G. applanatum makes a magnificent canvas for artists to etch upon.

Flavor, Preparation, and Cooking: A bitter mushroom tea is made from boiling the conks in a fashion similar to that described for *Fomes fomentarius* on page 227. The dried conk can be burned, emitting a pleasant, insect-repelling smoke. The diuretic and strongly antimicrobial tea (Suay et al. 2000) shows potential as a purgative for intestinal parasites and as a treatment for bacterial infections.

Mycorestoration Potential: Given its widespread range and perennial nature, this species can be seen as one of the sentinels of forest ecosystems, and a fantastic recycler of dead trees. A prodigious producer of spores, this mushroom can be amplified by loggers using spore-infused chain-saw oils. Active against staph and coliform bacteria, this species may have broad-spectrum antimicrobial properties. Its role in decomposing trees in forests throughout the world underscores its importance as a fungal leader in wood recycling. A mottled white rotter, this mushroom often occurs in tandem with turkey tails *(Trametes versicolor)*, suggesting to me that using these 2 species together could help restore toxic habitats where wood is abundant.

Comments: Often used as a "canvas" by artists, the large conks of G. *applanatum* have whitish pores that stain brown when bruised, allowing for line drawing. I have seen huge, beautifully scored specimens, depicting pastoral and sylvan scenes, enthroned as the centerpiece in the living room of the custodial temporary owner, mycologist David Arora. Artists can etch upon fresh specimens using a sharp needle, or they can dry the conks and then burn images onto them using a hot etching tool. Be careful when handling the conks, as your fingerprints in brown can appear on the delicate, white-pored underlayer, developing like a photographic print in a darkroom. A majestic species for landscapers, the perennial conks can be used as natural tables or shelves. David Arora also suggests that this mushroom, once harvested, is useful for making seats for stools.

Gilbertson and Ryvarden (1986–87) note that many wind-thrown aspens in Colorado sport this mushroom at their base. Consequently, we might theorize that this mushroom may be a facultative parasite—meaning that it opportunistically grows on stressed trees, leaving healthy ones alone. In my mind, this empowers the forest through the forces of natural selection.

Ganoderma lucidum (Wm. Curtis: Fries) Karsten and Allies

Common Names: Reishi (Japanese for "divine" or "spiritual mushroom"); ling chi, ling chih, ling zhi (Chinese for "tree of life mushroom"); mannentake (Japanese for "10,000-year mushroom"); mushroom of immortality; the panacea polypore.

Taxonomic Synonyms and Considerations: Once published under the name *Boletus lucidus* W. Curt., in Curtis's *Flora Londiensis* (published 1817–28), his plate of this mushroom has been chosen as the

▲ **FIGURE 263**

Regal mushrooms, *G. lucidum* and allies are annual polypores. The white leading edge of the cap margin grows rapidly in warm temperatures.

lectotype—the image upon which this species is based. In 1821, as taxonomy evolved, Fries moved it to *Polyporus*—*Polyporus lucidum* W. Curt: Fr. Typically showing a preference for warmer climates, *G. lucidum* grows on numerous hardwoods (oaks, elms, beech, birch, alder, maple) and is thought (Gilbertson and Ryvarden 1993–94) to live upon dead or dying Picea (Norway and Engelmann spruce). Its closest allies, *G. carnosum*, *G. resinaceum*, and the more distant *G. capense*, are separated by tenuous taxonomic features. Questions concerning taxonomy become increasingly important as novel constituents are isolated; in some cases, patents are pursued specific to this species, so if the mushrooms studied are not correctly identified, the species-specific patents may be invalidated. Hong and Jung (2004) published a recent DNA analysis of the clades surrounding this species complex, providing a good baseline for the further elaboration of species.

One historic and notable attempt to distinguish the North American from the Far Eastern types can be found in a Japanese-language article published by R. Imazeki (1937) titled "Reishi and *Ganoderma lucidum* that grow in Europe and America: Their Differences." Currently, the best treatises discussing the taxonomy of these polypores are Gilbertson and Ryvarden's monograph, *North American Polypores*, volumes 1 and 2 (1986–87) and Zhao's *The Ganodermataceae in China* (1989). The spore size of *G. lucidum* is more restrictive than the inclusive range of 13–17 μm in length by 7.5–10 μm in width characteristic of *G. oregonense* and *G. tsugae*. Nevertheless, Gilbertson and Ryvarden did not consider this feature to be more significant than habitat when delineating these 3 taxa in their "Key to Species." Placing emphasis on habitat may also be a dubious distinction, since these species produce fruitbodies on nonnative woods when cultivated. *G. lucidum* thrives in culture at higher temperatures, distinct from the cold-loving *G. oregonense* and *G. tsugae*, which do not produce chlamydospores in culture. Interfertility studies with some collections reveal that *G. curtisii* (Berk.) Murr. may merely be a yellow form of *G. lucidum* common to the southeastern United States. (See Adaskaveg and Gilbertson [1986, 1987] and Hseu and Wang [1991].) I expect that these taxa will soon be delineated using the repertoire of DNA mapping techniques.

G. oregonense is a much larger mushroom than *G. lucidum* and is characterized by a thick pithy flesh in the cap. Also, *G. oregonense* favors colder climates, whereas *G. lucidum* is found in warmer regions. (No occurrences of *G. lucidum* in the Rocky Mountain and Pacific Northwest regions have been reported, to the best of my knowledge.) *G. curtisii*, a species not recognized by Gilbertson and Ryvarden but acknowledged by Zhao (1989) and Smith-Weber and Smith (1985), grows in eastern North America and is distinguished from others by the predominantly yellowish cap as it emerges. These reishis, *G. lucidum*, *G. capense*, *G. carnosum*, *G. curtisii*, *G. oregonense*, *G. resinaceum*, and *G. tsugae*, represent a constellation of closely related species, probably stemming from a common ancestry. In Asia, *G. lucidum* has a number of unique allies. Most notably, a black-stalked *Ganoderma* species, *G. japonicum* Teng (= *Ganoderma sinense* Zhao, Xu et Zhang, colloquially known as "zi zhi") is considered to be a reishi.

◄ **FIGURE 264**

Ancient portrait from China by Chen Hungsho (1599–1652), of Hou Chang holding *G. lucidum*.

Introduction: The Japanese call this mushroom *reishi* or *mannentake* (10,000-year mushroom), whereas the Chinese know it as *ling chi, ling chih,* or *ling zhi* (mushroom [herb] of immortality). Renowned for its health-stimulating properties, this mushroom is depicted more often in ancient Chinese, Korean, and Japanese art than any other. Traditionally associated with royalty, health and recuperation, longevity, sexual prowess, wisdom, and happiness, it has been depicted in royal tapestries, often portrayed with renowned sages of the era. For a time, the Chinese even believed that this mushroom could bring the dead to life when a tincture made from it was laid upon the deceased's chest.

The use of *G. lucidum* spans more than two millennia. The earliest mention of ling chi occurred in the era of the first emperor of China, Shih-huang of the Ch'in Dynasty (221–207 B.C.E.). Thereafter, depictions of this fungus proliferated throughout Chinese literature and art. In the time of the Han Dynasty (206 B.C.E.–220 C.E.), while the imperial palace of Kan-ch'uan was being constructed, ling chi was found growing on timbers of the inner palace, producing nine "paired leaves." So striking was this good omen that emissaries were sent far and wide in search of more collections of this unique fungus. Word of ling chi thus spread to Korea and Japan, whereupon it was elevated to a status of near-reverence.

Description: Similar to a conk or kidney in shape, the cap of this woody-textured mushroom measures 5–20 cm in diameter and has a shiny surface that appears lacquered when moist, dulling when dried. (Reishi are sometimes so magnificent when fresh that growers lacquer the dried mushrooms to restore their original wet luster and save them as art pieces.) The cap can be a dull red to reddish brown, sometimes nearly black in color, and browns when touched. Whitish pores on underside. Areas of new growth whitish, darkening to yellow brown and eventually reddish brown at maturity, often with zonations of concentric growth patterns. Spores dispersed from the underside collect on the surface of the cap, imparting a powdery brown appearance when dry. Stem white to yellow, eventu-

ally darkening to brown or black, eccentrically or laterally attached to the cap, usually sinuous, and up to 10 cm in length by .5–5.0 cm thick. Spores brown, 7–12 by 6–8 μm.

Distribution: This mushroom is widely distributed throughout the world, from the Amazon through the southern regions of North America and across much of Asia. This mushroom is less frequently found in temperate than in subtropical regions, and its clade is common in the Gulf States region of the southeastern United States and throughout the Midwest, primary on hardwoods.

Natural Habitat: A cosmopolitan species, this saprophytic mushroom colonizes the widest range of hardwoods of any mushroom species I know of besides oysters. With varieties that live on oaks, elms, beeches, and even palms, the genetic diversity of *G. lucidum*'s is rich with so many ecotypes and varieties that careful study is required to clarify the differences between races or subspecies. This species is also reported as growing from spruces. In Australia, Chile, and in some Asian countries, some researchers believe that this mushroom behaves parasitically toward palm and tea trees, in contrast to the view held by most European and American mycologists that this mushroom is

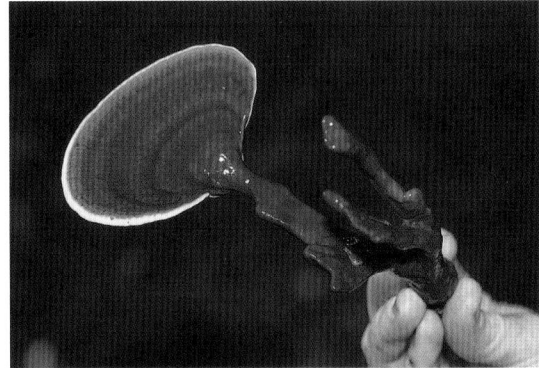

▲ **FIGURE 265**

When moist, the surface of reishi has a lacquered appearance. Whenever I see this mushroom, I am in awe of its beauty.

⚊ **FIGURE 266**

This short log inserted into a sand-filled nursery pot produces reishi mushrooms about 6 months after inoculation with plug spawn.

a saprophyte. The differences in viewpoint may reflect regional variation between the mushrooms or differences between species that have not yet been taxonomically delineated. An emerging view is that these classically saprophytic mushrooms are "facultative parasites," meaning that this species acts as an opportunistic infection, springing into action only when the host tree is stressed or diseased from other causes.

Type of Rot: White butt and root rot of hardwood, and occasionally conifer trees.

Fragrance Signature of Mycelium: Musty, mealy, fungoid, similar to that of a classic polypore.

Natural Method of Cultivation: Adaptive to many of the techniques described in this book, this aggressive species grows well on logs, stumps, or buried blocks of sawdust spawn. Outdoors, I prefer to cultivate this mushroom on logs, either using a technique like the standard method for growing shiitake employing plugs, or making short logs and, subsequent to inoculating with plug spawn, placing them into garden pots. To use this method, place a small amount of sand or gravel in the base of the pot, insert the plugged log, and add more sand or gravel around the upright log for support.

⚊ **FIGURE 267**

Reishi fruiting from an alder log 1 year after inoculation with plug spawn.

⚊ **FIGURE 268**

A *G. lucidum* laying yard. The advantages of ground contact are moisture and constant temperatures; the disadvantage is direct contact with biological competitors.

▲ **FIGURE 269**

G. lucidum fruiting from a bundle of inoculated sticks in China.

The sand or gravel helps to stabilize temperature and buffers moisture transfer. The log and pot can be watered to induce fruiting after 6 months of incubation, or when mycelium shows on the ends of the cut faces. (See page 178.)

Season and Temperature Range for Mushroom Formation: Typically a summer to early fall species. Temperature: 60–95°F.

Harvest Hints: When the fungi are within the fruiting range, and once the cap margin has thinned and has not reemerged for several weeks, the mushrooms have usually finished their life cycle. Occasionally, they can spontaneously regrow. With experience, cultivators develop a keen sense for the nuances of their strains and the conditions that indicate that the mushroom has reached maturity. The diameter of the log, tree, or stump on which they were grown directly influences the size of the conks.

Nutritional Profile: See page 202.

Medicinal Properties: Although directly active as an antimicrobial (Suay et al. 2000), according to most studies reishi mushrooms do not act directly as a tumoricidal against cancers as do many other mushroom species (Ooi et al. 2002). Reishi primarily functions as a

▲ **FIGURE 270**

Dr. Andrew Weil, a longtime proponent of medicinal mushrooms, stands amongst *G. lucidum* fruiting in one of my growing rooms. All the mushrooms came from one culture, and if they touch, they will fuse together and continue to grow.

biological response modifier, stimulating production of macrophages (often a consequence of the effects of interleukins-1, -2, -6, -10), activating the host's production of natural killer cells, T cells, and tumor-necrosis factors. More than 100 distinct polysaccharides and 119 triterpenoids from this species have been isolated, from both the mycelia and the fruitbodies (Zhou and Gao 2002). Of those that have been identified thus far, many triterpenoids and polysaccharides demonstrate immunomodulatory properties.

Triterpenoids are steroid-like compounds that inhibit cholesterol synthesis, allergenic response, and histamine release. These compounds are thought to be more concentrated in this species' fruitbodies than in the mycelium, a fact of significance for those choosing medicinal mushroom extracts for relief from bronchitis, asthma, and allergies (Hirotani and Furuya 1986; Han et al. 1998; Zhu et al. 1999). Lanostanic-type triterpenoids from spores of *G. lucidum* have been shown to limit the in vitro growth of meth-A and LLC tumor cell lines (Min et al. 2000) and cervical HeLa cells (Zhu et al. 2000).

Liu and others (2002) found that germinating or fractured spores produced more antitumor agents than dormant spores. Gao and fellow researchers (2002) discovered a new cytotoxic lanostanic triterpene aldehyde from the fruitbody of *G. lucidum* showing activity similar to that in a study by Min and others (2000). An ethanol fraction isolated from spores strongly stimulated the activity of T-lymphocytes (Bao et al. 2002). In response to hot water extracts of reishi mushrooms preserved in ethanol, versus saline controls, natural killer cell activity was significantly augmented when cancer cells were co-cultured with human spleen cells (Ohtomo 2001). Slivova and colleagues (2004) reported that *G. lucidum* inhibited breast cancer cell adhesion, reducing motility and migration of highly metastasized cancer cells.

Studies by Wang and others (1997) ascertained that the primary antitumor effects of *G. lucidum* are from biological response modification of the host. Reishi's polysaccharides caused a 5- to 29-fold increase in the tumor-necrosis factors, interleukins-1 and -6, and a substantial augmentation of T lymphocytes. *G. lucidum* has also been shown to help restore T-cell function in the spleen of gamma-irradiated mice (Chen et al. 1995). Lieu and others (1992) reported that polysaccharides of *G. lucidum* significantly inhibited the growth of leukemia (U937) cells. The antioxidant properties of reishi have been well established (Chang and But 1986; Chen and Zhang 1987; Wang et al. 1985; Yang et al. 1992; and Lee et al. 2001). Zhu and others (1999) found that several triterpene fractions, higher in the fruitbodies than in the mycelium, scavenge superoxide anions, interrupt-

ing the associated chain reaction of free radicals, thus providing a strong antioxidant effect. Similarly, Lee and others (2001) found that reishi inhibits hydroxyl radicals and prevents oxidative damage from the effects of cancer chemotherapies. A unique beta-glucan from the mycelium enhanced the production of nitric oxides from macrophages but decreased other free radicals and the collateral harm they cause to healthy cells (Han et al. 1998; Li et al. 2000; Zhou and Gao 2002). This mechanism was further elucidated by Kawakami and others (2002), who showed that tumor necrosis factors (alpha TNFs) were released by macrophages 8 hours after exposure to derivatives of mushroom polysaccharides targeting cancerous cells, followed 4 hours later by a burst of nitric oxide, which then killed the diseased cells.

Constituents—including lanostanic triterpenoids—from the fruitbodies of this remarkable species have been shown to be anti-inflammatory (Ukai et al. 1983) in the treatment of arthritis (Stavinoha et al. 1990, 1996; Lin et al. 1993; Mizuno and Kim 1996; Lee et al. 2001). In one study, reishi extracts compared favorably with prednisone but had few if any negative side effects (Stavinoha et al. 1990). In a small clinical study involving 33 patients, an aqueous extract of this mushroom inhibited platelet aggregation and gave positive results in treating atherosclerosis (Tao and Feng 1990). Another limited clinical study (Gau et al. 1990) of 5 HIV-positive hemophiliac patients likewise showed no adverse effect on platelet aggregation from extracts of *G. lucidum*, which was of concern due to the high adenosine fractions found in this mushroom. *G. lucidum* may prove useful for treating inflammation of the brain (Stavinoha 1997). Significant results were obtained recently in a clinical study using reishi components in the treatment of prostate inflammation (Small et al. 2000).

Concurrent with the well-known anti-inflammatory properties of *G. lucidum* is the production of interleukins-2, -6, and -8, which are typically associated with an inflammatory response of the immune system. This apparent contradiction—an immune enhancer being an anti-inflammatory—may be further explained by the fact that the effects of reishi can be bidirectional at different dosages. Bidirectionality of the anti-inflammatory and immunostimulatory effects, as measured by cytokine production, was found to be dose dependent when using polysaccharides from the closely related G. *tsugae* in a study by Gao and others (2000). The possible inflammatory influences may be ameliorated by the production of the steroidal triterpenoids, which are typically anti-inflammatory (Stavinoha et al. 1996). The end result of many studies is that G. *lucidum* is an anti-inflammatory agent and yet an immunity enhancer. D. Kim and others (1999) found that ganoderenic acid A in G. *lucidum* was a potent inhibitor of beta-glucuronidase, an enzyme closely related to liver dysfunction, and that ganoderenic acid A may be helpful for those developing cirrhosis from hepatitis.

Lin et al. (1995) determined that the water extract of fruit bodies of G. *lucidum* induced free-radical scavenging activity. Han and others (1998), Zhou and Gao (2002), and Li and others (2000) concurred that reishi polysaccharides potentiate the release of nitric oxide while enhancing the scavenging of free radicals by peritoneal macrophages, thus making them less inflammatory while enhancing interleukin, natural killer cell activity, and tumor necrosis factors.

The studies mentioned in the preceding paragraph underscore that reishi may play an important role in minimizing the effects of aging by reducing damage from oxidative stress associated with free radicals. Cao and Lin (2002) found that polysaccharides from this mushroom regulate the maturation of function of dendritic cells, critical for immune response, while Zhang and other researchers (2002) isolated yet another bioactive glucose-galactose-mannose sugar that enhances lymphocyte activity and immunoglobulin. Future research may better explain the unique, complex actions of this species and its diverse constituents.

Consumption of reishi helps respiration, since this species enhances the oxygen-absorbing capacity of the alveoli in the lungs, thereby enhancing stamina, not unlike ginseng (Chang and But 1986). Research by

Andreacchi and others (1997) demonstrated that a crude ethanol extract of G. *lucidum* increased coronary flow due to vasodilatation, with a corresponding decrease in diastolic blood pressure and no change to heart rhythm. Although called a blood-vessel/coronary dilator, patients should be aware that there are concerns about its use prior to surgery, as it might cause excessive bleeding (Andreacchi 1995). More recently, research suggests this mushroom restricts tumor angiogenesis.

Research in Seoul by Dr. Byong Kak Kim showed that extracts of this mushroom prevented the death of lymphocytes infected with HIV *and* inhibited the replication of the virus within the mother and daughter cells (Kim et al. 1994).

From this mushroom, Murasugi and others (1991) isolated and characterized the gene responsible for manufacturing a novel immunomodulating protein ("Ling Zhi 8"). The LD_{50}, an inverse measure of toxicity, shows this species has low toxicity even at relatively large doses (Chen and Miles 1996), making it a strong candidate for immunotherapy.

This mushroom also shows promise fighting chronic fatigue syndrome (CFS) by enhancing endurance (Aoki et al. 1987; Yang and Wang 1994).

Flavor, Preparation, and Cooking: Typically extracted in hot water for teas, tinctures, syrups, and soups. My family enjoys making a tea from fresh, living specimens, breaking them into pieces, boiling them in water for an hour, and then steeping for 30 minutes. The tea is then reheated, strained, and served without sweeteners. If a daily regimen of reishi tea is followed, as little as 3 to 5 g per person is a customary dose.

Mycorestoration Potential: Of the many mushrooms used throughout history, this species is the most admired mushroom in Asia. Having many beneficial properties (which were once folkloric but are now substantiated by science), reishi can be used for mycorestoration on many levels. Foremost, I see this species as a centerpiece in the reinvention of mushroom farms as healing arts centers. As an immune potentiator with anti-inflammatory properties, G. *lucidum*

◄ **FIGURE 271**

Co du Trong grew these *G. lucidum,* fruiting from sterilized sawdust, near Ho Chi Min City, Vietnam. After mixing rubber tree sawdust with bran and other materials, the substrate was brought to 212°F for 5 hours and, when cooled, inoculated with pure culture spawn.

brings into focus the fact that immune systems can be stimulated without inflammation, a concept that would seem to be an oxymoron to conventional immunologists. Foresters are faced with a similar dilemma. Thought to be a parasite by some and a saprophyte by many, this warm climate–loving mushroom and its closest allies grow on trees throughout the world. The many varieties offer ecotypes that can address issues specific to the regions in which they occur. This mushroom, although well explored medicinally, has been the subject of surprisingly little research into its mycoremediative properties. I hope readers and researchers will focus on this species in order to uncover more of its mycorestorative interactions.

► FIGURE 272

Reishi drying in the sun. Exposing mushrooms to sunlight stimulates the conversion of ergocalciferols into vitamin D2.

Comments: This mushroom is as beautiful as it is powerful. I was once visited by Dr. Joo Bang Lee, the famous Grandmaster of the Hwa Rang Do martial arts, who spent his early life in a Buddhist temple in Korea. When he entered into our growing rooms and saw the thousands of reishi, this sage paused and stood transfixed. Images from his youth, he said, filled his mind and his emotions surged as he looked upon the mushrooms with quiet reverence. I feel the same way every time I am around reishi. An intelligent being, reishi is one of nature's greatest displays of grace and

beauty. It invites quiet contemplation and earns my deep respect. I feel connected to this species.

Given the wide number of ecotypes, finding strains appropriate rot the wood types in your area is not difficult. Red, yellow, blue, and black varieties have been reported. With modern DNA tools, the clades of *Ganoderma* are now being discerned, and some strains formerly reported as *G. lucidum* are now known to be other species. See Hong and Jung (2004).

Grifola frondosa (Dicks: Fr.) S. F. Gray

Common Names: Maitake ("dancing mushroom"), kumotake ("cloud mushroom"), hen-of-the-woods, the dancing butterfly mushroom, mushikusa.

Taxonomic Synonyms and Considerations: Synonymous with *Polyporus frondosus* Dick ex. Fr. Closely allied to *Polyporus umbellatus* Fr. (also known as *G. umbellata* Pers.: Fr.), which has multiple caps arising from a common stem, a lighter color, and a more fragile texture. The primordia of *G. frondosa* are rich, dark gray brown to gray black in color, whereas those of *G. umbellata* are light gray. On a macroscopic level, these two mushrooms are easily distinguished from one another by their forms. On a microscopic level, the spores *G. umbellata* are substantially larger and more cylindrically shaped than the spores of *G. frondosa*.

Introduction: This mushroom is a delicious, soft-fleshed polypore with excellent nutritional properties. Of the polypores currently being studied, *G. frondosa* is attracting considerable attention from the pharmaceutical industry, especially in Japan, Korea, and increasingly in the United States. Several causal compounds appear to be at play, most notably the beta-glucans, especially the D-fraction constituents.

The transformation maitake undergoes from gray mounds, to brainlike balls, to labyrinthine folds and petals, and finally to extended leaflets upon maturity is another display of fungal elegance. We grow maitake indoors on alder sawdust (*Alnus rubra*) supplemented with organic oat bran, and we have developed several novel fruiting strains from mushrooms collected in the wild. Maitake is one of my favorite mushrooms. My body hungers for the taste of this mushroom, which I think is a reflection of its inherent beneficial properties. I like combining maitake, reishi, shiitake, and others to make an immune-enhancing mushroom tea. For more information on the cultivation of this mushroom, see my book *Growing Gourmet and Medicinal Mushrooms* (2000a).

▲ **FIGURE 273**

Ebikare Isikhuemhen happily holds a "hen" *(G. frondosa).*

Description: A large, fleshy polypore, dark gray brown when young, becoming lighter gray in age. (Some varieties fade to a light yellow at maturity.) Fruitbody is composed of multiple, overlapping caps, 2–10 cm in diameter, arising from branching stems, eccentrically attached and sharing a common base. Young fruitbodies are adorned with fine gray fibrils. The pores on the underside of the caps are white. Spores white, 6–7 by 3.5–5 µm.

Distribution: Grows in northern temperate, deciduous forests. In North America, primarily found in eastern Canada and throughout the northeastern and mid-Atlantic states. Rarely found in the northwestern and southeastern United States. Also indigenous to the northeastern regions of Japan and the temperate

▲ **FIGURE 274**

A beautiful "hen," ready for plucking, grows at the base of an oak tree.

▲ **FIGURE 275**

Dusty Yao finds *G. frondosa* fruiting from the root remnants of a tree near Mercersburg, Pennsylvania, in October. We found several beautiful hens in the area that day. Bicycles are a rapid and effective vehicle for hunting for *G. frondosa* in suburbia.

hardwood forests of China and Europe, where it was first discovered.

Natural Habitat: Found at the interface of ground and tree, near stumps, or at the base of dead or dying deciduous hardwoods, especially oaks, elms, maples, honey locust, black gum, beech, and occasionally larch. According to Gilbertson and Ryvarden (1986–87) and Overholts (1953), this mushroom has also been collected on or around pines and Douglas fir, although rarely so. Although found at the bases of dying trees and sometimes emerging from rotting roots, this mushroom is viewed by most mycologists as a saprophyte, exploiting tree tissue dying from other causes. Massive oaks, apparently healthy, often sport this mushroom at their base.

Type of Rot: White butt and root rot.

Fragrance Signature of Mycelium: Richly fungoid and uniquely farinaceous, sometimes sweet. To me, rye grain spawn of this fungus has a fragrance reminiscent of day-old fried corn tortillas. When the mushrooms begin to rot, a strong fishlike odor develops.

Natural Method of Cultivation: The inoculation of hardwood stumps or buried logs is recommended.

Given the size of the fruitbody and its gourmet and medicinal properties, this mushroom may well become the premier species for recycling stumps in hardwood forests. Although less frequently encountered, the occurrence of this mushroom on pines, Douglas fir, and larch is curious, confirming that some strains exist that could help recycle the millions of stumps dotting the devastated timberlands of the world. As forests decline as a result of acid rain, future-oriented foresters would be wise to explore strategies by which the dead trees could be inoculated and saprophytized by appropriate strains of maitake.

Those experimenting with stump culture should expect to wait 3 to 5 years before fruiting occurs. Strategies using high inoculation rates of plug spawn, supplemented with rope spawn, help colonize stumps. Stumps do not necessarily have to be "virgin." Maitake is well known for attacking trees already being parasitized by other fungi. Since maitake forms from buried wood, inoculations near the ground, just above the root zones, are recommended.

An easy way to grow maitake is to bury myceliated wood—either spawned logs or commercial maitake spawn blocks—in the spring, and then use a hoop frame and shade cloth to encourage fruitings in the fall.

Be forewarned that this mushroom often appears within the same 2-week period every year, unless there is drought, so you should make sure your daily walk takes you by your maitake patches. Too many times I have missed maitake fruitings, only finding them when they are well past their prime. Now I mark my calendar, October 1–15, but I begin checking in late September.

Season and Temperature Range for Mushroom Formation: In the fall, typically mid-September to late October in the northeastern United States. Temperature: 45–70°F.

Harvest Hints: I love the young brainlike forms, but this stage is usually only seen when cultivation is done indoors on sterilized sawdust. Best to harvest outdoor-grown mushrooms when the leaflets have extended but just before the edges brown, a sign of dieback.

Nutritional Profile: Our analysis of a 100 g serving shows the following: calories: 377; protein: 25.51 g; fat: 3.83 g; polyunsaturated fat: 1.12 g; total unsaturated fat: 2.08 g; saturated fat: 0.34 g; carbohydrates: 60.17 g; complex carbohydrates: 41.37 g; sugars: 18.80 g; dietary fiber: 28.50 g; cholesterol: 0 mg; vitamin A: 0 IU; thiamine (B1): 0.25 mg; pantothenic acid (B5): 4.40 mg; vitamin C: 0 mg; vitamin D: 460 IU; calcium: 31 mg; copper: 1.88 mg; iron: 7.6 mg; potassium: 2,300 mg; niacin: 64.80 mg; riboflavin: 2.61 mg; selenium: 0.056 mg; sodium: 14 mg; moisture: 4.75 g; ash: 5.74 g.

Medicinal Properties: This mushroom has antitumor properties (especially against breast, prostate, and colorectal cancers), as well as antidiabetic and antiviral properties (Nanba 1992, 1993, 1995, 1997; Yamada et al. 1990; Kawagishi et al. 1990; Ohno et al. 1984; Suzuki et al. 1984). In a nonrandomized clinical study of 165 advanced stage (III, IV) cancer patients, patients were either taking Maitake D-fraction with crude powder tablets only, or Maitake D-fraction crude tablets in addition to chemotherapy. "Tumor regression or significant symptom improvements were observed in 11 of 15 breast-cancer patients, 12 out of 18 lung-cancer patients, and 7 of 15 liver-cancer patients. When maitake were taken in addition to chemotherapy, these response rates improved by 12 to 28 percent" (Nanba 1997, 44).

Maitake beta-glucans increase tumor necrosis factors in human prostate cancer and are being explored as a treatment (Fullerton et al. 2000). Kodama and colleagues (2002), in a nonrandomized study, found that patients ranging from 22 to 57 years in age who had liver, breast, or lung cancer showed significant

◄ **FIGURE 276**

Outdoor fruitings of *G. frondosa* arising from blocks buried into rocky soil.

improvement (58 percent, 68 percent, and 62 percent, respectively) of immune-competent cells when maitake was combined with chemotherapy.

In vitro studies have shown that the 1,3-beta-D-glucans, the water-soluble fraction from the fruitbodies, stimulate cytokine production from macrophages, triggering an immune response (Kurashige et al. 1997; Adachi et al. 1994, 1998; Okazaki et al. 1995). By dry weight, the beta-glucans that are contained within the cell walls of a maitake can constitute 10 to 20 percent, while overall complex carbohydrates can constitute up to 41 percent of its mass. Structural in nature, these cell wall polysaccharides can decompose into several subcomponents, one of which is the well-known 1–6, 1–3-beta-D-glucan having a molecular weight of nearly 1 million. The denaturing of the water-soluble, high molecular weight polysaccharides to lighter subcomponents through digestion and/or moderate heat treatment (> 212°F) can enhance the bioavailability of beta-glucans, including the maitake-specific grifolan and its synergistic cousins. However, excess heat—over 302°F—can reduce (1–6) branched (1–3) beta-glucans into smaller subfractions (ranging from 6400 to 250,000 molecular weight) that, when isolated from one another, showed reduced or little activity (Mizuno and Zhuang 1995; Adachi et al. 1990). Hence, heat treatment between 212 and 250°F is well within the target range for making these compounds extractable and/or bioavailable without degradation into inactive constituents.

Ohno and others (1985) and Takeyama and colleagues (1987) isolated grifolan from mycelium grown in culture. As the density of the mycelium increases, leading to the eventual creation of a mushroom, so too, it is presumed, do the available beta-glucans. This group of polysaccharides also stimulates tumor necrosis factors (Ohno et al. 1995). Investigations into the production of nitric oxides from macrophages exposed to an extract of maitake mushrooms showed antitumor activity (Sanzen et al. 2001), results shared in common with *Agaricus brasiliensis* (Kawakami et al. 2002) and *G. lucidum* (Zhao and Gao 2002).

Sunlight and ultraviolet radiation (UVB) transform this mushroom's ergocalciferols into vitamin D2. Our research found that the vitamin D2 in maitake mushrooms grown and dried indoors was only 460 IU. When dried mushrooms were placed outside in the sun, the vitamin D2 soared to more than 21,000 IU (see page 207).

Ohnogi et al. (2004) filed a patent application the beneficial effects to humans of an aqueous extract of maitake for its role in regenerating nerves and improving neurological function.

Maitake has also been implicated as a possible treatment for diabetes (Kubo et al. 1994) by lowering and moderating glucose levels. A study by Manohar and colleagues (2002) on insulin-resistant mice showed that when a single dose from a maitake mushroom extract was introduced, circulating glucose lowered by 25 percent. Konno and others (2001) and Manohar suggest that maitake could aid in modulating glucose levels in diabetic patients. In 1988, Dr. Harry Preuss of Georgetown University announced investigations into the use of G. *frondosa* for treatment of type 2, non-insulin-dependent adult diabetes.

Flavor, Preparation, and Cooking: Toward the stem base, the flesh of this mushroom is thick and dense and is best when sliced. The upper petal-like caps are best when chopped. This mushroom can be prepared in many ways, delighting the connoisseur mycophagist. Simply slicing and sautéing à la shiitake is a simple method. This mushroom can also be stuffed with shrimp, sliced almonds, spices, and topped with melted cheese, then baked to golden brown. The late Jim Roberts of Lambert Spawn, a commercial mushroom spawn company, once fed me this dish featuring a 1-pound specimen that I devoured at one sitting, making for a very satisfying meal and a good night's sleep. Dried specimens can be powdered and used to make a refreshing tea.

Mycorestoration Potential: One of the best candidates for creating a medicinal and nutritious mushroom woodland landscape, maitake can be grown in your backyard. Through inoculations, maitake populations can be enhanced throughout the northeastern and Midwestern United States and much of Europe, within the supportive ranges of oaks, elms, honeysuckles, and beeches. Increasingly more common in suburbia than in the wild, this mushroom is following human's conquest and transformation of forests into neighborhoods. It loves to grow from the bases of aging trees or stumps, and from buried roots. Primarily known as a fleshy, delicious soft polypore, maitake may protect its host trees from invasion by aggressive parasitic fungi. I would not be surprised if some varieties of this mushroom are found to grow endophytically, protecting the trees from plague fungi.

Comments: Few mushrooms trigger my instincts like this one does. I love to grow it, to nurture its young forms that emerge looking like brains, with shooting fans that diverge and extend like leafy branches. This mushroom is not native to the region where I live. I try to make a trip to its native regions, usually Pennsylvania or New York, at least every other year. Friends who collect it search for it in groves of big oaks in neighborhoods, cemeteries, or parks. Very few collectors I have met have found this mushroom deep in the woods. It seems to have affection for human habitation and trees more than 40 years in age. (But if you scout neighborhoods for yards with large trees, be forewarned that you may be arrested for suspicious activity, especially in this day and age.) This is a great species, which, once it takes up residence, blesses its human partners with years of delectable delights.

Hericium abietis (Scopoli: Fries) S. F. Gray

Common Names: Conifer coral mushroom, comb tooth.

Taxonomic Synonyms and Considerations: So similar to *H. coralloides* and *H. ramosum* as to make it difficult to separate them from one another. I would not

be surprised if they are eventually found to be synonyms with diverse ecotypes. *H. abietis* is distinguished from *H. erinaceus* primarily by its preference for conifers, particularly spruce, hemlock, and Douglas fir, and by the fact that it produces loosely branched spines. *H. erinaceus*, however, is indigenous to hardwoods, has unbranched spines, and is ball-like in form.

Introduction: This easy-to-recognize mushroom is a favorite of Pacific Northwest mushroom hunters. Bright white in color, this mushroom often forms on the underside of downed conifers and produces a brown rot on the trees it decomposes. Differing from most mushrooms I know, this species can form from logs that have been denuded of their bark, suggesting to me that the mycelium penetrates deeply into the recesses of the host trees.

Description: A white, toothed mushroom, forming multiple branches that fork, diverge, and rebranch, and from which loosely arranged, needlelike spines hang. Emerging from a central tuberlike stalk, which is often short and easily missed when the mushrooms are plucked. Compared to mushrooms of similar mass, this

▲ **FIGURE 277**

Dusty Yao collects *H. abietis* in an old-growth forest near the slopes of Mount Rainier. This specimen was collected, cloned, and then consumed for a delicious dinner.

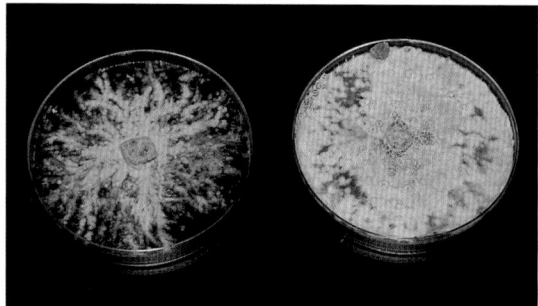

▲ FIGURE 278

The clone from the mushroom featured in the previous image grows under sterile culture in a nutrient-filled petri dish. Note the small clusters of baby mushrooms forming.

▲ FIGURE 279

H abietis fruits from Douglas fir wood chips and is genetically identical to the mushroom featured in figure 277. The time from collecting to cloning to fruiting was 4 to 5 months. Such cloning practices help preserve biodiversity by building a library of strains.

species is not as prolific a spore producer. I know of specimens of up to 20 pounds that have been collected; larger ones are likely. Spores white, 3–5 by 3–5 μm.

Distribution: Widely distributed across the temperate regions of the world, on conifers, mostly in the fall. Reported in the Pacific Northwest, and if the mushroom is indeed synonymous with *H. ramosum* and *H. coralloides*, its range could extend throughout the temperate conifer forests of the world.

Natural Habitat: On conifers, particularly hemlock, spruce, and Douglas fir.

Type of Rot: Reportedly producing a white rot of the heartwood, but reducing a tree to a brown mass. The mushrooms I have collected came from from downed, often barkless brown-colored decomposing trees.

Fragrance Signature of Mycelium: Sweet, appealing, increasingly fishy as the mushroom matures.

Natural Method of Cultivation: Using plug, sawdust, or rope spawn. Inoculated logs are best incubated a few feet off the ground and in the shade, so the downward forming mushrooms don't make ground contact. Once established, this mushroom can form annually for decades, depending upon the mass of the tree or stump.

Season and Temperature Range for Mushroom Formation: Primarily in the late summer through fall, fruiting well when temperatures cycle day to night within a 45–65°F range.

Harvest Hints: After maturity, the top dome begins to brown with the first signs of bacterial blotch. Best to harvest when the mushrooms are white.

Nutritional Profile: Our analysis of a 100 g serving shows the following: calories: 375; protein: 20.46 g; fat: 5.06 g; polyunsaturated fat: 0.83 g; total unsaturated fat: 1.85 g; saturated fat: 0.76 g; carbohydrates: 61.80 g; complex carbohydrates: 40.90 g; sugars: 20.90 g; dietary fiber: 39.20 g; cholesterol: 0 mg; vitamin A: 0 IU; thiamine (B1): 0.16 mg; pantothenic acid(B5): 7.40 mg; vitamin C: 0 mg; vitamin D: 57 IU; calcium: 8 mg; copper: 1.66 mg; iron: 6 mg; potassium: 2,700 mg; niacin: 11.80 mg; riboflavin: 2.26 mg; selenium: 0.091 mg; sodium: 4 mg; moisture: 6.69 g; ash: 5.99 g.

Medicinal Properties: Not yet known to this author.

Flavor, Preparation, and Cooking: Cut perpendicular to the downward spines and sauté over medium-high heat. The mushrooms release copious water. Season with rosemary or other herbs two-thirds of the way through the process, and continue cooking until the edges brown. Add white wine and then a touch of butter toward the end to enhance the crab- or fish-like flavor. If refrigerating, place into a paper sack, being careful that the mushrooms are not bruised in handling.

Mycorestoration Potential: Once established, *H. abietis* can produce from a dead conifer for many years, slowly digesting and disassembling a large tree into loose duff over one's lifetime. I see this species as one that is helpful to the elderly for enhancing brain and neurological function, and one that should be promoted in healing arts centers as part of self-healing programs. Parents could inoculate a large downed conifer for their children to enjoy throughout their lives.

Comments: This majestic species is eye-catching and is easily seen from great distances whether you are walking in the woods or driving along a country road. A longtime favorite of mushroom hunters, *H. abietis* has yet to be explored, even initially, for its medicinal properties. A candidate for aiding nerve regeneration, this mushroom is also likely to have antimicrobial properties similar to *H. erinaceus*. I view this mushroom as a brain food that increases intellectual acumen and strengthens the nervous system.

Hericium erinaceus (Bulliard: Fries) Persoon

Common Names: Lion's mane, monkey's head, bear's head, old man's beard, hedgehog mushroom, satyr's beard, pom pom, yamabushitake (Japanese for "mountain-priest mushroom").

Taxonomic Synonyms and Considerations: Formerly known as *Hydnum erinaceum* Fr. and sometimes cited as *Hericium erinaceum* (Fr.) Pers. *Hericium coralloides*

▲ **FIGURE 280**

Dusty Yao holds photographs from one of her fall mushroom hunts with her sister Liz, who holds *H. erinaceus*. We cloned this mushroom and then grew it, as featured here, in only 3 months from the day of collecting.

and *Hericium abietis* are similar species, but distinct in both their habitat preference and form. *H. coralloides* can also be cultivated on sawdust and differs from *H. erinaceus* in that its spines fork rather than emerging individually. *H. erinaceus*, when grown under culture and in high carbon dioxide environments, forms mushrooms bearing an uncanny resemblance to *H. abietis*, a fact few noncultivator mycotaxonomists realize.

Introduction: *H. erinaceus* is one of the few mushrooms imparting the flavor of shrimp or lobster when cooked. Producing a mane of cascading white spines, this mushroom can be grown on sterilized sawdust or bran or using the traditional log method first established for growing shiitake.

Description: Composed of downward-cascading, non-forking spines, this mushroom grows up to 40 cm in diameter in the wild. Typically white until aged and then discoloring to brown or yellow brown, especially at the top. Spores white, 4.5–5.5 by 4.0–4.5 μm.

Distribution: Reported from North America, Europe, China, and Japan. Of the *Hericium* species, this one is most abundant in the southern regions of the United States.

Natural Habitat: On dying or dead oak, walnut, beech, maple, sycamore, and other broadleaf trees. Found most frequently on logs or stumps.

Type of Rot: Reportedly a white heart rot.

Fragrance Signature of Mycelium: Rich, sweet, and farinaceous.

Natural Method of Cultivation: Inoculation of logs or stumps outdoors using sawdust or plug spawn à la the methods traditionally used for shiitake. This is one of the few mushrooms that produce well on walnut logs. Preferred woods for growing this mushroom include oak, beech, elm, and similar hardwoods. (The "paper" barked hardwoods such as alder and birch are not recommended.) Once inoculated, the 3- to 4-foot-long logs should be buried to one-third of their length into the ground in a naturally shady location. Walnut is comparatively slow to decompose due to its density, providing the outdoor cultivator with many years of fruitings. A heavy inoculation rate will shorten the gestation period.

Season and Temperature Range for Mushroom Formation: During the warm wet months. Temperature: 65–75°F.

Harvest Hints: Since the mushrooms are ball-like when young, they mature with a downward growth of the descending spines. The spore load of this mushroom is low compared to that of other species. When the mushrooms age, a soft rot begins on the crown, softening the interior brownish core with a rot. When these initial brown zones appear, it's time to pick.

Nutritional Profile: Our analysis of a 100 g serving shows the following: calories: 375; protein: 20.46 g; fat: 5.06 g; polyunsaturated fat: 0.83 g; total unsaturated fat: 1.85 g; saturated fat: 0.76 g; carbohydrates: 61.80 g; complex carbohydrates: 40.90 g; sugars: 20.90 g; dietary fiber: 39.20 g; cholesterol: 0 mg; vitamin A: 0 IU;

▲ **FIGURE 281**

Phan Ngoc Chi Lan stands between two rows of *H. erinaceus* in Vietnam.

thiamine (B1): 0.16 mg; pantothenic acid (B5): 7.40 mg; vitamin C: 0 mg; vitamin D: 57 IU; calcium: 8 mg; copper: 1.66 mg; iron: 6 mg; potassium: 2,700 mg; niacin: 11.80 mg; riboflavin: 2.26 mg; selenium: 0.091 mg; sodium: 4 mg; moisture: 6.69 g; ash: 5.99 g.

Medicinal Properties: Traditional Chinese medicinal practitioners prescribe this species for stomach ailments and for prevention of cancer in the gastrointestinal tract. Dr. Mizuno, of Shizuoka University, isolated acidic derivatives in this mushroom that are strongly effective against hepatoma cells. He further identified 5 distinct polysaccharides with potent antitumor properties that extended the life spans of patients (Mizuno 1995).

Other researchers patented an extraction process that isolates nerve growth stimulant (NGS) factor—compounds now known as erinacines and hericiones (Kawagishi et al. 1991). A novel erinacine, erinacine Q, isolated from liquid culture, is one precursor to this family of erinacines (Kenmoku et al. 2002). Erinacines stimulate neurons to regrow and rebuild myelin, and so they may possibly be significant in treating senility and Alzheimer's disease, repairing neurological trauma, increasing cognitive abilities, and perhaps improving muscle/motor response pathways, which would be helpful for those suffering from nerve degeneration, such as that occurring in muscular dystrophy. Kawagishi (2002) noted that lion's manes' low molecular weight compounds pass through the blood-brain barrier intact. Kolotushkina and colleagues (2003) further substantiated the neurological benefits from extracts of this mushroom through its myelin-generating influence on nerve and cerebellar glia cells in vitro. Ohnogi and others (2004) filed a patent on the beneficial effects of an aqueous extract of *H. erinaceus* for its medicinal, nerve-regenerating properties through human consumption.

Having strong antimicrobial properties against some fungi (*Aspergillus* species and *Candida* species) and bacilli, this mushroom is an important species for those wanting to have nerve-regenerating medicinal mushrooms growing close by.

Flavor, Preparation, and Cooking: I like frying this mushroom over high heat, covered so flavor is retained, until the fingerlike teeth are singed brown and become crispy. Next I like to add tamari or soy sauce and allow it to simmer for a few minutes with chopped onions, adding butter at the end. Strains vary substantially in their flavor profiles, from sweet to tart. When butter is added to the mushrooms, especially near the end of the cooking time, the seafoodlike flavor comes to the forefront.

Mycorestoration Potential: For those living in hardwood ecosystems, especially in subtropical regions, *H. erinaceus* can be instrumental in recycling fallen trees, giving bouquets of mushrooms for years as soil duff is created. Clusters of mushrooms can reach 10 or more pounds, appearing annually or biannually. As part of a medicinal mushroom forest, this species is an important asset. For coniferous woodlands, its close relatives *H. abietis* and *H. coralloides* are equally good candidates.

Comments: Once reserved for the palates of the royal families, this delectable mushroom not only has unique medicinal properties but also is also popular for its distinctive seafoodlike flavor. Primarily a hardwood saprophyte, *H. erinaceus* is widely distributed through-

out the world. I suspect that this species may be a "smart food," increasing the intelligence of those people who consume it. Families with histories of nerve disease, Alzheimer's, Parkinson's, or muscular dystrophy or similar susceptibility to nerve degeneration may want to establish a multigenerational tree or log that can not only provide them with delectable dinners for decades but also significantly and positively affect their health and that of their generations to come.

This mushroom is tolerant of subtropical temperatures and grows easily on sterilized sawdust sourced from many tree species, from oak to acacia to rubber. For more information on techniques for cultivating this and other medicinal mushrooms, please consult my earlier book *Growing Gourmet and Medicinal Mushrooms* (2000a).

▲ **FIGURE 282**

Dr. Andrew Weil holds a hardwood (oak) log hosting emerging *H. erinaceus* while traveling in China in 1984.

Hypholoma capnoides (Fries) Quelet

Common Names: The clustered woodlover, the smoky brown clustered woodlover, smoky gilled Hypholoma, Elsie's edible.

Taxonomic Synonyms and Considerations: *H. capnoides* is known by many as *Naematoloma capnoides* (Fr.) Karst. A sister species to *H. capnoides* is *H. fasciculare* (Hudson ex Fr.) Kummer (=*Naematoloma fasciculare* [Fr.] Quelet), a mushroom well worth knowing since it is poisonous! These 2 mushrooms are sometimes difficult to tell apart until they are upturned and the gills are examined. *H. capnoides* has smoky brown gills, whereas *H. fasciculare* has gills that are bright greenish yellow to dingy yellow in age. Furthermore, *H. fasciculare* is extremely bitter in flavor, whereas *H. capnoides* is mild. Once you know the differences, these species can be separated without difficulty. *Hypholoma*, *Psilocybe*, and *Stropharia* are the primary genera of the Strophariaceae family, which I featured in my first book, *Psilocybe Mushrooms and Their Allies* (1978).

Introduction: *H. capnoides* is an aggressive conifer stump decomposer. One precaution: *H. capnoides* is not a mushroom for those unskilled in mushroom identification. Several poisonous mushrooms resemble this species and inhabit the same ecological niche. This danger is entirely avoided by simply comparing spore print colors and honing your identification skills. Always be absolutely certain of the identification of a mushroom before ingesting it. For those not knowing the differences between, for instance, *H. capnoides* and *Galerina* species, the latter group having rusty brown spores, this is not a mushroom to grow. Additionally, *H. capnoides* has a poisonous sister—*H. fasciculare*—which although not likely to be deadly can sicken those who eat it. *H. fasciculare* has an olive-greenish cast and is often brighter than *H. capnoides* (see figure 286).

Description: Cap orange to orangish yellow to orangish brown to dull brown, 2–7 cm broad at maturity. Convex with an incurved margin, soon expanding to broadly convex to almost flattened, occasionally

▲ **FIGURE 283**

A young, emerging fruiting of *H. capnoides*. Note mature specimens off to the side.

▲ **FIGURE 284**

H. capnoides grows on dead wood, primarily conifers, and is an aggressive saprophyte competing well against several parasitic fungi. I have observed that stumps and logs having this mushroom do not readily host *Armillaria* species, which are devastating root-rot parasites. Some conifer stumps can support species fruitings of hundreds of mushrooms each season over a decade.

possessing a blunted nipple or umbo at the center of the cap. Cap margin often adorned with fine remnants of the partial veil, soon disappearing. Surface smooth, moist, and lacking a separable gelatinous skin (pellicle). Gills attached, soon seceding, close, white at first, soon grayish, and eventually smoky grayish

◀ **FIGURE 285**

H. capnoides has smoky gills when mature.

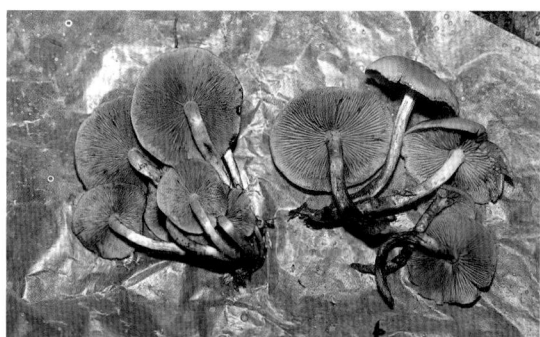

▲ **FIGURE 286**

Here, *H. fasciculare,* a gastrointestinally poisonous mushroom with olive-greenish gills when mature lies beside *H. capnoides,* and edible mushroom, which has smoky brown gills when mature.

purple brown in age. Stem 5–9 cm long, enlarged at the base, covered with fine hairs. Partial veil cortinate, sometimes leaving a faint annular zone, becoming dusted purple brown with spores on the upper regions of the stem. Usually growing in clusters. Spores are purple brown to purplish black, 6–7 by 4–4.5 µm.

Distribution: Widely distributed across North America, particularly common in the western United States. Also found throughout the temperate regions of Europe and probably throughout similar ecological zones of the world.

Natural Habitat: A lover of conifer stumps and logs, especially Douglas firs, ponderosa and lodgepole pines, hemlocks, western firs, spruces, and, more rarely, but significantly, on aged and rotting cedars and redwoods. (Redwoods and cedars are notoriously resistant to fungus rot; saprophytic mushrooms seldom grow on downed trees and stumps of these species.) I often find this temperate mushroom, along with other interesting relatives, in "beauty bark" used for landscaping around suburban and urban buildings. Although this species has not been reported on alder in the wild, I have successfully grown it on sterilized wood chips of *Alnus rubra.* Grows as far south as Arizona on ponderosa pines and north into the upper fringes of the boreal forests.

▲ **FIGURE 287**

Galerina autumnalis, a deadly poisonous mushroom, is easily distinguished from *Hypholoma* mushrooms by spore color: *Galerina* mushrooms produce rusty brown spores, whereas *Hypholoma* species produce purplish brown to black spores. A mistake in identification, easily avoided by noting spore color, can be fatal. *Be careful.*

Type of Rot: White rot, pulping conifers into loose, long "bleached" fibers.

Fragrance Signature of Mycelium: A fresh, sweet, forestlike, pleasant fragrance, similar to its cousin *Stropharia rugoso annulata.*

Natural Method of Cultivation: This aggressive species is one of the best for recycling millions of conifer stumps left from logging. This mushroom is a good species for inoculating stumps with rope, plug, or spore spawn because it produces long, forking, silky white, penetrating rhizomorphs. Since its stem butts regrow with such vigor, mycophiles throughout the temperate regions of the world can propagate this species without needing laboratories. For infusing chain-saw oils, I especially favor using spores of this mushroom. One can gather enough mushrooms during the rainy fall season from a single stump to inoculate hundreds more. Plug spawn—generated from stem butts or from commercial sources—gives you the most options for mycorestoration. We have had great success using plug spawn for inoculating burlap sacks filled with raw wood chips. Much of the knowledge I have gained from growing the woodland *Psilocybe* mushrooms such as *P. cyanescens* and *P. azurescens* is applicable to growing this species. (Please refer to *The Mushroom Cultivator* [Stamets and Chilton 1983], *Growing Gourmet and Medicinal Mushrooms* [Stamets 2000a], and see pages 289 to 292.)

Season and Temperature Range for Mushroom Formation: Typically during the fall rainy season in the northern temperate to sub-boreal regions of the world. Temperature: 50–60°F.

Harvest Hints: Mushrooms should be harvested when the caps are convex. Outdoors, this mushroom forms clusters, often with several dozen mushrooms arising from a common base. Bunches that hosted dozens of fruitbodies and weighed up to 4 pounds have been collected.

Nutritional Profile: Not yet known to this author.

Medicinal Properties: Not yet known to this author. Given this species' woodland habitat and aggressiveness in combating competitors, I think *H. capnoides* should be carefully examined for its antibacterial, antiviral, and immunity-enhancing medicinal properties.

Flavor, Preparation, and Cooking: Nutty and excellent in stir-fries, cooked like shiitake. On a subjective scale of 1 to 10, I would give it a 6. Most of my friends rate it as low as 4 and as high as 8 with one claiming it's her favorite edible mushroom and so giving it a 10. When cooked until crispy, this mushroom imparts to the palate a wonderfully nutty flavor.

Mycorestoration Potential: *H. capnoides* can be used for protecting against blights, for recycling stumps, and

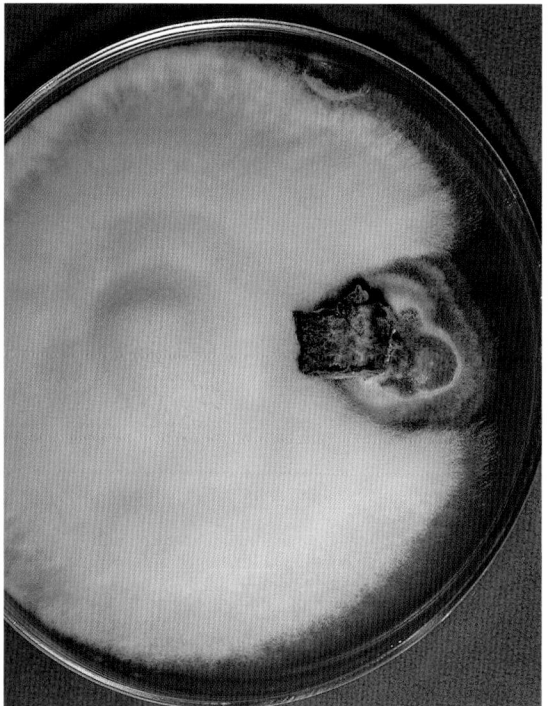

△ **FIGURE 288**

Mycelium of *H. capnoides,* a forest-friendly saprophyte, overrunning a culture of *Armillaria mellea,* a blight fungus that devastates thousands of acres of forests.

for enzyme production. Mycoforesters should carefully consider the judicious use of this fungus, favoring indigenous strains for forestalling blights (see figure 288). Chapman, Xiao, and Meyers (2004) found that its close relative, *H. fasciculare*, significantly reduced *Armillaria* blight in logged areas in experiments in British Columbia. I believe that *H. capnoides* can help mycoforesters lead logged habitats more rapidly down the path toward ecological recovery. For mycoremediation, this species is likely to be very powerful in breaking down an assortment of toxic wastes. I hope readers will further explore this species for its mycorestorative potential. The woodlover is a powerful ally with many interesting cousins, the noble *Psilocybe* species not least among them.

▲ **FIGURE 289**

H. capnoides fruits on a dead snag in the old-growth rain forest of the Olympic Peninsula.

Comments: *H. capnoides* grows well on conifer stumps and logs. Once stumps are inoculated using rope spawn, plug spawn, or sawdust spawn (implanted into fresh cuts or sandwiched on the top of the stump), rhizomorphs soon form from the sites of inoculation. Large-diameter stumps and logs produce crops every season for years, perhaps decades, usually accompanied by blossoming plant communities whose shade-producing canopies stimulate mushroom production and further enhance nutrient flow cycles. A further advantage of this mushroom is that it digests bark of many fir tree species.

I have found this mushroom several times on aged redwoods and cedars—in fact, this is the only edible mushroom I have found on these slow-to-rot trees. As such, this species should be closely examined for its unique enzymes—it may possess unharnessed mycoremediation properties.

Hypholoma sublateritium (Fries) Quelet

Common Names: Cinnamon cap, brick top, red woodlover, kuritake (Japanese for "the chestnut mushroom").

Taxonomic Synonyms and Considerations: Synonymous with *Naematoloma sublateritium* (Fr.) Karsten, a name now retired in deference to *Hypholoma*. This mushroom is a sister species to the edible *H. capnoides* and the poisonous *H. fasciculare*.

Introduction: Clusters of this species are often found on dead hardwoods, especially stumps, logs, and soils rich in wood debris. *H. sublateritium* is an excellent candidate for recycling stumps from the Midwest to the northeastern United States, in central Europe, and in central Asia.

Description: Cap 2–7 cm broad, hemispheric to convex, expanding with age to broadly convex, and eventually plane in age. Cap tan to brown to brick red, darker toward the center, and lighter yellow near the margin. Margin incurved at first and covered with cottony remnants of the partial veil, soon straightening, and eventually uplifted at maturity. Flesh relatively thick, bruising yellowish. Gills close, bluntly attached to the stem, pallid at first, becoming dark purple gray when mature. Stem 5–10 cm long by 5–10 mm thick, solid, covered with fibrillose veil remnants below the annular zone. The stem bases are often adorned with silky white rhizomorphs. This species often forms large clusters. Spores purplish black, 6–7.5 by 3.5–4 µm.

Distribution: Growing in eastern North America, Europe, and temperate regions of Asia (Japan and Korea).

Natural Habitat: Fruiting in the late summer and fall, primarily on the stumps of oaks and occasionally chestnuts, but not on firs, in direct contrast to its sister species *H. capnoides.*

Type of Rot: White rot, reducing logs into a pithy, whitish tuft of cellulosic fibers.

Fragrance Signature of Mycelium: Pleasant, rich, woodsy, sweet, reminiscent of the refreshing fragrance from a newly rained-upon forest, similar to the scent of the king Stropharia *(Stropharia rugoso annulata).*

Natural Method of Cultivation: Hardwood logs can be inoculated with dowel or sawdust spawn and laid horizontally side by side. Untreated sawdust can be used to bury the logs to one-third their diameter. Oak, chestnut, and perhaps the stumps of similar hardwoods can be inoculated using any one of the methods described in this book. Plug, rope, and spore spawn are effective methods for inoculating logs and stumps. Sawdust spawn can be broadcast through wood chips for creating outdoor mycological landscapes (see figures 195 to 197).

Season and Temperature Range for Mushroom Formation: Appearing from August through October in northern and eastern North America, northern and central Europe, and temperate regions of Asia. Temperature: 45–70°F.

Harvest Hints: Harvest before the gills become brown from maturing spores, and when the cobwebby partial veil that extends from the stem to the cap edge is stretched but not broken, typically when the caps are hemispheric to convex.

▲ **FIGURE 290**

The beautiful *H. sublateritium* fruiting from a block of sterilized sawdust.

▲ **FIGURE 291**

H. sublateritium fruiting from a log raft 6 months after inoculation with sawdust spawn, which was packed between scarified logs. These logs produced, without additional care, for 8 years.

⋀ **FIGURE 292**

H. sublateritium likes to fruit in dense families, often clustered, making it easy to pick.

⋀ **FIGURE 293**

A cluster of *H. sublateritium* emerging from a buried alder log inoculated with plug spawn.

Nutritional Profile: Our analysis of a 100 g serving shows the following: calories: 356; protein: 22.89 g; fat: 3.77 g; polyunsaturated fat: 1.51 g; total unsaturated fat: 1.78 g; saturated fat: 0.27 g; carbohydrates: 57.64 g; complex carbohydrates: 37.84 g; sugars: 19.80 g; dietary fiber: 34.70 g; cholesterol: 0 mg; vitamin A: 0 IU; thiamine (B1): 0.20 mg; pantothenic acid (B5): 14.60 mg; vitamin C: 0 mg; vitamin D: 42 IU; calcium: 65 mg; copper: 1.38 mg; iron: 25 mg; potassium: 2,500 mg; niacin: 79.60 mg; riboflavin: 1.35 mg; selenium: 0.128 mg; sodium: 35 mg; moisture: 8.35 g; ash: 7.35 g.

Medicinal Properties: Not yet known to this author. The properties of this mushroom remain largely unexplored, although mushroom hunters from Asia to the eastern United States seek the species.

Flavor, Preparation, and Cooking: Strongly flavored, it can be used in stir-fries or baked. Many recipes can incorporate cinnamon caps.

Mycorestoration Potential: As a possible defense against blight fungi, this species should figure as one of the premier mycorestoration candidates that can be sourced from nature. I suspect this species also can fight *Armillaria* root rot blight much in the same way as its cousins *H. fasciculare* and *H. capnoides* (Chapman,

⋀ **FIGURE 294**

H. sublateritium growing in the garden next to corn. Corralling a garden with logs inoculated with this mushroom is one of many ways to use this species.

Xiao, and Meyers 2004). Since this mushroom produces cordlike rhizomorphs that penetrate deeply into wood, an indication of mycelial fortitude, this species should be explored for its mycoremediative applications.

Comments: I once inoculated alder logs with *H. sublateritium* sawdust spawn, which then fruited for 8 years (see figure 291). The logs decomposed to the point of a loose pithy soft pulp, and yet mushrooms still were forming from the disheveled fibers. Such a long fruiting life and extreme decomposition suggests a digestive enzymatic system more powerful than that of most. I have grown this mushroom in sawdust mulch added to gardens.

The bright yellow margins in young specimens and reddish dark zone in the center of the caps, combined with spore color, gill tones, and the cobwebby veil, set this mushroom apart. However, *H. fasciculare*, a sister species sharing the same habitats, causes severe gastrointestinal pain and illness when consumed. *H. fasciculare* has purplish spores and a cobwebby veil but olive-greenish dark-colored gills. Many field guides describe these mushrooms in detail. If you plan to grow cinnamon caps, know the species well so you can avoid accidental poisoning (see figure 286).

Hypsizygus ulmarius (Bulliard: Fries) Redhead

Common Names: The elm oyster mushroom, shirotamogitake (Japanese for "white elm mushroom").

Taxonomic Synonyms and Considerations: Looking like an oyster mushroom, this species was once called *Pleurotus ulmarius* (Bull. ex Fr.) Kummer, later *Lyophyllum ulmarium* (Bull.:Fr.) Kuhner, and most recently *Hypsizygus ulmarius* (Bull.: Fr.) Redhead. Unique microscopic features qualify its placement into the genus *Hypsizygus. H. ulmarius* and *H. tessulatus* are closely related, living in the same ecological niche. *H. ulmarius* produces brown rot, while *Pleurotus ostreatus* produces a white rot—a major difference that aids in separating them taxonomically.

▲ **FIGURE 295**

H. ulmarius fruiting on a cottonwood in the Sol Duc River valley in the rain forest on the Olympic Peninsula.

▲ **FIGURE 296**

H. ulmarius fruiting in an experimental garden bed.

Introduction: A relatively rare and sometimes large mushroom that usually grows singly or in small groups on elms and beeches, the morphology of *H. ulmarius* closely parallels that of an oyster mushroom, but *H. ulmarius* is far better in flavor and texture. Increasingly popular in Japan, *H. ulmarius* has just recently made its entrance into the marketplace with rave reviews. Some marketers call them "king oysters," which is somewhat confusing because *Pleurotus eryngii* is better known by that name.

Description: Mushrooms hemispheric to plane, sometimes umbillicate, uniformly tan, beige, grayish brown, to gray in color, sometimes with faint streaks, and measuring 4–15 cm. (This mushroom can become quite large.) Cap margin in-rolled to incurved when young, expanding with age, even to slightly undulating. Gills decurrent and close. Stem eccentrically attached, thick, tapering, and curved at the base. Usually found singly, sometimes in groups of 2 or 3, rarely more.

Distribution: Throughout the temperate forests of eastern North America, Europe, and Japan. Probably widespread throughout similar climatic zones of the world.

Natural Habitat: A saprophyte on elms, cottonwoods, beech, maple, willow, oak, and occasionally on other hardwoods.

Type of Rot: Brown rot.

Fragrance Signature of Mycelium: Sweetly oyster-like with a flourlike overtone, not like anise, but pleasant and refreshing.

Natural Method of Cultivation: Inoculation of partially buried logs or stumps. I suspect that this mushroom will probably grow in outdoor beds consisting of an equal mixture of hardwood sawdust and chips, much like the garden giant. We have had great success pairing this species with a number of garden vegetables (see pages 192 to 196).

Season and Temperature Range for Mushroom Formation: Temperature: 50–70°F.

Harvest Hints: Harvest when the mushrooms are still convex to broadly convex, before they expand to plane.

Nutritional Profile: Not yet known to this author.

Medicinal Properties: I know of no published studies on the medicinal properties of *H. ulmarius*. Unpublished anecdotal reports suggest it is highly anticarcinogenic. Much of the research has been done in Japan. Traditional Chinese medicinal practitioners recommend it for treating stomach and intestinal diseases. The fact that this mushroom does not spoil as nearly as quickly as many *Pleurotus* species suggests antibacterial properties.

Flavor, Preparation, and Cooking: The same as for most oyster mushrooms. One of my preferred preparations is stir-frying in olive oil with a pinch of rosemary and sliced almonds.

Mycorestoration Potential: An ally for gardens and forests, this mushroom may operate like an endophyte, benefiting the plant with which it associates. As a brown rotter, its production of cellulases may be directed at breaking down cellulose (paper) products and toxins as varied as dioxins and wood preservatives such as chromated copper arsenate (CCA).

Comments: *H. ulmarius* is an excellent edible whose texture and flavor rank it, in my opinion, above all other oyster-like mushrooms, with the possible exception of *Pleurotus eryngii*. This mushroom is friendly to many garden vegetables, unlike its cousin *Pleurotus ostreatus*, which inhibits some plants. Also, *H. ulmarius*'s spore load is substantially less than that of most *Pleurotus* species. This mushroom is one I hope to see move to the forefront of the group of species used to benefit farms and gardens. It is likely to have anti-nematode properties, strong antimicrobial activity, and benefit the immune system. I also think this species is a good candidate to explore for lowering levels of low-density lipid (LDL) cholesterol; given that *Pleurotus ostreatus* has these properties, *H. ulmarius* may also have them.

Inonotus obliquus (Pers.: Fr.) Pilat

Common Names: Chaga, clinker polypore, clinker fungus, cinder conk, tschaga (Russian), tschagapilz (German), kabanoanatake (Japanese), black mass.

Taxonomic Synonyms and Considerations: *Polyporus obliquus* Pers.:Fr. = *Poria oblique* (Pers.:Fr.) Pilat

Introduction: A predominant canker common on birch and rarely on elm, beech, and hornbean. This mushroom has attracted interest for centuries, especially from eastern Europeans and Eurasians, and the northeastern North American First Peoples. Known also for its medicinal properties, this is one of the fire-starting mushrooms, along with *Fomes fomentarius*, that

FIGURE 297

Chaga, the sclerotium of *I. obliquus,* on a birch tree in Quebec, Canada.

allowed for the portability and convenient creation of fire from flint-spark tools. Chaga's availability in the boreal forests gave our ancestors ready access to fire-starting materials, allowed migration into northern regions, and helped our ancestors survive harsh winters.

Description: A flat-pored fungus developing on the surface or below the outer surface of dead trees. Pore surface brittle, dark reddish brown, with circular pores. Flesh bright yellowish brown, corklike. Forming a perennial sclerotium, dark to near black in color, on living trees. Spores transparent to light brown, 9–10 by 5.5–6.5 μm.

Distribution: Circumpolar, widespread throughout boreal deciduous forests, primarily in birch forests.

Natural Habitat: Growing on living and dead birch (*Betula* species), elm (*Ulmus* species), beech (*Fagus* species), and hornbeam or ironwood (*Ostrya* species) trees and stumps, causing a white heart rot. Sclerotia usually found on living birch, beech, elm, ash, and rarely on ironwood; the sporulating surface appears once the tree dies.

Type of Rot: White heart rot of living birch.

Fragrance Signature of Mycelium: Sweet, fungoid.

Natural Method of Cultivation: Inoculation of dying or dead trees, stumps, or logs. Similar to the method shown for reishi in inoculated logs (figure 192). Birch, beech, ironwood, and oak are the best woods for growing chaga, although, as with other mushrooms, many other tree species will probably support growth.

Season and Temperature Range for Sclerotium Formation: During the late spring to early fall. Temperature: 55–70°F.

Harvest Hints: The sclerotium is hard and can grow to nearly the size of a cantaloupe. In most cases, it can be removed from the tree only with considerable effort, often requiring a serrated knife. The spore-producing hymenium—a flattened multipored sheath—is usually difficult to find as it forms remote from sclerotium.

Nutritional Profile: Not yet known to this author.

Medicinal Properties: Researchers have identified an array of medicinally interesting compounds, including antitumor compounds; water-soluble and water insoluble hetero-polysaccharides; protein-bound polysaccharides; lanostane triterpenoids, including inositol (a vitamin B), betulin and analogues; and ergosterol peroxides. More proteins are in the mycelia than are in the fruitbodies. Hypoglycemic effects were measured from the ethanol-soluble fraction (Mizuno et al. 1999). Approved as an anticancer drug (befungin) in Russia as early as 1955 and reportedly successful in treating breast, lung, cervical, and stomach cancers (Hobbs 1995). Ryzmowska (1998) found that the water extract of *I. obliquus* inhibited the growth of cervical cancer cells in vitro. Burczyk et al. (1996) noted that some of this mushroom's constituents had a limiting effect on cell divisions of cancerous cells. Mizuno and colleagues (1996) and Kahlos and others (1996) noted that crude fractions from this mushroom showed antiviral activity against HIV and influenza. Shin and other researchers (2000) and Kahlos and others (1984, 1987, 1990; Kahlos and Hiltunen 1987, 1988) have extensively analyzed this species for its chemical constituents, finding suites of lanostanic triterpenoids, triterpenes, and betulinic acids, many of which are bioactive. Producing a unique melanin complex, these mycelial derivatives demonstrate antioxidant and genoprotective properties (Babitskaya et al. 2002). Recent research from Japan (Ohtomo 2001) shows that this mushroom, like many polypores, has strong immunomodulatory activities, regulating cytokine and interleukin response pathways, and stimulating macrophage and natural killer cell production. Chaga concentrates betulin from the bark of birch trees, just as *Taxomyces andreanae* concentrates taxol from the bark of the Pacific yew tree. Kahlos and others (1996) found that the external black skin of chaga sclerotia has 30 percent betulin, while the internal portions contain fungal lanostanes. This study suggests that teas would be better made from whole chaga, with the outer layer intact. Betulin, sourced from birch bark and/or chaga, has also shown promise in treating malignant melanoma, completely inhibiting tumors implanted in mice and causing apoptosis of cancerous cells (Pisha et al. 1995; Duke 1999).

Clearly, this species is yet to be fully explored by medical scientists for its complete repertoire of antimicrobial and immune-enhancing properties. Likely, many of the precursors responsible for its betulin production have antiviral properties. See also the description of the medicinal properties of *Piptoporus betulinus*.

Flavor, Preparation, and Cooking: Boil in water to create a bittersweet tea. Mushrooms prepared for fire starting have their outer skin removed and then are pounded in order to separate the hyphae, which are then dried into flammable wooly fabric.

Mycorestoration Potential: Possible candidate for preventing disease from more devastating blight organisms. Years ago, I spoke to a chestnut arborist from Quebec who had been able to heal blight-infected trees suffering from *Cryphonectria parasitica* using chaga. His method was simple and credible. He stated that he crushed chaga sclerotia into a powder,

⌃ **FIGURE 298**

Chaga can be ground into a powder and made into an immune-enhancing tea. Chaga tea has been made for centuries in eastern Europe and Eurasia.

➤ **FIGURE 299**

A massive *I. dryophilus* fruits at the base of this otherwise healthy oak. Is it a saprophyte, a parasite, an endophyte, or all of the above? I think this mushroom might prove to be a good medicinal.

adding water to make a thick paste. He placed this chaga paste directly into the lesions caused by chestnut blight and wrapped the wound with gauze to keep the paste in place. Over the next 2 years, the wounds healed over and the trees became blight resistant. I also wonder if *I. obliquus* can behave endophytically like *Fomes fomentarius*, helping to keep rapid forest-destroying diseases at bay.

Comments: Forming annually but lasting perennially, this distinctive fungus causes a black cankerous mass (sclerotium) on birch trees. This mushroom attracted the attention of Eurasians centuries ago (Maret 1991). Used traditionally for treatment of tuberculosis, ulcers, and digestive, heart, and liver cancers, the brittle chaga was stripped of its black outer mass, boiled, and used as a tea. About 3 to 5 g of dried mushroom per pot of tea was used. Once a tea was made, the boiled mass could be pounded into a poultice and applied to prevent infection and help repair cellular damage. Historically, this mushroom also enjoyed a reputation as an analgesic with anti-inflammatory properties.

The chaga sclerotium's compact size allowed for the portability of this natural medicine in ancient times. Chaga was a valuable asset in the pharmacopoeia of premodern peoples as a natural antibiotic, anti-inflammatory, and immunopotentiator, and as a practical fire-starter mushroom. (See also figure 47.)

Laetiporus sulphureus (Bull.:Fr.) Murr.

Common Names: Sulphur tuft; chicken-of-the-woods.

Taxonomic Synonyms and Considerations: For many years, chicken-of-the-woods was thought to be just one species, *Polyporus sulphureus* (*Laetiporus sulphureus*), but now taxonomists recognize that a constellation of species revolve around this bright-colored polypore. The classic *L. sulphureus* is yellow pored, grows on hardwoods, particularly oaks, and occurs in the eastern United States. The conifer-loving western North American form, *L. conifericola* Burdsall and Banik, grows on conifers and is also yellow pored (see figure 67). The eastern and Midwestern North American sulphur tuft, which is white pored and usually grows on the ground around oaks, is *L. cincinnatus* (Morgan) Burdsall.

This cluster of bright sulphur to sulphur orange shelf fungi differ in their geographical ranges, host tree preference, pore color, and genetic markers. To amateurs, they are virtually indistinguishable. The common name sulphur tuft encompasses these species concepts.

Introduction: Known as one of the most recognizable mushrooms, chicken-of-the-woods grows primarily in the summer months in North America. Visible at great distances, this mushroom is commonly found

➤ **FIGURE 300**

L. sulphureus fruiting on a beech stump in Kentucky.

on downed logs criss-crossing canyons along streams and rivers. When the fall rains begin, this mushroom melts into a loose mass before being reabsorbed into its habitat.

Description: An annual shelf-shaped mushroom, forming singly or with overlapping clusters, up to 2 feet in diameter, sometimes with a rudimentary pseudo stem, more often stemless. Brightly colored from yellowish to reddish orange above, and with yellowish pores below. The flesh is white, more pliable toward the margin, and chalkier toward the interior. Spores whitish to hyaline, ovoid to ellipsoid, 5–8 by 4–5 μm.

Distribution: *L. sulphureus* grows in eastern North America and western Europe. *L. conifericola*, its sister species, is found west of the Rocky Mountains, showing a preference for conifers. Sulphur tufts of all kinds are widely distributed throughout the world.

Natural Habitat: First appearing in late spring and growing through early fall, growing primarily on oaks (*Quercus* species). The conifer-loving *L. conifericola* prefers hemlocks.

Type of Rot: Brown cubical butt rot of living or dead trees with whitish, feltlike mycelium occupying the cracks as it digests wood.

Fragrance Signature of Mycelium: Pleasant, reminiscent of butterscotch.

Natural Method of Cultivation: Inoculating stumps or logs with plug or rope spawn. If stumps are checkered with fissures, then I recommend inserting plug spawn directly into the cracks for fast growth.

Season and Temperature Range for Mushroom Formation: Primarily a summer mushroom, fruiting at 60–80°F, maturing late in the season.

Harvest Hints: People differ in their reactions to eating this mushroom. An opinion shared by many is that edibility dramatically declines as the mushrooms age. Young, rapidly developing mushrooms with white margins are the best for eating. If you have enough of the leading edge of the cap margin, I recommend eating just that and discarding much of the remainder. However, I recently had one young cluster, barbecued, which cut and tasted like chicken. If you baste with teriyaki sauce, sprinkle the mushrooms with

▲ **FIGURE 301**

L. sulphureus fruiting on a conifer.

water to increase their moisture, add herbs and spices, and then bake in foil, the palatability of chicken-of-the-woods approaches respectability. However, under-cooked or aged mushrooms can cause consumers unpleasant gaseous gastrointestinal experiences. One mycologist I know served a chicken-of-the-woods soup to party guests, and it sickened all who ate it. This is an all-too-common type of story. I suspect that resident bacteria are the real culprits, unnoticed by most collectors. If you wish to consume this mushroom, choose the youngest, freshest specimens and cook them well. I recommend grilling them before adding to soups, since the temperatures achieved on the grill are much higher than those of boiling water (or soup).

Nutritional Profile: Not yet known to this author.

Medicinal Properties: Suay and other researchers (2000) found that extracts made from the mycelium of *L. sulphureus* had strong effects against the growth of staph bacteria *(Staphylococcus aureus)* and moderate effects against *Bacillus subtilis*. Sulphur tuft mycelium also demonstrated significant activity against the entero-bacterium *Serratia marcescens*, a microbe responsible for urinary tract infections and other infections in people suffering from burns, cuts, or cystic fibrosis.

Flavor, Preparation, and Cooking: Having a nutty fragrance and sometimes a pleasant chickenlike flavor, this mushroom is a favorite of summer mushroom hunters, especially when other edible mushrooms are scarce. I prefer to eat very young specimens, or the leading edge of new growth, discarding the tougher and drier interior regions in a bucket for later spore retrieval.

Mycorestoration Potential: I have never seen a "blight" of chicken-of-the-woods, despite its aggressiveness, nor have I seen the co-occurrence of this species with the *Armillaria* honey mushrooms, a fact that suggests to me that this species could prevent the migration of blights. That it also grows on utility poles treated with chromated copper arsenate (CCA) suggests that its powerful enzymes could be harnessed for breaking down recalcitrant toxins. Once in place, this mushroom grows for many years, its life span limited by the mass of the host tree.

Comments: Chicken-of-the-woods is one of the easiest mushrooms to recognize—delightfully obvious to mushroom hunters accustomed to straining their eyes in search of fungal delights. Causing a brown rot of living and dead trees, species in the taxonomic cluster of sulphur tufts are not rampant parasites; they are localized to trees already stressed or damaged from other causes and grow saprophytically long after the trees have died.

In culture, the mycelium of chicken-of-the-woods easily becomes airborne as cells fragment into short chains—perhaps offering an alternative method for spreading colonies when trees fall and break apart.

Lentinula edodes (Berkeley) Pegler

Common Names: Shiitake, golden oak mushroom, black forest mushroom, black mushroom, oak wood mushroom, Chinese mushroom, shiangu-gu or shiang ku (Chinese for "fragrant mushroom"), donku (Japanese), pasania (Japanese).

Taxonomic Synonyms and Considerations: Berkeley originally described shiitake mushrooms as *Agaricus edodes* in 1877. Thereafter, this mushroom has been placed variously in the genera *Collybia, Armillaria, Lepiota, Pleurotus,* and *Lentinus.* Most cultivators are familiar with shiitake as *Lentinus edodes* (Berk.) Singer. The genus *Lentinula* was originally conceived by Earle in the early 1900s and resurrected by Pegler in the 1970s, defining taxa formerly placed in *Lentinus.* Both genera are characterized by white spores, centrally to eccentrically attached stems, often-serrated gill edges, and a distinct preference for woodland environments. The genera differ primarily in microscopic features. Species of the genus *Lentinula* are monomitic; that is, they lack dimitic hyphae in the flesh, and have cells fairly parallel and descending in their arrangement within the gill trama. Members in the genus *Lentinus* have a flesh composed of dimitic hyphae and have highly irregular or interwoven cells in the gill trama. *L. boryana* is a subtropical species closely related to *L. edodes.*

Introduction: Shiitake mushrooms are a traditional delicacy in Japan, Korea, and China. For at least a thousand years, shiitake mushrooms have been grown on logs outdoors in the temperate mountainous regions of Asia. To this day, shiitakes figure as the most popular of all the gourmet mushrooms. Only in the past several decades have techniques evolved for its rapid-cycle cultivation indoors on supplemented heat-treated sawdust- and straw-based substrates.

Cultivation of this mushroom is a centerpiece of Asian culture, having employed thousands of people for centuries. We may never know who first cultivated shiitake, but the first written record can be traced to Wu Sang Kwuang, born in China during the Sung Dynasty (960–1127 C.E.). He observed that, by cutting logs from trees that harbored this mushroom, one could entice more mushrooms to grow when the logs were "soaked and striked" (see figure 136). In the early 1900s, the Japanese researcher Dr. Shozaburo Mimura published the first studies of the process of inoculating logs with cultured mycelium (Mimura 1904, 1915). Once inoculated, logs begin producing mushrooms 6 months to 1 year later and continue fruiting for years, largely dependent upon the tree type and diameter.

Description: Cap 5–25 cm broad, hemispheric, expanding to convex and eventually plane at maturity. Cap dark

◄ **FIGURE 302**

L. edodes is the most popular wood-decomposing cultivated mushroom in the world.

▲ **FIGURE 303**

L. edodes flushing from eucalyptus logs in Brazil.

brown to nearly black at first, becoming lighter brown in age or upon drying. Cap margin even to irregular, inrolled at first, then incurved, flattening with maturity and often undulating with age. Gills white and even at first, becoming serrated or irregular with age, bruising brown when damaged or with age. Stem fibrous, centrally to eccentrically attached, and tough in texture. Flesh bruises brownish.

Distribution: Limited to the Far East; native to Japan, Korea, and China. Not known from North America or Europe. With the continued deforestation of the Far East, the genetic diversity of this species appears increasingly endangered.

Natural Habitat: This mushroom grows naturally on dead or dying broadleaf trees, particularly the shii tree (*Castanopsis cuspidata*), *Pasania* species, *Quercus* species, and other Asian oaks and beeches. Although occasionally found on dying trees, shiitake is a true saprophyte, exploiting only necrotic tissue.

Type of Rot: White, mottled rot.

Fragrance Signature of Mycelium: Grain spawn has a smell similar to that of crushed fresh shiitake, sometimes slightly astringent and musty. Sawdust spawn has a sweeter, fresh, and pleasing odor similar to that of the fresh mushrooms.

Natural Method of Cultivation: On hardwood logs, especially oak, sweet gum, poplar, cottonwood, alder, ironwood, beech, birch, willow, eucalyptus, and many other broadleaf woods. The denser hardwoods produce for as long as 6 years. The more rapidly decomposing hardwoods have approximately half the life span, or about 3 years. The fruitwoods are notoriously poor for growing shiitake. Although shiitake naturally occurs on dead oaks and beeches, the purposeful cultivation of this mushroom on hardwood stumps in North America has had limited success thus far.

Logs should be incubated without making ground contact, since native soil microbes can compete with the shiitake mycelium. Covering the logs with a tarp, burlap sacks, or shade cloth can aid colonization by limiting the damaging effects of humidity fluctuation. Incubate in shade.

For more information on the cultivation of shiitake on logs, see Fujimoto (1989), Przybylowicz and

▲ **FIGURE 304**

L. edodes fruiting on oak logs in Washington State.

◀ **FIGURE 305**

L. edodes fruiting from an alder log.

Donoghue (1988), Leatham (1982), Komatsu et al. (1980), Kuo and Kuo (1983), and Harris (1986).

Season and Temperature Range for Mushroom Formation: From the spring through early fall. Temperature: 50–80°F.

Nutritional Profile: Our analysis of a 100 g serving shows the following: calories: 356, protein: 32.93 g; fat: 3.73 g; polyunsaturated fat: 1.30 g; total unsaturated fat: 1.36 g; saturated fat: 0.22 g; carbohydrates: 47.60 g; complex carbohydrates: 31.80 g; sugars: 15.80 g; dietary fiber: 28.90 g; cholesterol: 0 mg; vitamin A: 0 IU; thiamine (B1): 0.25 mg; pantothenic acid (B5): 11.60 mg; vitamin C: 0 mg; vitamin D: 110 IU; calcium: 23 mg; copper: 1.23 mg; iron: 5.5 mg; potassium: 2,700 mg; niacin: 20.40 mg; riboflavin: 2.30 mg; selenium: 0.076 mg; sodium: 18 mg; moisture: 9.61 g; ash: 6.13 g.

This species can be a prodigious producer of vitamin D when fresh mushrooms are exposed to sunlight, especially when the gills face the sun during drying. (See page 207.) Indoor-grown mushrooms, when dried in the dark, contain very little vitamin D2 but are high in ergosterols and ergocalciferol, the UV-activated precursors to vitamin D2. For the elderly, or for those living in northern climates who stay indoors, 10 g dry weight or 100 g wet weight of shiitake exposed for 1 day to direct sun could provide nearly enough vitamin D to meet the entire requirement for a week.

Medicinal Properties: The most popular and best-studied medicinal mushroom, shiitake has remained a focus of research since the late 1960s. Lentinan, found in shiitake mushrooms, is a heavy molecular weight polysaccharide (around 500,000 molecular weight), free of nitrogen. According to Mizuno (1995), lentinan has no direct cytotoxic properties but is instrumental in activating a host-mediated response. Macrophages respond to lentinan and, in turn, stimulate lymphocytes and other immune cell defenses. Lentinan is a protein-free polysaccharide, in comparison to flammulin, a protein-rich polysaccharide found in enokitake *(Flammulina velutipes)*. Ng and Yap (2002) found that lentinan is orally active, and they suggest its use as a vaccine in the prevention of tumor development. Sia and Candlish (1999) found that a shiitake extract enhanced the production of normal white blood cells, leading to phagocystosis.

Another extract of shiitake mushroom mycelium, *Lentinula edodes* mycelium (LEM) is an orally active, protein-bound polysaccharide. A water-soluble, lignin-rich fraction from LEM (JLS-18) has been found to have 70 times the in vitro antiviral activity of LEM and activates natural killer cells, T cells, macrophages, and interleukin-6 (Yamamoto et al. 1997). Both LEM and the JLS-18 fraction have strong antitumor properties. Gu and Belury (2005) found that, when comparing 5 of our cold water and ethanol extracts from the living mycelia of *Hericium erinaceus, Grifola frondosa, Ganoderma lucidum,* and *L. edodes,* the *L. edodes* extract caused significant but selective apoptosis to

▲ **FIGURE 306**

Dr. Andrew Weil holds a harvest of our organically grown shiitake.

melanoma cells in vitro without causing harm to non-tumorigenic healthy cells.

In a study by Ghoneum at Drew University, 11 cancer patients with advanced malignancies were treated with an active hemicellulose compound derived from *L. edodes* fermented mycelium and showed significant improvement. Ghoneum (1995) found that arabinoxylane, a derivative of the fermentation of rice by shiitake, turkey tail *(Trametes versicolor)*, and the split-gill mushroom *(Schizophyllum commune)*, increased human natural killer cell activity by a factor of 5 in 2 months. Arabinoxylanes are an enzymatic consequence of the digestion of rice by living mycelium. Composed of xylose and arbinose sugars, they have diverse medical benefits (Hawkins 2001; Mondoa and Kitei 2001). Arabinoxylanes from mushroom-fermented rice also have antiviral effects (Ghoneum 1995, 1998).

That shiitakes yield antiviral compounds has been well documented. An extract from shiitake mycelium has been shown to be effective against herpes simplex virus type 1 (Sarkar et al. 1993). A water-soluble lignin derivative limited HIV replication in vitro and stimulated the proliferation of bone marrow cells (Suzuki et al. 1990). Clinical trials with lentinan in the treatment of HIV patients showed inhibitory activity (Gordon et al. 1998). However, Abrams (2002) found no significant advantage in using lentinan in treating AIDS patients. A serine proteinase inhibitor has been recently isolated from the fruitbodies of shiitake (Odani et al. 1999). This mushroom has also been suggested as a treatment for chronic fatigue syndrome (T. Aoki et al. 1987) and as an overall tonic.

Shiitake has broad antibacterial properties. Antibacterial tests have proven positive against several microbes (Hirasawa et al. 1999). In one study, lentinan was shown to be effective at preventing septic shock (Tsujinaka et al. 1990). Hatvani (2001) found that the cell-free extracts from the liquid fermentation of mycelium significantly inhibited growth of the yeast *Candida albicans* and the bacteria *Streptococcus pyogenes, Staphylococcus aureus,* and *Bacillus megaterium.* Hirasawa also reported activity against *Streptococcus* species.

Flavor, Preparation, and Cooking: Superb fresh or dried, shiitake can be enjoyed in a wide variety of dishes. A traditional Japanese soup recipe calls for slicing the mushrooms and placing them in a preheated chicken broth complemented with chopped green onions and sometimes miso. Shiitake are steeped in this soup broth for a few minutes and served hot. The flavor of slightly cooked shiitake is tart and totally different from the flavor imparted by thorough cooking. Chinese restaurants usually rehydrate dried shiitake and simmer them in the broth of stir-fries. I have found that the flavor is best preserved if the pan is covered during cooking, which minimizes the loss of the aromatic flavors through evaporation (fragrance carries flavor away from the mushrooms). Although others have noted the type of substrate affects flavor, I don't feel confident in stating that this is always the case.

Our family regularly consumes shiitake. Our favorite method of cooking them is to sauté mushrooms that have been torn, not cut. By tearing the mushrooms, cells are pulled apart along cell walls, preserving the flavor within. The stems are first cut off and then the mushrooms are pulled apart, starting from the cut stem base. Olive oil or a similar light oil is added to the wok or frying pan and placed over high heat. Once the oil is hot, add the torn mushrooms, *cover the pan,* and stir frequently. As the mushrooms cook, a more meatlike flavor emerges. Chopped onions, chopped walnuts, sliced almonds, and other condiments can be added as desired. This preparation can be used as a base for many dishes; add the stir-fried shiitake to steamed rice, fish, pasta, chicken, or vegetables. Cultivators like the shelf stability of fresh shiitake, which are less susceptible to spoilage than most other mushrooms. A spicier flavor emerges as shiitake dry.

Mycorestoration Potential: In the recycling of wood debris from forest thinning, this mushroom has huge market appeal and is adaptive to a variety of woods in numerous climates. Furthermore, the enzymes

L. edodes flourishes from one of our mushroom kits. Independent of the number of mushrooms that form, the yield remains about the same. The more mushrooms that are produced, the smaller they are; while conversely, the fewer mushrooms that form, the larger they are. A 5-pound sawdust kit produces about 1 pound of fresh mushrooms. The spent kit can be used as spawn for inoculating logs, or it can be repeatedly sterilized and reinoculated, giving rise to more fruitings, though shrinking each time until expiring.

secreted by this mushroom break down polycyclic aromatic hydrocarbons (PAHs), polychlorinated bi-phenols (PCBs), and pentachlorophenols (PCPs). Hatvani and Mecs (2003) demonstrated that shiitake mycelium could be used remove toxic metals and decolorized industrial dyes from contaminated effluents. Growers of shiitake can harness the enzyme-saturated wood, post harvest, for use in mycofiltration and/or mycoremediation. We have been successful taking spent shiitake sawdust blocks (after fruitings have subsided) and repacking the mycelial sawdust into burlap sacks for mycofiltration. After becoming saturated with effluent and microbes, the burlap sacks often fruit. I encourage shiitake farms growing mushrooms on sawdust to explore the possibilities of recycling their mycelial blocks in burlap for mycofiltration

applications. The subsequent reemergence of the mushrooms is a bonus unexpected by most cultivators.

Comments: I love this mushroom, its emergent forms, and its many flavor dimensions. This is my mainstay gourmet mushroom, one that I never tire of eating despite having access to hundreds of pounds per week. If mushrooms are grown in conditions where humidity fluctuates, the caps often crack, increasing the strong, spicy flavor. Many shiitake varieties—ecotypes—are being cultivated, which differ in their preferences for wood types, time before fruiting, and seasonal preferences. Although the mycelium can be grown on a wide variety of carrier materials, only a few strains will produce, for instance, on cereal straws. Whether the goal is to create mycofiltration membranes, break down toxins, remove heavy metals, or produce food crops or medicines, this species has enough attributes to give mycorestorationists many options.

Macrolepiota procera (Scop.:Fr.) Singer

Common Name: Parasol mushroom.

Taxonomic Synonyms and Considerations: Previously known as *Lepiota procera* (Scop. Ex. Fr.) S.F.G., this species has been the subject of several researchers who have worked to distinguish it from other large *Lepiota*-like species, such as *L. rachodes* Vittad. Vellinga and others (2003) divide these species in 2 major clades (and genera), although they share many macroscopic similarities. A poisonous mushroom, *Chlorophyllum molybdites*, looks similar but differs from the white-spored parasol mushroom described here by producing greenish spores.

Introduction: One of the most majestic of all edible mushrooms, the parasol mushroom can achieve mammoth sizes, commonly more than 1 foot in diameter and standing nearly 2 feet tall. This species is easily naturalized in your backyard and, once resident, appears annually. On our property, satellite colonies have spontaneously appeared in the years after our first flush from a spawned patch. Of all the mushrooms I have grown, this one is my favorite for outdoor cultivation because of its statuesque form, deliciousness, and visibility. This mushroom is one of nature's greatest artistic forms, gracing landscapes with delectable dignity.

Description: Cap 5–50 cm in diameter, spheroid at first, then convex, expanding to broadly convex, and eventually plane in age. Surface smooth when young, soon breaking into darkened wooly to fibrous patches emanating from the center of the cap, which sports a dark-colored umbo. Flesh white or reddish but not staining when cut or bruised. Gills free, close, white at first, becoming dingy-colored with age. Stem relatively slender, 10–60 cm long and 1–2 cm thick, swelling into a bulbous base, adorned with fine brownish scales upon underlying white surface. Partial veil thick, white, floccose; separating and falling into a membranous, persistent ring on the upper regions of the stem. The membranous ring is often movable. Spores white, ellipsoid, 12–18 by 8–12 μm.

Distribution: Widely distributed, common throughout the temperate lowlands of Europe and North America, and naturalizing in non-native regions when planted outdoors. Reported from eastern Canada through New England and south to Mexico, west to Michigan, and from south Arizona westward to southern California. Probably more widely distributed than the literature indicates.

Natural Habitat: A lover of human habits and habitats, parasol mushrooms frequent trails, yards, gardens, and woodlands where the ground is exposed to dappled sunlight and with a minimum undercanopy. This "edge-runner" mushroom loves the margins between field and forest habitats, with a particular affinity for conifers, aspens, and oaks. It loves mulch, and I sense that grass and its thatch encourage primordia to form and develop.

Type of Rot: White rot.

Fragrance Signature of Mycelium: Musty.

Natural Method of Cultivation: I have been successful with 2 methods. The first is to make a 4-inch-deep bed of sawdust and wood chips, using commercial sawdust spawn from Fungi Perfecti at the rate of 5 pounds per 100 square feet. Once spawn has been broadcast, the bed overgrows with grass, which is then cut several times in a season. Subsequent scatterings of wood chips are introduced in the late spring and midsummer. Placing this mushroom in moist, shallow grassy depressions sloping toward

▲ **FIGURE 308**

The stately *M. procera* first fruited from a spawned patch and then spread, with satellite colonies erupting in multiple locations on our property. They grow very quickly. See the next photo for these same mushrooms just a day later.

▲ **FIGURE 309**

One day later, the cap expands, the partial veil becomes a membranous ring on the stem, and the gills flare as white spores are released. A poisonous look-alike, *Chlorophyllum molybdites,* can be deadly but has greenish spores, not white, which makes it easy to distinguish from this mushroom.

watersheds with good exposure to the sun encourages fruitings. Harnessing spores and stem butts for inoculations can greatly expand a few mushrooms into hundreds.

Another fun and entomologically curious method, in which I borrow from my experiences with *Lepiota rachodes* (= *Chlorophyllum rachodes*), is to inoculate thatch ant mounds (*Formica* species), which ants build from conifer needles, thatch, and wood debris. The nests become infused with white mycelium within a few months of inoculation. The mounds fruit a year or more later and are most productive when the ants abandon the nest. In the process, the ants spread mycelium which creates satellite colonies.

Season and Temperature Range for Mushroom Formation: Fruitings typically peak in the fall, from mid-September, extending to December in the more southern regions, fruiting when temperatures hover between 55–70°F.

Harvest Hints: As is the case with most mushrooms, younger ones are better, but older ones retain good flavor even if comparatively drier and not as delectable. Younger mushrooms tend to have few maggots, and the closed caps allow them to be stuffed for cooking.

Nutritional Profile: Chang and Hayes (1978) report that this mushroom has 20 percent crude protein, less than 4 percent fat, 69 percent total carbohydrates, 7 percent glycoproteins, 7 percent fiber, and 12 percent ash.

Medicinal Properties: The mycelium of this mushroom exudes extracellular antibiotics that are effective against *Staphylococcus aureus* and marginally inhibitory to *Enterococcus faecium* (Suay et al. 2000). This author does not know of other medicinal properties.

Flavor, Preparation, and Cooking: We like this mushroom baked, sliced and sautéed, or barbecued. Cook with olive oil. If needed, add a little water to the hot pan when you add the mushrooms; cover the pan while steaming, and then sear the mushrooms until

▲ **FIGURE 310**

Several dozen mushrooms sprouted in front of our laboratories, making for many *M. procera* feasts.

▲ **FIGURE 311**

Parasol mushrooms often come up in groups, usually synchronized in their growth. Such flushes as seen here "captivate" and "bemushroom" all those who discover them.

▲ **FIGURE 312**

M. procera sawdust-chip spawn is spread over a thatch ant mound. Expecting a ferocious reaction, I was surprised when the ants calmly went about distributing the mycelium, unperturbed by my invasive action. Curious.

▲ **FIGURE 313**

Two days later, all the spawn has been incorporated into the nest as part of its architecture, leaving behind the large chips that were in the sawdust-chip spawn. We await the fruitings—how long? Nature knows.

the fringes are browned. Add onions, garlic, and rosemary or other spices near the end of cooking. I also enjoy stuffing young specimens of this versatile mushroom: dice the stems or caps of other mushrooms, mix them with nuts, seasonings, breadcrumbs, or eggs, stuff into the inverted sphere-shaped parasol caps, and bake them in a preheated oven at 375°F for 30 to 45 minutes. Adding soy sauce, tamari, or teriyaki sauce when serving makes my taste buds stand up and shout for more.

Mycorestoration Potential: Few delicious mushrooms grow in grassland thatch mixed through with complex forest debris. I recommend using this mushroom as an ally at the edges of forests, fields, streams, ponds, or estuaries. Also, as an ant-friendly species, this mushroom can be joined with ant colonies to encourage their populations, speeding up the recycling of brush while stimulating growth of these delicious mushrooms. Its full range of antimicrobial properties is yet to be determined. I suspect that this species has great but presently unknown potential.

Comments: The parasol's reputation as an excellent edible extends back centuries. The earliest botanical books that depict mushrooms often feature the parasol as one of the most prominent species, along with shaggy manes, chanterelles, morels, and porcini. As a thatch decomposer, this mushroom has the underexplored ability to adapt to complex habitat mixtures. Its interactions with insect communities, particularly ants, warrant further exploration, since this mutualistic relationship also benefits ecosystems and hungry mycophiles. I wait each fall with anticipation, wondering how massive the next fruiting will be. By feeding the mother mycelium with occasional influxes of debris (wood chips, sawdust, cut grass), we ensure that the beds do not die out. Without renewal, the parasol patch's life span is just a few years before the species moves on. Since each mushroom can generate up to 10 g of spores (which is approximately 10 billion), satellite colonies are common.

Morchella angusticeps Peck and allies

Common Names: The black morel, the conic morel, Peck's morel.

Taxonomic Synonyms and Considerations: Morels remain in a taxonomic morass. The gene sequences of the morel, an Ascomycete, seem more variable than those of better-restricted mushroom species, particularly the gilled and polypored Basidiomycetes. In my cultivation efforts, I have seen a gamut of forms fruiting from a single mushroom patch, convincing me that the morphology of morels is influenced by environmental factors, particularly habitat. Some of the best ongoing work is being reported by Michael Kuo (2004); updates can be viewed on his website at www.mushroomexpert.com, where 14 species are described. A study by O'Donnell and others (2003) found 28 distinct species, 24 of them being endemically unique. *Morchella* are divided into 2 major clades: the yellow-tan-grays and the black morels.

Introduction: In this book, I focus on the morels growing in burns, with which I have had the most experience. These mushrooms—some of the first organisms to appear after a fire—include *M. angusticeps*, the classic black North American morel, and *M. atrotomentosa*, the black-footed morel. Current researchers seek to find taxonomically significant differences between morels.

Morels, *M. angusticeps* and *M. atrotomentosa*, both frequent burned habitats in the first year after a fire. I have cloned morels from burn sites, grown out the mycelium to create spawn, started a bonfire, and, once cooled, mixed in the spawn. Morels then pop up in the spring, but in many cases they look different from one another, even from the same burn pile. This variation in forms shows how mutable the phenotype can be.

I most commonly find black morels in fire pits, and what may be sister species growing amongst landscaping bark, in gardens, near construction sites, and beneath fir trees. Morels growing in these varied habitats all have black ridges, although I do not know if they are different species.

If morel taxonomists were to cultivate morels and fruit them, they would likely be more lenient in their efforts to strictly separate these species by morphology or habitats. I have long been curious about the differences in DNA that are expressed when one phenotype is elicited from the background habitat versus another. Most cultivators know that habitat makeup influences form while few taxonomists have tuned into this relationship. This ongoing debate among taxonomists can be confusing to amateurs, but it may be comforting for you to know that even the experts are still confused. You can help sort it out by sharing your experiences with the leading scientists studying this subject. Michael Kuo has an online submission form at

➤ **FIGURE 314**

The black morel, a complex containing *M. angusticeps* and *M. conica*.

www.mushroomexpert.com for those who wish to send in specimens from their bioregions.

Description: A honeycombed, ribbed species with edges that darken with age. The cap is typically conical in shape, measuring 2–8 cm wide by 2–8 cm high. Stem white, hollow, with a granular texture, measuring 5–12 cm long by 2–4 cm thick. White mycelium is attached to the base of the stem, which is often swollen.

Distribution: Widely distributed throughout the temperate regions of the world.

Natural Habitat: Black morels are common in the spring in burned areas (primarily in the first year after burning), in newly laid landscaping bark, in gardens, near construction sites, and less frequently in conifer forests. I often hear of morels fruiting adjacent to discarded gypsum board or adjacent to cemented walkways.

Type of Rot: Brown rot, I suspect, from my observations of this species decomposing sawdust.

Fragrance Signature of Mycelium: Pleasant, like crushed fresh morel mushrooms.

Natural Method of Cultivation: The method that has produced the most consistent results for me has been to use sawdust spawn mixed into a burn site. I make use

FIGURE 315

Yellow morels, a complex encompassing *M. esculenta* and *M. deliciosa.* The specimens featured here came from an old apple orchard in Washington State.

◄ **FIGURE 316**

This yellow morel *(M. esculenta)* emerged from this sclerotium-like tuber. This mushroom was growing under mature Douglas fir trees near Olympia, Washington. Morels grow from these types of subterranean masses, which can be variable in form.

of a bonfire site—usually a 4- to 10-foot-diameter fire pit—and inoculate it with spawn during the late summer to early fall. After the fire has cooled, I rake out the ground and evenly spread handfuls of spawn over the singed earth. A bag of sawdust spawn weighing 5 pounds effectively inoculates 100 square feet. Additional native burned earth can be laid upon the spawn to a depth of 2 to 4 inches. I also leave chunks of burned wood, which help provide shade. (The morels often form sideways in the shadows.) In March, or 2 months before the natural morel season, I put up a shade cloth to limit sun exposure and increase the number of morels surviving to maturity. Morel spawn can be mixed into equal proportions of sand, ash, gypsum, gravel, and natural (nonpetroleum) charcoal for a similar effect. Success depends upon numerous factors, the strain being the first and foremost.

Once your first flushes come up, stem butts from cultivated *Morchella* are recommended for respawning. Morel spawn grows very quickly, produces a grayish and eventually brownish feltlike mycelium, and then disappears as it coalesces into sclerotia, the "eggs" that sprout into mushrooms. I have found that adding a light sprinkling of native grasses can increase fruiting, and the grasses appear to benefit from contact with morel mycelium.

Season and Temperature Range for Mushroom Formation: Fruiting in the early to late spring, when day to night temperatures fluctuate from near freezing to moderate temperatures. Ecotypes can vary substantially. Temperature: 45–65°F.

Harvest Hints: Black morels are best picked as the ridges darken and spores are released.

Nutritional Profile: 20 percent protein, 5 percent fat, 9 percent fiber, 64 percent carbohydrates.

Medicinal Properties: Morels contain volatile, toxic hydrazines, which are denatured in cooking. Breathing the fumes of cooking morels can be dangerous, perhaps deadly. Duncan and other researchers (2002) isolated a novel galactomannan that enhanced

▲ **FIGURES 317**

Damein Pack, lead grower at Fungi Perfecti, inoculates a bonfire site with morel sawdust spawn.

▲ **FIGURE 318**

When spawned in the summer to fall, black morels usually pop up the following spring.

macrophage response with a molecular weight of nearly 1,000,000 Daltons. To my knowledge, this is the first study showing morels activate a positive immune response. Go morels!

Flavor, Preparation, and Cooking: A superb edible, this mushroom should be thoroughly cooked in a well-ventilated room, because some individuals are sensitive to breathing its volatile vapors or to eating incompletely cooked mushrooms. Morels marry well with cream and butter, making delicious white sauces. Hundreds of recipes using this mushroom have been published. (See cookbooks listed at www.fungi.com.)

Mycorestoration Potential: I believe that morels are a significant species in helping fire-damaged ecosystems recover. When summer forest fires sweep the landscape, the habitat is spottily sterilized. Morel mycelium spreads quickly in this biologically impaired environment, initially encountering little microbial competition. With the onset of rain, the soils' pH quickly neutralize, and, come spring, morels are some of the first organisms to arise from an otherwise barren landscape. As morels mature and swell with water, their scent attracts insects and small and large mammals seeking food. These animals bring seeds from adjacent lands, as well as life-breeding microbes, which they deposit in nitrogen-rich dung, further nourishing the land. We can harness the restorative power of this mushroom by deliberately spreading native black morels and pairing them with grasses and other plants, improving the rate of recovery.

Comments: Morels remain one of the most mysterious of all mushrooms and have captured the imagination of collectors, artists, writers, and scientists for centuries. Their distinct forms, fantastic flavors, and spring season appearances have helped to popularize them as a preferred gourmet mushroom. Mushroom hunters can depend on finding native morels, peculiar to certain habitats, in their bioregion.

The black morels are special in that they are one of the very few fleshy mushrooms to grow in burned landscape. Several cup fungi—orange peel (*Aleuria aurantia* and allies), mo-er (*Auricularia* species and *Peziza* species)—pop up, but these are thin, nonfleshy mushrooms. In a sense, morels lead the charge in habitat renewal. Although they fuel the food chain, I know of no formal studies on their role in recovering ecosystems.

J. L. Dahlstrom and others (2000) studied whether or not morels can form mycorrhizae with trees. Although these mushrooms do not form mycorrhizae in the classic sense, they observed that morel mycelium penetrated into the feeder roots of pine, larch, and Douglas fir seedlings and facultatively behaved like mycorrhizae by benefiting the hosts. Although clearly a

▲ FIGURE 319

This field of *M. esculenta* is a find of a lifetime and a dream come true!

saprophyte, morels might grow endophytically within the roots of many forest plants. If either model is true, this could better explain why morels suddenly appear in the ashy remains of forests where they had previously never been seen. Perhaps morel mycelium infiltrates the root zones of most conifer forests in western North America, remaining resident but unseen for years, only to emerge in its fleshy form after a fire.

For more information, I recommend Michael Kuo's book *Morels* (2005).

Pholiota nameko (T. Ito) S. Ito et Imai in Imai

Common Names: Nameko, slimy Pholiota.

Taxonomic Synonyms and Considerations: Many species in this genus sport a glutinous, slimy veil, coating the cap's surface and producing dull brown spore deposits. Three species are notably delicious. The most popular *Pholiota* is *P. nameko*—synonymous with *P. glutinosa* Kawamura. Formerly placed in *Collybia* and *Kuehneromyces*, this mushroom has a smooth cap and a thick glutinous veil covering it. A pair of scalier cousins, *P. squarrosoides* Peck and *P. squarrosa-adiposa*, are also edible and can be cultivated using the parameters outlined here. All produce rhizomorphs at the base of their stems that regrow upon replanting.

Introduction: Nameko is one of the most popular cultivated mushrooms in Japan, ranking close behind shiitake and enokitake. This mushroom has an excellent flavor and texture. Cultivators growing *Pholiota* mushrooms outdoors run the risk of misidentification with toxic wild species such as *Galerina* spp. or even *Hebeloma* spp. Although *Pholiota* mushrooms produce dull brown spore prints and the deadly *Galerina* species emit rusty brown spores, the color differences may be subtle to the untrained eye. A mistake in identification can be fatal. Other features also distinguish the differences between these genera. Those not knowing how to distinguish *Pholiota* from *Galerina*, *Hebeloma*, or other toxic mushrooms should avoid cultivating any

potential look-alikes. Refer to field guides for more information and consult a qualified mycologist if there is any chance of misidentification. That being said, this is one of my favorite mushrooms.

Description: Cap 3–8 cm, hemispheric to convex, and eventually plane. Surface covered with an orangish, glutinous slime, thickly encapsulating the mushroom primordia, thinning as the mushrooms mature. The slime quickly collapses, leaving a viscid cap. Cap surface smooth. Gills white to yellow, becoming brown with maturity. Partial veil glutinous-membranous, yellowish, adhering to the upper regions of the stem or

▲ **FIGURE 320**

The estimable Albert Bates of Mushroompeople holds one of our *P. nameko* kits. Once harvested, the mycelium in the kit is still alive and will regrow if placed onto newly cut wood. Additionally, the stem butts of the mushrooms can be used for growing more mycelium.

along the inside peripheral margin. Stem 5–8 cm long, equal, covered with fibrils and swelled near the base, to which clusters of whitish to golden rhizomorphs are attached. Spores cinnamon brown to dull brown, 4–7 by 2.5–3 μm.

Distribution: Common in the cool, temperate highlands of China and Taiwan, and throughout the islands of northern Japan. Not yet known from Europe or North America.

Natural Habitat: In the temperate forests of Asia, growing on stumps of broadleaf trees, especially oaks (*Quercus* species) and beech (*Fagus crenata*).

▲ **FIGURE 321**

Six months after inoculating, *P. nameko* fruits from partially buried alder logs spawned from a spent mushroom. Once established, fruitings of this mushroom recur for a long time.

Type of Rot: White.

Fragrance Signature of Mycelium: Musty, farinaceous, not pleasant.

Natural Method of Cultivation: On logs of broadleaf hardwoods, especially beech, poplar, alder, aspen, oak, eucalyptus, and probably many other woods. This mushroom does particularly well on stumps and on log rafts. I inoculated alder logs using sawdust spawn packed into logs scarred with a chain saw and laid wood chips atop them. The logs produced bountiful crops in the fall for more than 5 years, pulping the wood into long, fibrous, whitish strands. Stumps will fruit for much longer. Recommended inoculation methods include plug, sawdust, rope, and stem butt spawn.

Season and Temperature Range for Mushroom Formation: Fruits in the fall. Temperature: 50–70°F.

Harvest Hints: Pick the mushrooms while the caps are still covered with a veil, or soon after opening—a time when the caps are unusually convex in shape. You may want to wear gloves when picking these slippery mushrooms, since their slimy overcoat makes them difficult to grasp. Also, the glutinous slime acts like an adhesive, picking up any debris that you touch while picking mushrooms.

Nutritional Profile: Our analysis of a 100 g serving shows the following: calories: 364; protein: 33.65 g; fat: 3.91 g; polyunsaturated fat: 1.01 g; total unsaturated fat: 1.29 g; saturated fat: 0.17 g; carbohydrates: 48.36 g; complex carbohydrates: 29.26 g; sugars: 19.10 g; dietary fiber: 28.10 g; cholesterol: 0 mg; vitamin A: 0 IU; thiamine (B1): 0.28 mg; pantothenic acid (B5): 17.50 mg; vitamin C: 0 mg; vitamin D: 38 IU; calcium: 18 mg; copper: 1.6 mg; iron: 16 mg; potassium: 2,500 mg; niacin: 106 mg; riboflavin: 3.06 mg; selenium: 0.103 mg; sodium: 4 mg; moisture: 8.50 g; ash: 5.38 g.

Medicinal Properties: Not yet known to this author.

Flavor, Preparation, and Cooking: A *very* slimy mushroom, nameko is easily diced into miniature cubes and can be used imaginatively in a wide variety of dishes, from stir-fries to miso soups. Once the glutinous slime is cooked away, the mushroom becomes quite appetizing, having a crunchy texture and nutty/mushroomy flavor. This is a great edible. Although when lightly cooked this mushroom is pleasantly satisfying, I prefer the strong nutty flavor that results from thorough cooking.

Comments: This group of slimy *Pholiotas*—*P. nameko* and species in the *P. squarrosoides* complex—are superb allies for the connoisseur mushroom landscaper. Once they are skilled in the art of identification and can distinguish *P. nameko* from other *Pholiota* species and particularly the *Galerina* species, mycophiles can enjoy the fruits from this species for years after it has taken up residence. The deadly *Galerina autumnalis* (see figure 287) is also orange and has brown spores and a ring. *P. nameko* can share these general characteristics. A misidentification can have deadly consequences. Be aware that sometimes native *Galerina* mushrooms pop up in beds of wood chips that have been spawned with other mushrooms. Be careful!

One appealing feature of these *Pholiotas* is their tendency to produce large clusters, making them visible from afar. The same methods described for growing *P. nameko* can be applied for growing other *Pholiota* species, including *P. squarrosoides* and *P. squarrosa-adiposa*. All are delightfully visual and most are deliciously edible. For more information, consult David Arora's *Mushrooms Demystified* (1986) and Alexander Smith and L. R. Hesler's *The North American Species of Pholiota* (1968).

Piptoporus betulinus (Bull.:Fr.) Karst.

Common Names: Birch polypore, birch conk, kanbatake (Japanese).

Taxonomic Synonyms and Considerations: Also known as *Polyporus betulinus* (Bull.: Fr.) Fr., the birch polypore is easy to recognize. A related oak-degrading species, *Piptoporus soloniensis* (Dub.:Fr.) Pil, grows in the southeastern United States and central Europe. *Piptoporus quercinus* (Schrad.) Pilat grows in central Europe (but not in North America) on oaks, as its name implies, but is rare.

Introduction: One of the premier polypore mushrooms in the boreal forests of North America, Europe, and Eurasia, this species prefers birches and is one of the most common polypores found in such forests, along with *Fomes fomentarius* and *Inonotus obliquus*. Prehistoric humans found this mushroom useful for multiple purposes. When cut thinly and dried, the paperlike slices are flammable. They can also be used as a bandage to stop bleeding and prevent infection, and they have anti-inflammatory properties.

Description: Cap 2–30 cm broad, conk-shaped, convex expanding to broadly convex, and nearly plane; smooth, fleshy, with a distinct incurved margin and recessed, easily separated pore layer; sometimes mildly

◀ **FIGURE 322**

P. betulinus fruiting on a paper birch *(Betula papyrifera)*.

tough, but often with a pliable, leathery, smooth suedelike outer covering that can be dented when touched or contacted during growth. When cut, the flesh is white, marshmallow-like. Spores white in deposit, 5–6 by 1.5–2 μm.

Distribution: Throughout the temperate regions of the world wherever birch forests occur.

Natural Habitat: Exclusive to paper birch *(Betula papyrifera)* and yellow birch *(Betula alleghaniensis)*. Although inoculations have produced conks on pine, spruce, and poplar, this mushroom does not naturally occur upon them. Even though it is thought to be a parasite by conventional foresters, it is not very aggressive. The birch polypore is more common in birch forests that are climaxing than in younger ones.

Type of Rot: Brown cubical rot of the sapwood. White mycelial mats forming in the cracks of wood, often yellowish brown at interface.

Fragrance Signature of Mycelium: Rich, fungoid, pleasant.

Natural Method of Cultivation: Preferred method is plug spawn of dead trees or stumps, particularly birch; although alder can be used, it is less successful.

▲ **FIGURE 323**

Dr. Christian Ratsch photographs *P. betulinus* in Germany.

Season and Temperature Range for Mushroom Formation: Primarily a summer mushroom, but forming from May through November. Temperature: 50–90°F.

Harvest Hints: Best harvested midsummer to early fall. The spore-bearing pores can be harvested, air-dried, and then used as a spore inoculum. Mushroom fruitings from previous years, usually bug ridden, sometimes persist for several years after formation.

Nutritional Profile: Not yet known to this author.

Medicinal Properties: This fungus is useful for stopping bleeding, preventing bacterial infection, and acting as an antimicrobial agent against intestinal parasites. Capasso (1998) postulated that the famous 5,000-year-old ice man (nicknamed "Otzi") found on the slopes of the Alps in the fall of 1991 may have used this fungus to treat infection from intestinal parasites *(Trichuris trichiura)*. The novel antibiotic piptamine has been isolated from this fungus (Schlegel et al. 2000). Suay and others (2000) tested the extracellular metabolites from this species grown in culture, finding them effective in retarding the bacterium *Bacillus megaterium*. I suggest this mushroom as a candidate for studying its capacity to fight anthrax *(Bacillus anthracis)*.

Kanamoto and others (2001) noted that betulinic acid derivatives from *P. betulinas* showed unique antiviral activity against HIV, blocking viral reproduction. After I sent proprietary mycelial extracts of my strain of this mushroom to researchers as part of the U.S. Biodefense program, scientists associated with the U.S. Army Medical Research Institute of Infectious Diseases (USAMRIID) and the National Institutes of Health (NIH) reported to me that my extracts of this polypore selectively killed the virus without harming healthy human cells. (They used the vaccinia virus and cowpox, instead of smallpox, for reasons of obvious safety concerns.) In January 2004, I filed a patent on this novel discovery.

Manez and colleagues (1997) found that selected triterpenoids from this species reduced chronic dermal inflammation. Kamo and others (2003) also noted its anti-inflammatory effects. Pisha and other researchers (1995) found, in studies using mice, that betulinic acid, a pentacyclic triterpene, was specifically toxic to melanoma without causing adverse effects to the host. Pisha and others (1995) found that betulinic acid facilitated apoptosis of melanoma. This compound has been further evaluated for the treatment or prevention of malignant melanoma.

Kawagishi and others (2002) reported on novel enzyme-inhibiting properties of this mushroom, factors affecting tumor development.

Flavor, Preparation, and Cooking: When young, the birch polypore can be thinly sliced, boiled, and added to soups, imparting a pleasant, mild flavor. Once plucked from its tree host, the mushroom has a shelf life of only 2 to 4 days before souring when stored at room temperature.

Mycorestoration Potential: Potentially protective from invasive, more aggressive parasites. Attracts numerous beetle species, which can subsequently attract birds, including woodpeckers. An interesting species with complex interactions with other forest organisms, this mushroom has mycorestorative properties that have not yet been explored. If this species initially acts like an endophyte, forestalling more pernicious blights, then birds, particularly woodpeckers, may help trees survive by inoculating them with the birch polypore. Now that the birch-growing *Fomes fomentarius* is a known endophyte and it is recognized that woodpeckers transfer its mycelium, the case can be made that woodpeckers may be involved in a complex mutualistic relationship with this fungus. I hypothesize that this may also be true of birds' relationships with *P. betulinus*.

Comments: This species is one of the most prominent and frequently seen mushrooms on birch. Discovered with the famous 5,000-plus-year-old ice man found in the Alps, *P. betulinus* and its uses transcend cultures and millennia. Forest peoples have long recognized this species as an aid for survival. The birch polypore

will continue to be an important medicinal species in our fungal armamentarium for fighting viral, bacterial, and other immunological diseases. I believe that its genome should be carefully surveyed for new antimicrobial and especially viral agents.

Pleurotus eryngii (De Candolle ex Fries) Quelet sensu lato

Common Names: The king oyster, boletus of the steppes, cardoncello (Spanish).

Taxonomic Synonyms and Considerations: Synonymous with *Pleurotus fuscus* (Batt.) Bres. Bresinksy and others (1987) commented upon varieties specific to ecological niches and, although the mushrooms may appear morphologically identical, the distribution of these ecotypes is quite distinct. A variety called *P. eryngii* var. *nebrodensis* (Inz.) Sacc., white in form, has an excellent flavor, a stout form, and long shelf life. Some refer to this mushroom as a new species, *Pleurotus nebrodensis*. This "nebrodensis" was thought be indigenous to Spain until Venturella (2002) determined that a new taxon, *P. eryngii* var. *elaeoselini*, merited publication, and that others had inappropriately

◄ **FIGURE 324**

A wild *P. eryngii* attached to the base of a carrot plant.

applied the name *P. nebrodensis* to this mushroom. Another mushroom thought to be a variety of *P. eryngii* (*P. eryngii* var. *ferulae*), found on the host *Ferula communis*, a giant fennel, was determined by Urbanelli and others (2002) to be a genetically distinct species. In nature, these varieties are terrestrial, growing saprophytically in association with members of Umbilliferaceae, the carrot family. This cluster of varieties and/or closely related species produces some of the most delectable of all the oyster mushrooms.

Introduction: The king oyster, *P. eryngii*, is by far the best-tasting oyster mushroom, well deserving of its name. Popular in Europe and becoming increasingly so in Asia, this stout, thick-fleshed mushroom is one of the largest species in the genus. Preferring hardwoods, this mushroom is easy to grow. Although it produces on cereal (wheat) straws, the yields are not as substantial as those of *P. ostreatus* and *P. pulmonarius* on this same material at the same rate of spawning unless supplements are added. This mushroom prefers wood to straw as a substrate for fruiting.

Description: Cap 3–12 cm in diameter, at first convex, expanding with age, becoming funnel shaped, with the margin typically in-rolled, extending with age. Stem 3–10 cm in length, central, thick, tapering downward. Gills fairly distant, thin, grayish, and decurrent. Mushrooms light brown to whitish in color, depending upon varieties and growth conditions. Forming individually or in small groups. Cultivated mushrooms achieve a greater stature and overall size than ones collected in the wild.

Distribution: Throughout southern Europe, North Africa, central Asia, and southern Russia. Not yet reported as native to North America.

Natural Habitat: This mushroom is a saprophyte—some postulate a facultative parasite—on dying and dead *Eryngium campestre* and other *Eryngium* species, members of the carrot family. Zervakis and others (2001) isolated 5 distinct strains of *P. eryngii* from other terrestrial plants, including *Ferula*

communis, *Cachrys ferulacea*, *Thapsia garganica*, and *Elaeoselinum asclepium*.

Type of Rot: White.

Fragrance Signature of Mycelium: Rich, sweet, and classically oysteresque but without overtones of anise.

Natural Method of Cultivation: Outdoors, on partially buried logs and on stumps inoculated with plug, sawdust, or rope spawn. Outdoors this mushroom tends to form at the interface of stump and ground. Mycelium can be run on cereal straws but fruitings are sparse in comparison to those on wood-based substrates.

Season and Temperature Range for Mushroom Formation: Growing in the late summer to early fall in Spain and across southern Europe to Italy, in the Mediterranean region. Temperature 60–70°F.

Nutritional Profile: Not yet known to this author.

Medicinal Properties: Sano and others (2002) found that ethanol extracts of this mushroom had anti-allergenic and anti-inflammatory effects when tested

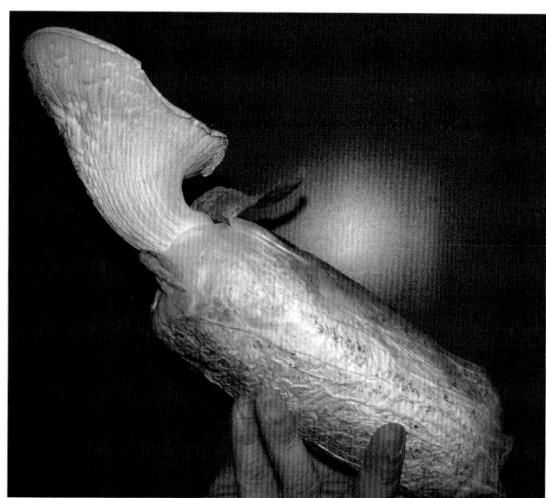

▲ **FIGURE 325**

P. eryngii var. *nebrodensis,* a white variety of the king oyster, is increasing in popularity as a premier gourmet mushroom.

on mice exposed to oxazolone to induce an allergic reaction. Yaoita and others (2002) isolated a novel sterol, potentially having anti-inflammatory activities, from this mushroom.

This mushroom may regulate lipid metabolism in the liver, preventing the hyperaccumulation of low-density cholesterol, as has been reported for *P. ostreatus*. However, to the best of my knowledge, no studies have yet been done on the cholesterol-lowering effects of this species.

Flavor, Preparation, and Cooking: Stir-fried until the edges become crispy and golden brown, the king oyster is far superior to *P. ostreatus* and *P. pulmonarius*. This mushroom has a strong, delicious nutty flavor, crunchy texture, and excellent sliceability, and it stores well under refrigeration. Its short gills and comparatively low spore load extend shelf life.

Mycorestoration Potential: Produces powerful enzymes, similar to those of *P. ostreatus*, that degrade a variety of toxins. Rodriguez and colleagues (2004) used the extracellular metabolites in vitro, isolating a new laccase from this mushroom to effectively degrade 2,4-dichorophenol (the base for Agent Orange, a source of dioxins; currently used in weed killers and pesticides and thought to cause nerve damage and cancer) and benzo(a)pyrenes (cancer-causing agents from petroleum products). Because this mushroom does better in a terrestrial setting than do most species from the genus *Pleurotus*, *P. eryngii* might be better applied to the degradation of surface soils containing these types of toxins, especially in the construction of mycofiltration buffers or overlaid as mycoremediation membranes.

Comments: As a fungal ally, the king oyster satisfies several important needs. Its reputation as a gourmet mushroom has made it one of the most popular in the world. Gary Lincoff, renowned mycologist and author of *The National Audubon Society Field Guide to Mushrooms* (2000), stated that during a mycophiles' tour of Europe, culinary professionals traveling with

him rated this mushroom one of the highest. New strains such as *P. eryngii* var. *nebrodensis* are even more flavorful and becoming increasingly popular in Asia. That this species can thrive terrestrially, in contact with complex soil microbes, suggests that it might be more useful than *P. ostreatus* for in-ground placement in areas where destruction of toxins is required, and where *P. ostreatus* would face microbial obstacles. For a mushroom to degrade toxins and yet retain its deliciousness is a unique combination of talents. However, we do not yet know whether this species bioaccumulates heavy metals or other toxic substances from the soils in which it grows. For this reason, I recommend that you do not ingest mushrooms from toxic lands until analytical studies have established their safety.

Pleurotus ostreatus (Jacquin ex Fries) Kummer

Common Names: The oyster mushroom, oyster shelf, tree oyster, straw mushroom, hiratake (Japanese), tamogitake (Japanese).

Taxonomic Synonyms and Considerations: First described in 1774 from Austria by Jacquin, and in 1871 by Kummer, no type specimens survived. *Pleurotus ostreatus* represents a huge complex of subspecies, varieties, and strains. Since no type collection exists, the species concept has been reconstructed based on newer isotypes. Petersen and Krisai-Greilhuber (1996), with help from Vilgalys and Sun (1994), came to the rescue by establishing a new reference collection, an epitype, which was then used for interfertility studies. There are 2 major ecotypes in *Pleurotus*—ones from North America (brown forms) and ones from Europe (blue or brown forms). The closely related *P. populinus* and *P. pulmonarius* can be difficult to distinguish from *P. ostreatus*. For an excellent description of the history, taxonomy, and delimiting features of *Pleurotus*, see the University of Tennessee's Mycology Lab

△ FIGURE 326

The classic oyster mushroom *(Pleurotus ostreatus)*, fruits from a cut face of a log.

△ FIGURE 328

Grey and pink oyster mushrooms fruiting from a straw-stuffed chair.

◄ FIGURE 327

Oyster mushrooms fruiting on a dead tree.

website, http://fp.bio.utk.edu/mycology/Default.htm, which is updated as new data becomes available.

Introduction: Of all the cultivated mushrooms in the world, *P. ostreatus* is the easiest to grow. This species adapts to such a spectrum of substrates as to boggle the mind. So many plant products, from paper, straw, wood, seeds—even water hyacinths—are appetizing to oyster mushroom mycelium. Oyster mycelium can digest 5 pounds of wood, reducing it to less 50 percent of its mass, in a few months. The by-products of this decomposition—water, carbon dioxide, enzymes, alcohols, carbohydrates, and mycelial mass—benefit the food chains of other organisms. Oyster mushroom cultivation can help alleviate poverty and hunger by recycling waste materials of little economic value and turning them into nutritious and medicinally beneficial products. The oyster mushroom's many by-products can also be harnessed to help heal environments contaminated by a wide assortment of pollutants.

Description: Cap convex at first, expanding to broadly convex, eventually flat and even upturned in age, 5–20 cm or more in diameter. White to yellow to grayish yellow to tan, rarely with pinkish tones, to lilac gray to gray brown. Cap margin smooth to undulating like an oyster shell. Color varies according to the strain, lighting, and temperature conditions. Stems are typically eccentrically attached to the cap. Flesh generally thin. Some strains form clusters; other strains produce individuals. Spores whitish in deposit, ellipsoid, 7–9 by 3.5–5 μm.

Distribution: Many varieties are distributed throughout the hardwood and fir forests of the world.

Natural Habitat: Common on broadleaf hardwoods in the spring and fall, especially cottonwood, oak, alder, maple, aspen, ash, beech, birch, elm, willow, and poplar. One ecotype, occurring on conifers (*Abies* species), has been variously described as a variety of *P. ostreatus*, or as a separate but closely related species, *P. pulmonarius*. Oyster mushrooms have been reported from other conifer species but are not as common as those found on deciduous woods. Although sometimes found on dying trees, *P. ostreatus* is thought to be a saprophyte; but it can become a facultative parasite when the host is stressed, capitalizing on the newly generated dead tissue. Occasionally occurring on composting bales of straw, and in Mexico on the waste pulp from coffee production (see figure 332). The most abundant fruitings of this species occur in low valleys and riparian habitats.

Type of Rot: White.

Fragrance Signature of Mycelium: Sweet, rich, pleasant, distinctly aniselike and almost almondlike.

Natural Method of Cultivation: Outdoors on logs or stumps inoculated with spawn from spores, or with sawdust, corncob, dowel-plug, or rope spawn. We have also had good experiences using spent straw (used to produce oyster mushrooms), which was then sandwiched between bales of straw or burlap sacks. Pagony (1973) reported that, on average, more than 1 pound of mushrooms per year was harvested from inoculated poplar stumps for more than 3 years. Of the 200 poplar stumps, ranging in size from 6 to 12 inches in diameter, which were inoculated in the spring, *all* produced mushrooms by the fall of the following year. As expected, hardwoods of greater density, such as oak, take longer to produce but sustained flushes for many more years. Girdling and using rope spawn in the cut groove is an effective method of inoculation.

Season and Temperature Range for Mushroom Formation: Most prolific in spring; strains in western North America that grow in the fall tend to produce late in the season. Temperature: 40–75°F.

Harvest Hints: Oysters are best harvested before the caps flatten and heavily sporulate, usually when the mushrooms are convex in shape. Most strains in the wild produce only a few mushrooms in clusters. I prefer cluster-producing strains, because a sizable harvest of these young, succulent mushrooms can be picked at once. As spores mature, edibility declines and perishability increases.

Nutritional Profile: Our analysis of a 100 g serving shows the following: calories: 360; protein: 27.25 g; fat: 2.75 g; polyunsaturated fat: 1.16 g; total unsaturated fat: 1.32 g; saturated fat: 0.20 g; carbohydrates: 56.53 g; complex carbohydrates: 38.43 g; sugars: 18.10 g; dietary fiber: 33.40 g; cholesterol: 0 mg; vitamin A: 0 IU; thiamine (B1): 0.16 mg; pantothenic acid (B5): 12.30 mg; vitamin C: 0 mg; vitamin D: 116 IU; calcium: 20 mg; copper: 1.69 mg; iron: 9.1 mg; potassium: 2,700 mg; niacin: 54.30 mg; riboflavin: 2.04 mg; selenium: 0.035 mg; sodium: 48 mg; moisture: 6.73 g; ash: 6.74 g.

▲ **FIGURE 329**

Oyster mushrooms fruiting from inoculated logs placed over an irrigation slough. Here, wedges were cut out, the cavities were packed with sawdust spawn, and the wedges were reinserted.

▲ **FIGURE 330**

Oyster mushrooms fruiting from rope. This rope spawn is ideal for inoculating notched stumps.

▲ **FIGURE 331**

David Brigham inserts rope spawn of oyster mushrooms into a groove made by a chain saw.

Medicinal Properties: Studies (Gunde-Cimerman 1999; Gunde-Cimerman and Cimerman 1995; Gunde-Cimerman and Plemenitas 2002; Bobek et al. 1998) show that *P. ostreatus* and other closely related species naturally produce isomers of lovastatin (3-hydroxy-3-methylglutaryl-coenzyme A reductase), a drug approved by the FDA in 1987 for treating excessive blood cholesterol. Constituents similar to lovastatin are present in higher numbers in the caps than in the stems and are more concentrated on the mature gills, especially in the spores. One model showed that plasma cholesterol turnover was significantly enhanced by 50 percent, with a corresponding 25 percent decrease in the liver compared to the controls (Bobek et al. 1995). This accelerated plasma turnover of cholesterol resulted in an overall reduction beyond baselines. This family of compounds may explain the often-reported cholesterol-reducing effects of many woodland mushrooms (Bobek et al. 1998, 1999).

When mice were implanted with sarcoma 180 and oyster mushrooms constituted 20 percent of their daily diet, the tumors were inhibited by more than 60 percent after 1 month compared to the controls (Ying 1987). In another study, when rats were fed a diet composed of 5 percent oyster mushrooms and administered dimethylhydrazines to induce tumors, fewer tumors formed than in the controls. In this study, Zusman and others (1997) found that when rats were given corncobs colonized by oyster mushrooms, they were significantly protected from chemicals that otherwise induced colon cancer, reducing incidence from 47 percent to 26 percent; corncobs without mycelium provided no protection. A lectin from the fruitbodies of oyster mushrooms, when injected into mice, showed potent activity against implanted tumors of sarcoma 180 and hepatoma H-22. Bobek and Galbavy (2001) have identified a novel beta-glucan, pleuran, which has antioxidant effects and may help prevent cancers from metastasizing.

Wang and Ng (2000) identified an ubiquitin-like protein from oyster mushrooms that inhibits HIV-1 reverse transcriptase activity, causing cleavage of transfer RNA. This unique form of ubiquitin appears to govern cell division, inhibiting cells that are infected with HIV. Piraino and Brandt (1999) have also identified an ubiquitin from *P. ostreatus* that is useful as an antiviral.

In 2004, we grew several hundred pounds of oyster mushrooms for the first clinical study using medicinal mushrooms in the United States. This small clinical study, entitled "Antihyperlipidemic Effects of Oyster

Mushrooms," is being led by Dr. Donald Abrams, chief oncologist of San Francisco General Hospital. As described on the clinical trial pages of the National Institutes of Health (Abrams 2004), "The primary goal of this study is to evaluate the short-term safety and potential efficacy of oyster mushrooms *(Pleurotus ostreatus)* for treatment of hyperlipidemia in HIV-infected patients who are taking Kaletra, a protease inhibitor (PI) that is commonly used in highly active antiretroviral therapy (HAART)."

An issue with protease inhibitors being used by AIDS patients is that this antiviral medicine interferes with lipid metabolism in the liver, causing the hyper-accumulation of low-density lipoproteins (LDLs) and leading to clogged arteries, heart disease, and disfigurement. Should oyster mushrooms help remodulate liver metabolism while patients are undergoing treatment with protease inhibitors, then this species may become a recommended adjunct to HIV therapy. Many other species of *Pleurotus* are also likely to contain these liver-remodulating compounds.

Not all properties of oyster mushrooms are beneficial. Workers at commercial oyster farms commonly report allergic reactions to the spores of *P. ostreatus*. Symptoms include fever, headache, congestion, coughing, sneezing, nausea, and general malaise (Kamm et al. 1991; Horner et al. 1993). Workers, who may at first tolerate contact with oyster spores, often develop increased sensitivity with continued exposure. Comparatively few individuals are allergic to oyster mushrooms after they have been cooked. For more information, consult Reshef and colleagues (1984) and Mori and others (1998).

Horner and others (1993) found that, in a comparative study of 701 patients, approximately 10 percent of Americans and Europeans showed an allergenic response to extracts of *P. ostreatus*, while *Psilocybe cubensis* triggered the highest allergenic response for Americans at 12 percent and 16 percent for Europeans.

Flavor, Preparation and Cooking: Oyster mushrooms have a mild, nutty flavor that becomes stronger the longer they are cooked. I prefer a general sauté in olive oil, browning the mushrooms' gills and cap edges, adding herbs (rosemary, thyme, oregano, and basil) toward the end. Alternatively, basting the gills with an herbal-tamari concoction and then broiling them for several minutes results in a sumptuous meal.

Mycorestoration Potential: Oyster mushrooms have a demonstrated ability to break down petroleum-based pollutants, particularly the polycyclic aromatic hydro-carbons, the core molecules within oil, diesel, pesti-cides, herbicides and many other industrial toxins. Their adaptive ability is nothing short of amazing. My work with the Battelle Pacific Northwest Laboratories in Sequim, Washington, showed that *P. ostreatus* produces powerful denaturing enzymes that could dis-mantle several recalcitrant industrial toxins, leading the way to habitat restoration (Thomas et al. 1999). These activities were later confirmed by the work of

▲ **FIGURE 332**

Dusty Yao holding a bucket of oyster mushrooms fruiting from coffee grounds. Coffee grounds from espresso are essentially steam pasteurized, so are perfect for growing oyster mush-rooms. The mushrooms break down the caffeine, an important effect that can help prevent caffeine toxicity from the runoff from coffee plantations.

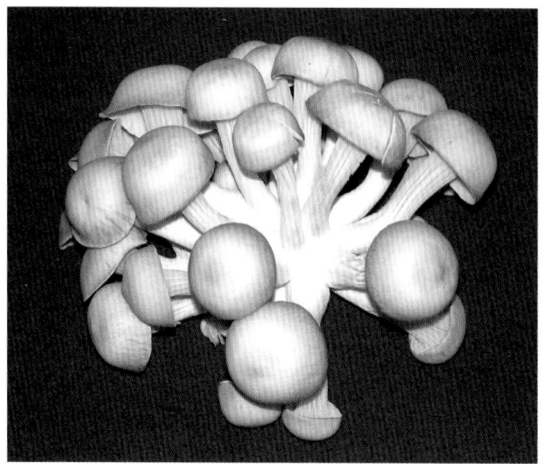

⬛ **FIGURE 333**

A cluster of a sporeless strain of *P. ostreatus*. This unique strain has nude gills, free of any hymenium or basidia (see page 14).

European researchers, including but not limited to Bhatt and others (2002), Cajthaml and others (2002), and Eggen and Sasek (2002), who also found that "spent compost" worked better at breaking down toxins than fresh, pure culture spawn, a discovery that has broad implications for the value-added use of the waste substrate coming from oyster mushroom farms. Novotny and others (2001, 2003) showed that oyster mushrooms also decolorize industrial dyes.

Also, noteworthy is that oyster mushrooms are selective in their absorption of heavy metals. Highly absorbent of mercury, oyster mushrooms concentrate up to 140 times the level of this heavy metal in the substrates upon which it is grown (Bressa et al. 1988). Tolerant of high levels of many heavy metals in its habitat, *P. ostreatus* does not appreciably hyperaccumulate cadmium or lead. Such dynamics allow oyster mushrooms to be grown in polluted environments having a complexity of toxic wastes. I would not advise eating mushrooms grown in these environments until they have been carefully analyzed for the heavy metals to which the oyster mycelium is exposed. For more

information of the mycorestorative properties of oyster mushrooms, see chapter 10.

Comments: If one mushroom can steer the world on the path to greater sustainability, fighting hunger, increasing nutrient return pathways in ecosystems, destroying toxic wastes, forestalling disease, and helping communities integrate a complexity waste streams, oysters stand out. Growing throughout the world, this ubiquitous mushroom species has adaptive abilities that are nothing short of amazing. From Africa to Asia to Europe to the Americas, ecological leaders are learning how to harness its powers. Oyster mushrooms are well positioned to lead the way for rebalancing vast waste streams that currently overload our ecosystems. This is the first mushroom I recommend beginning growers to try, and most succeed in their first attempts. For more information on the cultivation of oyster mushrooms, please consult *The Mushroom Cultivator* (Stamets and Chilton 1983) and *Growing Gourmet and Medicinal Mushrooms* (Stamets 2000a).

Psilocybe cubensis (Earle) Singer = *Stropharia cubensis* Earle

Common Names: San Isidro, golden top, cubie.

Taxonomic Synonyms and Considerations: First described as a *Stropharia* from Cuba in the scientific literature by Earle in 1904, this mushroom sports a well-developed membranous ring on its stem, a feature of former taxonomic significance in defining the genus *Stropharia*. Subsequent studies by Singer moved this species to the *Psilocybe* genus. Dr. Gaston Guzman has studied this species for more than 40 years, and his monograph on the subject is the most complete to date (Guzman 1983). (A major revision is nearly completed and will be available through www.fungi.com.)

Introduction: The best-known of the *Psilocybe* mushrooms, *P. cubensis* is a psilocybin-active mushroom, producing a form of psychedelic intoxication when ingested. Many of the techniques developed in the "underground" cultivation of this mushroom, long revered by native peoples and now by modern populations, have been adapted for the cultivation of gourmet and medicinal mushrooms. Currently the cultivation of this mushroom is illegal in the United States, but it is legal in many other countries, including England, Thailand, and the Netherlands. However, the legal status of this mushroom and other psychoactive *Psilocybe* species is often in a state of flux as governments struggle with legal definitions. Consult the appropriate legal statutes before pursuing cultivation.

I do not recommend this mushroom for use by the general public. However, in my opinion, this species and its relatives can be helpful for sparking creativity in artists, philosophers, theologians, mathematicians, physicists, astronomers, computer programmers, psychologists, and other intellectual leaders.

I personally believe that the computer and Internet industries and astrophysics have been inspired through use of this fungus, which has stimulated the imagination and fields of vision of scientists and shamans with complex fractals, hyperlinking of thoughts, and mental tools for complex systems analysis. Many users over thousands of years have elevated this and other *Psilocybe* mushrooms to the level of a religious sacrament.

Description: Cap 1–8 cm broad, conic to convex, expanding in age to plane, with or without an umbo. Cinnamon reddish brown when young, lightening with maturity, becoming yellowish to yellowish white, with the center regions remaining darker. Surface smooth, sometimes with flecks of scaly tissue, soon removed by rain. Flesh whitish, bruising bluish. Gills pallid when young, becoming purplish brown to purplish black to black with maturity. Partial veil membranous, falling from the margin as the cap expands, becoming a white sheathlike ring in the upper regions of the stem, with the veil sometimes bluish toned, and often covered on top with purplish spores. Stem equal

▲ **FIGURE 334**

Wild *P. cubensis*, here growing in Palenque, Mexico, has a long history of use by indigenous peoples. A so-called "magic" mushroom, this mushroom is legal in some countries and illegal in others. Before cultivating, please check the legal status of this mushroom. Be careful.

◄ **FIGURE 335**

P. cubensis is a stately mushroom, emanating an air of elegance and beauty, traits long admired by those who have grown and picked it.

in diameter, swelling toward the base, often adorned with fuzzy mycelium to which whitish rhizomorphs are attached. Spores purplish brown in deposit, ellipsoid, 11–17 by 8–11 µm.

 **FIGURE 336**

During the 1970s and 1980s, it became popular for college students to grow *P. cubensis* in grain filled jars, often in closets converted into mini-growing rooms.

▲ **FIGURE 337**

A hoop frame covered with black plastic loosely draped to just above the ground provides a humid and warm environment for outdoor cultivation during the summertime, as this garden demonstrates.

Distribution: Globally throughout the tropics and subtropics, especially in pastoral and grassland habitats.

Natural Habitat: Naturally occurring on cattle, oxen, yak, water buffalo, and elephant dung.

Type of Rot: White.

Fragrance Signature of Mycelium: Mealy.

Natural Method of Cultivation: Several methods work well for cultivating this mushroom outdoors. The most productive method I have witnessed is to use leached cow manure or button mushroom–style composts mounded outdoors among garden vegetables, or under shade or "bug-out" cloths. See *The Mushroom Cultivator* (Stamets and Chilton 1983). Some strains of this mushroom fruit well on straw, and to a lesser degree on wood, provided ample spawn is used. Stem butts with rhizomorphs still attached regrow with vigor when replanted into supportive habitats. For other methods of cultivation, see my earlier book,

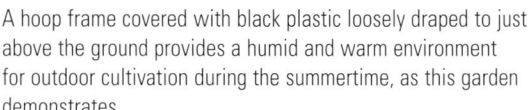

▲ **FIGURE 338**

Untarped once or twice a day for misting, mushrooms erupt as a unified flush from pasteurized leached cow manure. Many mushrooms can be grown in this fashion outdoors. Tightly woven white "bug-out" cloth can also be used, provided high humidity and moisture can be maintained. These mushrooms orient toward light and incoming airflow.

coauthored with Jeff Chilton *The Mushroom Cultivator* (1983).

Season and Temperature Range for Mushroom Formation: Growing during the rainy season in the

▲ FIGURE 339

These rapidly growing mushrooms can mature in a single day.

tropics and subtropics, often more common from May through September, although I have found them in Palenque, Mexico, in January and February. Temperature: 70–90°F.

Harvest Hints: Best harvested when the mushrooms are young, before the caps expand to plane, and before maggots consume them.

Nutritional Profile: Not yet known to this author.

Medicinal Properties: Western scientists are now conducting the first explorations into the medicinal properties of psilocybin and its analogues since the early studies at Harvard ("The Harvard Psilocybin Project") by Timothy Leary and Richard Alpert assisted by Ralph Metzner (Weil 1963). The fallout from the 1960s cultural revolution had the net effect of forestalling research into the mushroom's psychological benefits. Recently, this mushroom and its active constituents psilocybin and psilocin began to be re-explored for their use as an aid to psychoanalysts and psychiatrists in treating patients struggling with alcoholism, drug addiction, trauma, autism, and end-of-life issues. The National Institutes of Health have approved and continue to consider the use of psilocybin and psilocybin-containing mushrooms for treating psychological disorders. Medical practitioners are now

recognizing what shamans have known throughout history: with proper guidance, the psilocybin mushroom can benefit patients struggling with psychological difficulties and seeking spiritual enrichment.

Despite miscategorization as an illicit drug, this mushroom and other psilocybin species have no addictive potential. In fact, users feel repulsed with frequent use, often claiming that the experiences were powerful and meaningful but not ones they wished to immediately repeat. An illicit "drug" that causes users to refuse frequent doses does not fit the standard definition of a drug with high abuse potential. Furthermore, tolerance accumulates, with effects dramatically declining over consecutive days of use.

The FDA has conditionally approved a small clinical study of psilocybin at the University of Arizona, supervised by Dr. Francisco Morenzo, to treat patients who suffer from obsessive-compulsive disorder, after he reported benefits from the supervised use of psilocybin in a controlled setting (Moreno and Delgado 1997). Other studies are planned for the use of psilocybin to treat cluster headaches. Dr. Charles Grob (2004) is studying whether the use of psilocybin can reduce end-of-life anxiety in stage IV cancer patients. For updates on this and related research, please see www.maps.org/research/psilo/azproto.html.

Toward the end of his life, a good friend of mine, Bill Webb of Big Sur, California, called me to say that for a few days his hearing had been restored as a result of ingesting this mushroom. He was emphatic in his insistence that I tell others because he believed this revelation was medically significant. Given psilocybin's enhancement of the visual and auditory senses and its role as a temporary neurotransmitter, I think this anecdotal report should be taken seriously. Psilocybin and its analogues may help treat neuropathy.

Flavor, Preparation, and Cooking: If *P. cubensis* mushrooms are boiled in a large volume of water, which is changed out several times and removed, they can be denatured of their psychoactive properties. For purposes of preserving their potency but ridding them

A primordial cluster of *P. cubensis* explodes from a mound of pasteurized leached cow manure. Here, asbestos-free vermiculite is placed upon the beds to retain moisture. Although vermiculite limits contaminants and helps mushrooms form, it is messy and requires extra effort to clean the vermiculite particles and dust from harvested mushrooms.

▲ **FIGURE 341**

A multi-canopied overgrowth of trees provides shade and helps prevent evaporation, buffering microclimates, stimulating mushroom growth, and decreasing the need for watering.

the neurotoxins VX, sarin, and soman, and the decomposition of munitions. A scavenger of phosphorus and nitrogen, this species secretes powerful enzymes that deserve further study.

Comments: Like many of the mushrooms described in this book that are restorative to the environment, this mushroom and its allies, under the guidance of a mature counselor, can be restorative to the soul. When I have taken these mushrooms, one theme surmounts all: the Earth is calling to us to be good shepherds, to wake up to our potential, to stop the destruction of the Earth's diversity of species and its habitats, telling us that we are one with the universe, not apart, that we are enmeshed in the continuum and that the positive power of goodness permeates the cosmos. Spiritual people of all religions, from Buddhists to Christians, can find that the experience affirms their religious beliefs. Those who have had psilocybin-facilitated religious journeys often state that words cannot adequately convey the meaning of their experiences.

Many who use these mushrooms recreationally are not prepared for the experience and so the positive benefits are not realized. Some individuals become paranoid or terrified due to the intensity of the encounters.

of pathogens, mushrooms harvested from dung or areas where coliforms and protozoa thrive should be boiled in water for 5 minutes. For more information, see my earlier book, *Psilocybin Mushrooms of the World* (1999b).

Mycorestoration Potential: A potential use is the destruction of phosphorus-bound toxins, including

Caution is strongly advised. For parameters that may help guide you to a positive experience, see *Psilocybin Mushrooms of the World* (Stamets 1999b).

Psilocybe cyanescens Wakefield and Allies

Common Names: Cyan, caramel cap.

Taxonomic Synonyms and Considerations: The name *Psilocybe* is ancient Greek for "bald head," referring to the smooth surface texture of the cap (a feature I increasingly empathize with as I grow older!). The genus *Psilocybe* has such close affinities to *Stropharia* and *Hypholoma* that, historically, the separation of these genera presented taxonomic difficulties. These genera are clustered within the family Strophariaceae, which also includes the more distantly related genus *Pholiota*. Alexander Smith (1949) proposed that the family might best be represented by only 2 genera: the genus *Pholiota* and the macro genus *Psilocybe*, which would also envelope species of *Stropharia* and *Hypholoma* (as *Naematoloma*). Current DNA studies by Rytas Vilgalys, at Duke University, show that *Pholiota* and *Psilocybe* are discretely different but the genus *Psilocybe* is divided between the bluing, psilocybin-containing *Psilocybes* and their shorter, nonbluing,

non-psilocybin-containing sister species represented by the constellation of taxa related to *P. montana*, *Stropharia*, and *Hypholoma*. Research will soon be published that will elaborate on these distinctions.

P. cyanescens has several close allies, including *P. azurescens*, *P. subaeruginosa*, *P. subaeruginascens*, and *P. aerugineomaculans*. Another *Psilocybe*, perhaps a new species, is appearing with increasing frequency during the winter months in the San Francisco Bay area (see figure 345). All species in this taxonomic cluster have chestnut- to caramel-colored caps, grow on wood chips, bruise bluish, sport rhizomorphs around the stem bases, and are potent psilocybin producers.

Similar to oyster mushrooms in the variety of materials this group of species can consume, *P. cyanescens* and its allies shows extraordinary aggressiveness: the rhizomorphs run fast; fork frequently; grip the substrate with tenacity; thoroughly permeate wood chips, sticks, and thatch from grasses; and digest a wide array of debris, including paper products.

Introduction: *P. cyanescens* and allies taught me how to propagate mushrooms outdoors and create mycological landscapes. Its affection for growing around and benefiting neighboring plants fascinates me. Although this mushroom can grow in barren wood

◁ **FIGURE 342**

The wood chip–loving *P. cyanescens* is characterized by wavy caps, silky stems, and bruising bluish.

◁ **FIGURE 343**

A private *P. cyanescens* mushroom patch.

◀ **FIGURE 344**

Mycelium on birch dowels—grown from stem butt spawn—grasps the wood with its lacy rhizomorphs. This spawn, when planted into more wood chips, grows mushrooms in 6 to 12 months, given conducive conditions. The following photograph depicts the mushrooms grown from this form of natural spawn.

chips, more frequently *P. cyanescens* and allies grow in environments between wood chips and grasslands. As "edge runners," these fungi project fans of exquisitely forming and divergent rhizomorphs, and rich soil appears in the wake of its mycelial waves. This group of species can be a powerful ally to your environment, and to the ecology of your mind.

Description: Caps are hemispheric at first, soon becoming convex, expanding to broadly convex and eventually plane in age, 2–10 cm in diameter. Caps chestnut to caramel colored, hydrophanous, fading to straw yellow in drying, sometimes with bluing tones or spots. Cap margins are incurved at first, covered with a cobwebby veil that attaches to the upper regions of the stem, expand with maturity, and often undulate with age. Gills are brown to dark brown, often mottled, and bluntly attached to the stem, typically with a thin, whitish outer edge. The stem is centrally attached to the cap, silky white to dingy brown near the base, often covered with fine whitish fibrils that can bruise bluish. The stem is cartilaginous, even, straight to sinuous, usually swelling and curved toward the base. Each stem base is usually fitted with a coarse, radiating array of white rhizomorphs. Although individual mushrooms can be found, the majority grows in groups or clusters.

Distribution: The species represented in this group of mushrooms are found throughout the temperate forests of the world, including but not limited to the coastal Pacific Northwest of North America; northeastern North America; the British Isles; and temperate zones of Europe, Eurasia, Asia, southern Argentina, Chile, South Africa, New Zealand, and Australia.

△ **FIGURE 345**

Buddha overlooks a *Psilocybe* garden. This species, not yet named, heralds now from the San Francisco Bay Area and probably is new, or least a newly imported species.

Natural Habitat: Primarily deciduous woods, especially in riparian areas. These mushrooms fruit in the cool, rainy seasons, primarily during the fall to early winter. They show a particular fondness for the decorative (nonbark) mulch used in landscaping around newly constructed buildings. In fact, they are common in urban and suburban areas and are actually rare in natural settings. Ideal locations for collecting this mushroom are in the landscaped property surrounding institutional and government facilities: libraries, utility companies, churches, colleges, computer internet service providers, and even courthouses and police stations! Other locations for finding these mushrooms are rhododendron, rose, and azalea gardens, often in city parks.

Type of Rot: White.

Fragrance Signature of Mycelium: Sweet and sour, rich to a point of being unpleasant, farinaceous, and reminiscent of spoiling corn.

Natural Method of Cultivation: Cultivation of this mushroom follows essentially the same strategy as that used for growing *Stropharia rugoso annulata*. Sawdust spawn is broadcast into hardwood or conifer (Douglas fir) chips that have been laid down outdoors in a partially shaded environment. The wood chips should be variable in size, ranging from $1/8$ to 4 inches in length.

Season and Temperature Range for Mushroom Formation: Temperature: 45–75°F.

Harvest Hints: Mushrooms are best harvested when young, before spore maturity and damage from insects occurs. Grasping the mushrooms from the stem base minimizes damage to the flesh, indicated by the bluing reaction. Clusters of mushrooms can be picked, and, once the stem butts are severed, the fruit bodies easily dry, shrinking to a small fraction of their original size. The fresh stem butts, replete with rhizomorphs, can be used for creating spawn on paper, cardboard, burlap, dowels, and wood chips.

⋀ FIGURE 346

If burlap bags stuffed with wood chips are inoculated with sawdust, dowel, or stem butt spawn, the mycelium fully colonizes in 4 to 6 months, typically over winter in the Pacific Northwest.

⋀ FIGURE 347

A species related to *P. cyanescens* is *P. azurescens*. Both enjoy wood chips actively being overgrown with grasses. The grasses collect dew, which streams down their stems to the ground, encouraging mushrooms to form.

Nutritional Profile: Not yet known to this author. This mushroom is normally consumed not for its nutritional value but for its mind-influencing properties.

Medicinal Properties: Not yet known. Two of the active constituents within this mushroom, psilocybin and psilocin (and their associated analogues), may prove to be nerve tonics. See discussion of *P. cubensis*.

▲ **FIGURE 348**

Another close relative to *P. cyanescens* is *P. azurescens,* which also loves grasses growing through wood chips.

Flavor, Preparation, and Cooking: Flavor is bitter, farinaceous, and distasteful and sometimes induces gagging. If made into a tea, it must be used within 1 or 2 days, since bacteria will proliferate. The tea can be frozen or ethanol can be added to raise the alcohol content to above 25 percent in order to prevent souring. Freshly harvested mushrooms can be dried and stored airtight in cool, dark conditions until ready for use. Freezing dried mushrooms is the method preferred by many. Any mushrooms showing deep blackish blue lesions—sites of bacterial contamination, usually localized to wounded tissue—should be avoided.

Mycorestoration Potential: I suspect that this group of species has enormous mycoremediation potential, but they have yet to be explored, largely because they are illegal in many countries. Since the compound psilocybin contains phosphorus and nitrogen, a bluing species in this group was tested and shown to be effective in scavenging these elements from toxins, dephosphorylating VX-like compounds, decomposing TNT-related munitions, and denaturing a wide variety of other industrial toxins. Enzymes and their decomposition pathways within this group of species have yet to be characterized.

Comments: A pioneering species, this mushroom is one of the first species to appear when wood is chipped. (I estimate that in western Washington and Oregon this species proliferates in 1 of every 4 delivered loads of wood chips created by the utility company's annual chipping of tree branches to prevent interference with power lines.) Many varieties of these mushrooms are edge runners, appearing along roads and trails and within ecological borders between deciduous and coniferous zones.

In temperate, low-elevation climates, if wood chips are placed around buildings for landscaping, this group of mushrooms can predominate. Mostly riparian species associated with alders, eucalyptus, blackberries, and brushy plants, these fast-growing, white-rot mushrooms pulp wood in a few years. Having your own *Psilocybe* mushroom patch in your backyard, whether you ingest these mushrooms or not, is a source of pride for many mushroom connoisseurs. These mushrooms grow well with many ornamental shrubs underneath deciduous and coniferous trees (but not cedars or redwoods), and if provided with annual debris fields of wood chips they may remain resident in your landscape for more than a lifetime. See also my earlier books *Psilocybin Mushrooms of the World* (1999b), *Growing Gourmet and Medicinal Mushrooms* (2000a), and *The Mushroom Cultivator* (Stamets and Chilton 1983).

Sparassis crispa Wulf ex Fries = *Sparassis radicata* Weir = *Sparassis herbstii* Peck

Common Name: Cauliflower mushroom.

Taxonomic Synonyms and Considerations: *S. crispa,* the cauliflower mushroom, is at the center of a constellation of species encompassing the species concepts of several other ecotypes, including *S. radicata,* *S. laminosa,* and *S. herbstii,* all of which look very similar to one another. The classic radicate stem ("tapering root") typical of *S. radicata* emanates from the root zone of conifers in western North America. Those forms found on oaks in the eastern United States and in Europe lack the long-tapered root stem, tend to have thicker folds, and are generally smaller. The hardwood-degrading type may be a different species, S. *herbstii*, or its synonym S. *spathulata.*

Introduction: Popular in northern Europe and the United States, S. *crispa* is one of the easy-to-identify edible mushrooms. Specimens can grow to enormous sizes (up to 70 pounds!), especially those native to the old-growth fir forests of northwestern North America. I suspect that the mycelium of this species is widespread throughout mature forests, especially where ground-water is near the surface, but is not noticed until the mushrooms appear.

Description: When young, whitish to yellowish white, looking like a cauliflower, then like a brain tightly covered linguine egg noodles, soon with ridges elongating into flattened, wavy leafletlike structures, diverging from the center; smooth, whitish to creamy yellow, with the margins darkening with age or during drying. Growing up to 2 feet (60 cm) across with a central stem often tapering into a long taproot, darkening toward the base. Cultivated specimens are much

smaller than their wild cousins, with mushrooms from the Pacific Northwest old-growth forests providing the largest specimens. Spores whitish in deposit, smooth, ellipsoid, 5–7 by 3–5 µm.

Distribution: Found throughout the temperate regions of Europe and northeastern and western North America

Natural Habitat: Growing from the ground interface at the base of conifer trees, primarily Douglas firs. In eastern North America and Europe growing at the base of oaks from late July through November, preferring older, late-succession forests. Also growing at the base

⋀ **FIGURE 349**

Cauliflower mushrooms like these S. *crispa* are superb edibles and, in my opinion, guard forest ecosystems from attack by quick-to-kill fungal blights, such as *Armillaria* root rot.

⋀ **FIGURE 350**

S. *crispa* fruiting from hemp rope. This rope can lasso notched stumps, inoculating them with mycelium of this *Armillaria*-fighting species. Once in place, the cauliflower mushroom can fruit for years.

of snags or stumps. Once it forms, this mushroom reappears like clockwork every year. I have little doubt that *Sparassis* species exist in mycelial form in the woods for many years before fruitbodies form.

Type of Rot: Brown root and butt rot.

Fragrance Signature of Mycelium: Foresty, rich, sometimes sickeningly sweet.

Natural Method of Cultivation: Dowel and rope spawn can inoculate stumps. Sawdust spawn or a stem butt from a wild harvested mushroom can be buried in the ground near a tree base, making contact with a newly inflicted root wound.

Season and Temperature Range for Mushroom Formation: From late summer through early winter. Mushrooms tend to stand for weeks before decomposing, usually after maturity and subsequent to heavy rains. Temperature: 45–65°F.

Harvest Hints: Best harvested when the leaflets have fully extended but before the edges brown from age. I prefer the smaller fruitbodies with tight folds, but then the moisture content must be reckoned with; the younger fruitbodies are heavy with water, and water content declines as the leaflike petals extend. Young, developing mushrooms can be brushed free of debris so that, as they mature, dirt does not become embedded deeply into the extending folds. Dig deeply to fully pull out the mushroom and its long, tapering taproot. (This root often regrows when replanted.) Once harvested, shake the mushroom upside down to remove debris and insects. Some collectors steep the mushroom submerged in ice water overnight to both wash the mushroom and drive out the bugs. The next day, the mushrooms are drained of water. Groups of petal-like folds can be sliced lengthwise and air-dried and/or frozen for future use. A cluster of this delicious species can be used to make several meals for a family.

Nutritional Profile: Not known to this author. I plan to analyze this species in the future.

Medicinal Properties: Contains the antibiotic sparassol, also present in some *Armillaria* species, that is antifungal. Other metabolites inhibit *Bacillus* bacteria. This mushrooms is a source for novel antitumor beta-glucans; Ohno and others (2000, 2002) found that its polysaccharides had protective effects in cancer-induced mice, increasing cytokine production, causing vascular dilation, and leading to increased leukocyte activity. In a study by researchers at the Tokyo University of Pharmacy and Life Science, healthy patients in a small control group given an *S. crispa* fraction then showed an increase in whole blood cytokines. This baseline study was followed by a study of 14 patients who were afflicted with various cancers and consumed the same product, with 9 showing improvement (Ohno et al. 2003). *S. crispa* looks promising as an immunotherapeutic agent.

Flavor, Preparation, and Cooking: A "leafy" fruit body best torn into pieces and then cooked in mass. *Sparassis* species are crunchy, with a nutty, mild, but enjoyable flavor. I prefer tearing the leaflets apart, cooking them in a wok with lots of vegetables, using the excess water from the mushrooms to steam the vegetables. The cauliflower mushroom is exceptionally good when dipped in beer-batter and deep-fried. Also excellent added to casseroles, soups, and stews.

Mycorestoration Potential: I view this species as a strong candidate for forestalling or preventing the invasion of devastating forest fungal blights such as honey mushrooms in the genus *Armillaria*. *S. crispa* outcompetes many *Armillaria* species but does not initially kill the tree. Like any good parasite, the fungus allows the tree to grow, feeding upon it slowly over time. (Some mycologists question whether or not this species is truly parasitic (Ammirati 2004). Over the course of its life, *Sparassis* puts up a territorial defense, preventing more destructive fungal parasites from entering the tree. Woodward and others (1993) identified 2 new antifungal metabolites from *S. crispa*. In the near future, we plan to inoculate the perimeter of a clear-cut populated by *Armillaria* to see if cauliflower mushrooms can be used as a mycological barrier preventing further invasion. Far better to have a woodland mature into an ancient forest and climax with cauliflowers than for a young one to be destroyed by *Armillaria*, making it prone to forest fires.

Consumers and those practicing mycorestoration should be aware that *S. crispa* uptakes arsenic from its habitat (Slejkovec et al. 1997), although not as efficiently as many other mushroom species.

Comments: Finding this majestic mushroom, one of the easiest to identify, usually means trekking into the forest, an adventure I love to undertake. I find it mostly in very wet, old-growth or mature second-growth forests where the trees are more than 70 years old. I encountered a fruiting of this mushroom on a stump made by loggers who had cut the tree half a century earlier. From this and my personal experiences, I think cauliflower mushrooms are long-term residents in maturing forests. Many mushroom buyers purchase these mushrooms, known for their commercial value as a gourmet edible, from collectors who also foray for chanterelles and matsutake. This mushroom's immune-enhancing and antifungal properties, combined with its excellent flavor and gargantuan size, meet all the qualifications of a mycorestorative ally.

Stropharia rugoso annulata Farlow apud Murrill

Common Names: Garden giant, king Stropharia, burgundy mushrooms, the wine cap.

Taxonomic Synonyms and Considerations: A distinct and not yet taxonomically controversial species, this mushroom has also been called *S. ferrii* Bres. or *S. imaiana* Benedix.

Introduction: Mushrooms from this species can be massive. I have grown specimens weighing 5 pounds apiece and measuring nearly 2 feet in diameter. A burgundy color gives the mushrooms a stonelike appearance. More than once I have thought the young emerging buttons were smooth rocks that had mysteriously appeared. To the touch they feel cold, smooth, and firm but not as hard as a rock. They grow at a remarkable rate, transforming from adolescence to adulthood between morning and dusk. The upper surface of the sheathlike ring is lined with gill plates, hence its name, *rugoso annulata*, which means "ridged ring." Known throughout Europe as a garden mushroom, this species thrives in newly disturbed soils rich in wood debris. A cosmopolitan species, *S. rugoso annulata* has other remarkable talents valuable for use in mycorestoration. One is its projection of thick, cordlike rhizomorphs, and the other is its ability to live in microbially complex soils infused with wood debris. Rich soil is left in the wake of the garden giant as it consumes debris. It seems friendly to most plants and is an especially good companion to corn (see figure 214). As is to be expected given its size, each garden giant can produce a prodigious outflow of spores.

▲ **FIGURE 351**

S. rugoso annulata growing in a mixture of soil and wood chip hosting grasses. *Stropharia* and *Psilocybe* mycelium love root masses of grasses, especially when infused with wood chips.

Description: Cap 4–13 cm, nearly ovoid at first, embedded in the ground, expanding to convex with deep incurved margins, soon expanding to convex to broadly convex to plane, sometimes with whitish, spore-dusted veil remnants adhering to the margin. Cap reddish brown at first, fading in age to straw color, sometimes with tones of purplish brown spores in streaks across it. Margin incurved at first, connected to the stem by a thick, membranous veil. Veil breaking with age to form a thick membranous ring radially split with gill-like ridges, usually darkened with spores. Toothlike veil remnants often seen at the time the ring separates from the cap. Stem thick, equal, enlarging toward the base, often bulbous, from which thick, white radiating rhizomorphs emanate. Stem base and stalk soon riddled with, sometimes hollowed out from, larvae. Spores are ellipsoid, smooth, purplish to brown in mass, and measure 11–13 by 7.5–8 μm.

Distribution: Common in the mid-Atlantic States including New York, New Jersey, and Massachusetts. Although first described from North America in 1922, this mushroom is known from Europe, New Zealand, and Japan. Given the export and trade in ornamental plants, it is difficult to know from where this species first heralded.

Natural Habitat: In hardwood forests and/or among hardwood debris or in soils rich in undecomposed woody matter, especially common in the wood-chip mulch used in outdoor urban or suburban plantings of ornamentals. Growing in disturbed grounds rich in broken sticks or wood chips. In Japan, the mycologist Yokoyama found a strain inhabiting rice straw. I have had scattered reports of the garden giant growing with cereal grasses. In central Europe, S. *rugoso annulata* has naturalized in cornfields where years earlier it was first introduced as a mulching species. We have had garden giant colonies in our garden for many years. As long as the debris fields are replenished, this species stays resident. Mushrooms from straw mulch have less density than those growing on wood chips.

▲ **FIGURE 352**

A flush of baby S. *rugoso annulata* fruits from cased, pasteurized wheat straw. Note the rhizomorphs channeling nutrients to primordia and adolescent mushrooms.

▲ **FIGURE 353**

This wood chip garden of S. *rugoso annulata* produced hundreds of mushrooms over several years. The yearly influx of new chips kept the bed alive. Three years after the last influx of wood chips, the fruitings ceased and the colony moved on. Note the ganglion of rhizomorphs attached to stem bases.

Type of Rot: White.

Fragrance Signature of Mycelium: A unique, forest-rich, perfumed fragrance is outgassed by the mycelium.

Natural Method of Cultivation: I like creating a mother patch—essentially a mycelial lens—an oblong pile of wood chips, shallow in depth, inoculated with garden giant sawdust spawn, and fully infused with

◀ **FIGURE 354**

Azureus Stamets holds a specimen of *S. rugoso annulata* estimated to weigh around 3 pounds. Car-stoppers, these mushrooms are best grown in the backyard, out of sight by drive-by mushroom hunters.

▲ **FIGURE 355**

S. rugoso annulata fruiting in unison in a mulch bed in New Zealand. Note the regal crowns with toothlike decorations below the caps. These will soon fall into a membranous ring, also known as an *annulus*. See next photo.

▲ **FIGURE 356**

The same garden giants as in the previous figure, 2 days later, in a majestic display of mycological beauty. Note how the membranous rings have now fallen from the caps and the characteristic purplish brown to black spores dust their upper sides.

mycelium. Once a bed of wood chips has become colonized by mycelium, usually after 6 months to 1 year, or alternatively when straw is colonized, usually in 3 to 6 months, the mycelium can be expanded by simple transplantation (see pages 190 to 192).

Season and Temperature Range for Mushroom Formation: Primarily late spring to early fall, climaxing in midsummer. Temperature: 50–90°F.

Nutritional Profile: 22 percent protein; 34 mg niacin per 100 g. In view of the scant information I have found, I plan to have this species analyzed and will post the results in future publications.

Medicinal Properties: Showing activity against coliforms, this mushroom exudes antibiotic metabolites that affect diverse bacterial populations, given that it competes so well in microbe-rich soils. The medicinal constituents of this species are largely unexplored.

Flavor, Preparation, and Cooking: Young buttons can be the size of softballs and are great for stuffing and broiling in the oven or grilling on the barbecue. Or for a fabulous meal, dice the stems and cook until the edges are browned, then add onions, garlic, and walnuts, and complement with rosemary and assorted spices. I like this mushroom well cooked. The garden giant declines in edibility as spores mature and the flesh thins. For a more crispy culinary experience, thinly slice the mushrooms and baste them with teriyaki sauce, frying or grilling them until the edges are dark brown and the flesh is tan.

Garden giants should not be eaten more than 2 or 3 days in a row because some people's digestive enzymes cannot handle the load. Make sure they are thoroughly cooked.

Comments: Of all the mushrooms in this book, this may be the premier species for habitat restoration because of its tolerance of complex habitats, love of disturbance, adaptive appetite, friendliness to important garden allies, and the ability of its stem butts to regrow. This species has a wonderful scent and other interesting properties that attract insects, from bees to flies. It is also highly attractive to earthworms. I think this mushroom can play a pivotal role in fortifying ecosystems, especially in gardens and along waterways.

A species that provides food for mammals, insects, and fish; digests garden, farm, and forest wastes; eats coliform bacteria; and enriches soil and stimulates plant growth is truly remarkable. Yet to date, we know little about the many roles *S. rugoso annulata* may potentially play in supporting the ecosphere.

Trametes versicolor (L:Fr.) Pilat

Common Names: Turkey tail, kawaratake (Japanese), yun zhi (Chinese).

Taxonomic Synonyms and Considerations: Synonymous with *Coriolus versicolor* (L.:Fr.) Quelet and formerly called *Polyporus versicolor* (L.:Fr.), so named for the multicolored zonations on the cap. Diverse ecotypes abound, and the mushrooms can be purplish, grayish brown to brown to blackish brown in tone. Mushrooms that can be confused with this species include the false turkey tail *(Stereum ostrea)*, and the hairy parchment mushroom *(Stereum hirsutum)*, which often have brown underbellies. *Lenzites betulina*, a gilled polypore, has an overall similarity in appearance when seen from afar.

Introduction: Turkey tail, a ubiquitous woodland polypore growing in forests across the planet, is one of the most common mushrooms in the world. Turkey tails

▲ **FIGURE 357**

T. versicolor comes in many colors and has beautiful zonations on the caps. This is one of the most common mushrooms in woodlands throughout the world, here fruiting on an aspen log.

have long been revered in Asia and Europe for their medicinal properties. Their use extends back hundreds, and probably thousands, of years. Studies in the past 20 years have unveiled that the enzymes secreted by its mycelium are some of the most powerful toxin-destroying agents yet identified from a natural source. This species offers unique tools for healing both people and the planet from the ravages of pollution.

Description: An annual shelf polypore, with thinly formed flattened caps, often in large clusters of overlapping fruitbodies that are finely fuzzy on the upper surfaces, concentrically colored with distinct zones of brown, from buff to reddish brown, sometimes bluish, fading in age, rarely whitish to cream colored. Underside covered with a cream- to buff-colored pore layer. Flesh tough, pliable, cream-colored, with a thin black layer beneath the cap surface. Spores whitish in deposit, cylindrical but slightly curved, smooth, 5–6 by 2–2.5 μm.

Distribution: Global, widely distributed throughout the boreal, temperate, subtropical, and tropical regions of the world. Few mushrooms other than oysters can boast such a wide range.

An oak stump explodes with *T. versicolor* mushrooms. In essence, this stump is a powerfully medicinal platform, producing artful mushrooms with highly anticarcinogenic and mycorestorative properties.

Natural Habitat: Turkey tail grows on virtually all dead hardwoods and is the most common mushroom found in deciduous forests, but it can also grow upon conifers, particularly fir, spruce, pine, larch, juniper, and cypress (Gilbertson and Ryvarden 1986–87).

Type of Rot: White.

Fragrance Signature of Mycelium: Pleasant, smoothly sweet, and distinctly polypore-esque.

Natural Method of Cultivation: In logs, using the same method used to cultivate shiitake, or buried logs in pots, as for reishi. Plug, sawdust, or rope spawn can inoculate stumps. Stumps can host multiple species: a natural pairing with turkey tail is *Ganoderma*, particularly *Ganoderma lucidum* and *Ganoderma applanatum*. See figure 96.

Season and Temperature Range for Mushroom Formation: Forming annually and growing from late spring through early winter in widely diverse climates, from pines in Arizona to the boreal birch forests in the extreme north of North America. Temperatures: 50–90°F.

Harvest Hints: Since *T. versicolor* is highly vulnerable to insect attack, mushrooms should be picked early,

prior to heavy sporulation; in some regions this corresponds to midsummer to early fall. Mushrooms can be strung up and dried in the sun.

Nutritional Profile: Our analysis of a 100 g serving shows the following: calories: 369; protein: 10.97 g; fat: 1.51 g; polyunsaturated fat: 0.27 g; total unsaturated fat: 0.32 g; saturated fat: 0.06 g; carbohydrates: 77.96 g; complex carbohydrates: 76.06 g; sugars: 1.90 g; dietary fiber: 71.30 g; cholesterol: 0 mg; vitamin A: 0 IU; thiamine (B1): 0.07 mg; pantothenic acid (B5): 1.70 mg; vitamin C: 0 mg; vitamin D: 62 IU; calcium: 34 mg; copper: 0.65 mg; iron: 8.7 mg; potassium: 570 mg; niacin: 9.30 mg; riboflavin: 1.06 mg; selenium: 0.007 mg; sodium: 6 mg; moisture: 8.00 g; ash: 1.56 g.

Medicinal Properties: Probably the best-documented medicinal mushroom, wild strains of *T. versicolor* (= *Coriolus versicolor*) typically show remarkable vitality and aggressiveness in culture. The mycomedical activity is twofold: both as an antitumor compound, inhibiting growth of cancer cells, and in stimulating a host-mediated response, bolstering the immune system's natural killer cells (Garcia-Lora et al. 2001). Lin and others (1996) showed that *Coriolus versicolor* polysaccharides (CVP) enhanced the recovery of spleen cells subsequent to gamma irradiation. Recent studies at the New York Medical College suggest that ethanolic extracts of yun zhi show promise as an adjuvant therapy in treating hormone-responsive prostate cancer by slowing tumorigenesis (Hsieh amd Wu 2001). This species, or its derivatives, have been also been used to treat a wide variety of cancers (breast, lung, colon, sarcoma, and other carcinomas).

T. versicolor is the source of PSK, commercially known as krestin, and is responsible for several hundred million dollars of sales of this approved anticancer drug in Asia. PSK is derived primarily from mycelial cultures but can also be extracted from fruitbodies. Used clinically in the treatment of cervical cancer in conjunction with radiation therapy, PSK has helped substantially to increase survival rates. In clinical studies, patients afflicted with gastric cancer and treated with chemotherapy showed a decrease in cancer recurrence and an increase in the disease-free survival rate when conventional treatment was combined with a regimen using the protein-bound polysaccharide PSK from *T. versicolor* (Sugimachi et al. 1997; Nakazato et al. 1994). By all measures, this treatment protocol was clearly cost-effective. PSK reduces cancer metastasis (Kobayashi et al. 1995) and stimulates interleukin-1 production in human cells (Sakagami et al. 1993). PSK was also found to be a scavenger of free radical oxidizing compounds (superoxide anions) through the production of manganese superoxide dismutases (Kobayashi et al. 1993; H. Kim et al. 1999).

A highly water-soluble, low-cytotoxic polysaccharopeptide (PSP) isolated from this mushroom has been proposed as an antiviral agent inhibiting HIV replication, based on an in vitro study (Collins and Ng 1997). PSP is a classic biological response modifier (BRM), inducing gamma interferon, interleukin-2, and T-cell proliferation, differing chemically from PSK in that it has rhamnose and arbinose, while PSP has fucose (Ng 1998). Dong and others (1996, 1997) reported that a polysaccharide peptide (CVP) and its refined form (RPSP) not only have antitumor properties but also elicit an immunomodulating response by inhibiting the proliferation of human leukemia (HL-60) cells while not affecting the growth of normal human peripheral lymphocytes. Yang and others (1992) also found that a smaller polypeptide (SPCV for smaller polypeptide from *Coriolus versicolor*, 10,000 molecular weight) significantly inhibited the growth of leukemia cells. Kariya and colleagues (1992) and Kobayashi and others (1994a, 1994b) showed that the protein-bound polysaccharides of *T. versicolor* express activity mimicking superoxide dismutase (antioxidating). Lin and others (1996) showed that CVP enhanced the recovery of spleen cells subsequent to gamma irradiation. Ghoneum (1995, 1998) conducted 2 small clinical studies in the United States using arabinoxylane, a product from fermenting *T. versicolor*, *Lentinula edodes* (shiitake), and *Schizophyllum commune* on rice; it showed a dramatic 5-fold increase in natural killer cell activity within 2 months of treatment. PSK has also been found to be a strong antibiotic, effective against *Escherichia coli*, *Staphylococcus aureus*, *Pseudomonas aeruginosa*, *Candida albicans*, *Cryptococcus neoformans*, *Listeria monocytogenes*, and other microbes pathogenic to humans (Sakagami and Takeda 1993; Mayer and Drews 1980; Tsukagoshi et al. 1984; Sakagami et al. 1991). Both PSK and PSP are present in the mycelium and can be extracted from fermented cultures.

A report by Ikekawa (2001) showed that an extract (PSK) was ineffective as an antitumor agent with implanted sarcoma 180 tumors, while aqueous extracts from fleshier mushrooms like shiitake (*Lentinula edodes*), enokitake (*Flammulina velutipes*), and oyster (*Pleurotus ostreatus*) elicited a strong host-mediated response, leading to significant regression of tumors. Ikekawa further states, "although extracts of *Trametes versicolor* was approved as an anti-cancer drug in Japan and that of *P. linteus* in Korea, they are not very active in oral administration (p.o.) experiments" (Ikekawa 2001, 293). In contrast to reishi, aqueous extracts of *T. versicolor* and PSK appear to be ineffective in controlling sarcoma 180, but they may be effective against other forms of cancer, or when whole fractions, not isolated fractions, are employed. PSK has been used to normalize immune function in patients with chronic rheumatoid arthritis (Hobbs 1995).

In studies of our turkey tail mushroom extract by Fisher and other researchers (2003) at St. Mary's Hospital in San Francisco, several strains of carcinoma cell lines were cultured and given serial dilutions. Clear cytotoxicity and apoptosis against carcinomas were demonstrated without harming healthy cells in a dose-dependent manner. I am also working with a group of researchers use of our *T. versicolor* mycelium-based extract in a multiyear clinical trial with breast cancer patients through the University of Minnesota Medical School in collaboration with Bastyr College of Medicine. The first year will be focused on in vitro tests, followed several years of in vivo (patient) studies (Standish et al. 2004)

The mycelium and the fruitbody—composed of compacted, differentiated mycelium—also produce antimicrobial compounds, natural defenses preventing rot. The traditional, historical use in teas and soups seems well warranted as an antimicrobial additive to the human diet.

Flavor, Preparation, and Cooking: Tough and leathery, the fruitbodies are extracted in water by boiling in soups or in teas; this mushroom has been used for centuries as a natural medicine.

▲ **FIGURE 359**

T. versicolor fruits from a conifer log in the old-growth forest of the Olympic Peninsula.

Mycorestoration Potential: In the forest, the turkey tail only fruits on dead trees and is thought to be a saprophyte. When I grew several strains in a culture dish, turkey tail mycelium overran the parasitic honey mushroom *(Armillaria mellea)*, the parasitic cauliflower *(Sparassis crispa)*, and the woodlover *(Hypholoma capnoides)*. I envision that this species, perhaps better than others, could be used to prevent blight fungi from sweeping through forests by setting up a competing mycelial perimeter of inoculated stumps. If a sufficient band of stumps in a clear-cut, for instance, could be inoculated with turkey tail mycelium using rope or plug spawn in advance of the march of *Armillaria* mycelium, then the blight might be stopped in its path. Turkey tails also work well for mycorestoration strategies using burlap sacks filled with sawdust or wood chips, which are useful for filtering metals, organophosphates, polycyclic aromatic hydrocarbons (PAHs), and microbes *(Escherichia coli, Listeria monocytogenes, Candida albicans,* and *Aspergillus* species) from effluents. Once charged with water-enriched nutrients and placed in a conducive setting, the burlap sacks fruit with mushrooms. These mushrooms, if proven to be free of compromising toxins, could conceivably be used for immunotherapy.

T. versicolor is the focus of a spectacular array of studies showing that the secondary metabolites from its mycelium—its laccases and oxidized lignin peroxidases—are highly effective in breaking down PAHs including antracines (Johannes et al. 1996; Field et al. 1992), pyrenes, flourene (Sack and Gunther 1993), methoxybenzenes (Kersten et al. 1990), and styrene (Milstein et al. 1992). These compounds could also aid in the destruction of dimethylmethylphosphonates (nerve toxins) and even the bleaching of pulp (Katagiri et al. 1995). Arica and other researchers (2003) found that the heat-killed mycelium of *T. versicolor* could selectively absorb mercuric ions from aquatic systems, making this mushroom potentially useful for mycofiltration of mercury.

This species responds with different enzyme expressions depending upon the nutritional or toxic profile of the contacted habitat. Black and Reddy (1991) isolated 6 genes from this fungus that were responsible for lignin peroxidase production, and Iimura and Tatsumi (1997) identified 2 novel genes whose expression was induced by exposure to pentachlorophenol (PCP). As more researchers use this fungus to degrade toxic wastes, more gene expressions and enzymatic pathways are likely to be discovered: its genome is worthy of more studies.

To the best of my knowledge, mycoremediation efforts using *T. versicolor* against any of the above toxins do not directly infringe on any outstanding species-specific patents. (Check with a qualified intellectual property attorney before embarking on commercialization. Please abide by all patent laws.)

Comments: Few mushrooms have as many properties beneficial to humans and the environment as this one. How long humans have used this mushroom remains a mystery. Turkey tails were an ingredient in early folk remedies in Asia, the Americas, and Europe. Many cultures throughout the world have incorporated the multicolored fruitbodies as works of art, wearing or carrying the mushrooms in earrings and necklaces, on clothing, and even on ancient weaponry.

The number of ecotypes of this mushroom is astonishing—native strains can be found in virtually all woodland habitats, allowing for the amplification of native strains to help locally specific mycorestoration efforts. I find this mushroom fascinating for its broad range of abilities to aid humans as we struggle to live in harmony with our environment.

➤ **FIGURE 360**

This turkey tail strain is one I cloned from a broken branch from my apple tree. Actually, when my kids were toddlers, they jumped on Dr. Andrew Weil while he was laying in our hammock that was supported by two apple trees. One of the branches broke and Andy hit the ground with a thud and with gleeful kids on board. I threw the branch into the bushes. Turkey tail mushrooms fruited which I cloned, giving rise to the mushrooms seen here. This strain shows strong activity against prostate and other cancers. Such is the way science meanders forward.

The End . . .

of the beginning of the mycorestoration revolution . . .

GLOSSARY

agar: A product derived from seaweed. Valued for its gelatinizing properties and commonly used to solidify nutrified media for sterile tissue culture.

agarics: Mushrooms with gills.

anamorph: The state of fungi, typically asexual, whereby conidia or unspecialized hyphal cells are expressed, characteristic of many Ascomycetes.

anastomosis: The fusion of hyphal cells followed by an exchange of cellular contents between 2 mycelial networks.

annulus: A ring, collar, or cellular skirt forming on the stem, typically originating from a portion of the partial veil.

appressed: Flattened.

ascus, asci: A saclike cell typical of the class Ascomycetes, usually containing 6 or 8 spores. Most cup fungi and morels (*Morchella*) belong to this group.

basidia: The clublike cells that give rise to 4 (more rarely 2 or 6) spores.

Basidiomycetes: The class of fungi that bear spores upon a clublike cell known as a basidium. Pore, gill, tooth, and jelly fungi (*Auricularia* spp.) belong to this class.

biological efficiency: The percentage measurement of the yield of fresh mushrooms from the dry weight of the substrate: 100 percent biological efficiency is equivalent to saying that from a substrate with a moisture content of 75 percent, 25 percent of its mass will yield fresh mushrooms having a moisture content of 90 percent.

block: A term used in mushroom culture to refer to the cube-shaped mass of sawdust substrate contained within plastic bags. Once the mycelium has grown through the substrate, the plastic can be stripped off and the mycelium holds the mass together. Blocks can be used individually or collectively to build "walls" of mushroom mycelium.

brown rot: A condition caused by the degradation of cellulose by fungi. It leaves the substrate a brown color largely due to undecomposed lignin. Solid blocks of wood are used for testing whether a fungus causes brown rot or white rot.

capitate: Having a swollen head.

carpophore: The fruiting body of higher fungi.

casing: A layer of water-retaining materials applied to a substrate to encourage and enhance fruitbody production.

cheilocystidia: Variously shaped sterile cells on the gill edge of mushrooms.

clamp connection: A small, semicircular, hollow bridge laterally attached to the walls of 2 adjoining cells and spanning the septum between them. See figure 73.

collyboid: Resembling mushrooms typical of the genus *Collybia*—groups of mushrooms clustered together at the base and having convex to planar caps.

conidia: A uninucleate, exteriorly borne cell formed by constriction of the conidiophore.

conidiophore: A specialized stalk arising from mycelium upon which conidia are borne.

context: The internal flesh of the mushroom, existing between the differentiated outer layers of the mushroom.

coprophilic: Dwelling on and having an affection for manure.

cystidia: Microscopic sterile cells arising from the gill, cap, or stem.

deciduous: Used to describe trees that seasonally shed their leaves.

decurrent: The attachment of the gill plates to the stem of a mushroom wherein the gills partially run down the stem.

dikaryotic: The state wherein 2 individual nuclei are present in each fungal cell.

deliquescing: The process of autodigestion by which the gills and cap of a mushroom melt into a liquid. Typical of some members in the genus *Coprinus*.

dimitic hyphae: Fungal flesh typified by 2 kinds of hyphae.

dimorphic: Having 2 forms, often used to describe a species that has a conidial and ascomycetous state.

diploid: A genetic condition wherein each cell has a full complement of chromosomes necessary for sexual reproduction, denoted as 2N.

disc: The central portion of the mushroom cap.

eccentric: Off centered.

ellipsoid: Oblong shaped.

endospores: Spores formed internally.

evanescent: Fragile and soon disappearing.

farinaceous: Grainlike, usually in reference to the scent of mycelium or mushrooms.

fermentation: The state of actively growing microorganisms, usually in a liquid environment.

fibrillose: Having fibrils, or small "hairs."

filamentous: Composed of hyphae or threadlike cells.

flexuose, flexuous: Bent alternately in opposite directions.

flush: A crop of mushrooms collectively forming within a defined time period, often repeating in a rhythmic fashion.

fruitbody: The mushroom structure.

fruiting: The event of mushroom formation and development.

generative hyphae: The thin-walled, branched, and narrow cells that give rise to the spore-producing layers and surface tissues. Species typified by clamp connections will have clamps at the septa of the generative hyphae.

genotype: The total genetic heritage or constitution of an organism, from which individual phenotypes are expressed.

heterothallic: Having 2 or more morphologically similar pairs of strains within the same species. The combination of compatible spore types is essential for producing fertile offspring.

homothallic: Having 1 strain type that is dikaryotic and self-fertile, typically of mushrooms that produce 2 spores on a basidium.

hymenium: The fertile outer layer of cells from which basidia, cystidia, and other cells are produced.

hypha, hyphae: The individual fungal cell.

hyphal aggregates: Visible clusters of hyphae resembling cottony tufts of mycelium, often preceding but not necessarily leading to primordia formation.

hyphosphere: The microscopic environment in direct proximity to the hyphae.

karyogamy: The fusion of 2 sexually opposite nuclei within a single cell.

lageniform: Thin and sinuous.

lamellae: The gills of a mushroom, located on the underside of the cap.

lamellulae: The short gills originating from the outer peripheral edge of the cap but fully extending to the stem.

lignicolous: Growing on wood or a substrate composed of woody tissue.

lignin: The organic substance that, with cellulose, forms the structural basis of most woody tissue.

macroscopic: Visible to the naked eye.

meiosis: The process of reduction division by which a single cell with a diploid nucleus subdivides into 4 cells with 1 haploid nucleus each.

membranous: Being sheathlike in form.

micron: one millionth of a meter.

mitosis: The nonsexual process of nuclear division in a cell by which the chromosomes of 1 nucleus are replicated and divided equally into 2 daughter nuclei.

monokaryon: The haploid state of the mushroom mycelium, typically containing 1 nucleus.

monomitic: Fungal flesh consisting only of thin-walled, branched, and narrow (generative) hyphae.

myceliated: A term I made up to describe the condition whereby the mycelium has colonized or infused through a substrate.

mycelium, mycelia: Fungal network of threadlike cells.

mycology: The study of fungi.

mycophagist: A person or animal that eats fungi.

mycophile: A person who likes mushrooms.

mycophobe: A person who fears mushrooms.

mycorrhizal: A symbiotic state wherein mushroom mycelium forms on or in the roots of trees and other plants.

mycosphere: The environment in which the mycelium operates.

mycotopia: A term I coined to describe an environment in which fungi are actively used to enhance or preserve ecological equilibrium.

natural culture: The cultivation of mushrooms outdoors, benefiting from natural weather conditions.

nucleus: Concentrated mass of differentiated protoplasm in cells containing chromosomes and playing an integral role in the reproduction and continuation of genetic material.

oidia: Conidia (spores) borne in chains.

pan/panning: The dieback of mycelium. May be caused by a variety of factors, but primarily sudden drying after wetting.

parasite: An organism living on another living species and deriving its sustenance to the detriment of the host.

partial veil: The inner veil of tissue extending from the cap margin to the stem and at first covering the gills of mushrooms.

pasteurization: The rendering of a substrate to a state where competitor organisms are at a disadvantage, allowing mushroom mycelium to flourish. Steam or hot water is usually used; biological and chemical pasteurization are alternative methods.

phenotype: The observable physical characteristics expressed from the genotype.

photosensitive: Sensitive to light.

phototropic: Growing toward light.

pileocystidia: Sterile cells on the surface of the cap.

pileus: The mushroom cap.

pinhead: A dotlike form that develops into a mushroom. The pinhead is the earliest visible indication of mushroom formation.

pleurocystidium, pleurocystidia: The sterile cells on the surface of mushroom gills, distinguished from those sterile cells occurring on their outer edges.

primordium, primordia: The mushroom at the earliest stage of growth, synonymous with "pinhead." See figure 14.

radicate: Tapering downward. Having a long rootlike extension of the stem.

rhizomorph: A thick stringlike strand of mycelium. A rhizomorph can consist of 1 enlarged cell or many, usually braided.

rhizosphere: The space encompassing the rhizomorph or the zone around the roots of plants.

saprophyte: A fungus that lives on dead organic matter.

sclerotium, sclerotia: A resting stage of mycelium typified by a mass of hardened mycelium resembling a tuber and from which mushrooms, mycelia, or conidia can arise. Sclerotia are produced by both Ascomycetes and Basidiomycetes.

sector: Usually used to describe fans of mycelium morphologically distinct from the type of mycelium preceding and bordering it.

senescence: The state whereby a living organism declines in vigor due to age and becomes susceptible to disease, characterized by decreasing cell viability, decreasing cell divisions, and a loss in DNA copying ability.

septate: Cells with distinct walls.

septum, septa: Structural divisions between cells, i.e., cell walls.

skeletal hyphae: Coarse, inflated cellular network consisting of thick-walled, unbranched cells lacking cross walls. Skeletal hyphae give mushrooms a tough, fibrous texture, especially at the stem base. Except for the basal cell, they are typically clampless.

spawn: Any material impregnated with mycelium, the aggregation of which is used to inoculate more massive substrates.

species: A biologically discrete group of individuals that are cross-fertile and give rise to fertile progeny.

sporeless strains: Strains that don't produce spores. Sporeless oyster strains are highly sought after, given the health problems associated with growing these mushrooms indoors.

spore: A reproductive cell or "seed" of fungi, bacteria, and plants. In fungi, these discrete cells are used to spread fungi to new ecological niches and are essential in the recombination of genetic material.

sporocarp: Any fruitbody that produces spores.

sterilization: The rendering of a substrate to a state where all life-forms have been made inviable by means of heat (steam), chemicals, gas, UV radiation, pressure, molecular stressing, or radioactivity. Sterilization usually implies prolonged exposure to temperatures at or above the boiling point of water (212°F; 100°C) at or above atmospheric pressure.

stipe: The stem of a mushroom.

strain: A race of individuals within a species sharing a common genetic heritage but differing in some observable set of features which may or may not be taxonomically significant.

stroma: A dense, cushionlike aggregation of mycelium forming on the surface of a substrate; generally does not lead to fruitbody formation.

subhymenium: The layer of cells directly below the hymenium.

substrate: Straw, sawdust, compost, soil, or any organic material on which mushroom mycelium will grow.

super-pasteurization: Prolonged pasteurization utilizing steam, typically for 12 to 48 hours at or near 212°F (100°C) at or near atmospheric pressure. Super-pasteurization is a method commonly used to render sawdust substrates, in bulk, into a form usable for the cultivation of shiitake, oyster, reishi, and similar mushrooms.

taxon, taxa: A taxonomic unit, usually in reference to a species.

thermogenesis: The natural and spontaneous escalation of temperature occurring in substrates as fungi, bacteria, and other microorganisms flourish.

through-spawning: Mixing spawn evenly throughout the substrate.

top-spawning: Placing spawn as a layer on the top of a substrate.

trama: The internal layers of cells between the gills of mushrooms.

universal veil: An outer layer of tissue enveloping the cap and stem of some mushrooms, best seen in the youngest stages of fruitbody development.

variety: A subspecies epithet used to describe a consistently appearing variation of a particular mushroom species.

vector: The pathway through or carrier on which an organism travels or the pathway by which DNA is expressed.

veil: A tissue covering mushrooms as they develop.

white rot: A condition whereby a substrate is rendered light in color from the fungal decomposition of lignin (delignification), leaving cellulose largely intact. Solid blocks of wood are used for testing whether a fungus causes white rot or brown rot.

(For more information on mycological terminology, I highly recommend *Ainsworth and Bisby's Dictionary of the Fungi*, 9th edition, edited by P. M. Kirk, P. F. Cannon, J. C. David, and J. A. Stalpers, published by CABI Science International, Surrey, England; available through www.fungi.com.)

➤ **FIGURE 361**

The author holds a *Rhizopogon* species and *Ganoderma oregonense* deep in the old-growth forest. To communicate with the author, send all inquiries to info@fungi.com.

RESOURCES

Field Guides for Identifying Mushrooms

General Identification

Edible and Poisonous Mushrooms of the World by Ian R. Hall, Steven L. Stephenson, Peter K. Buchannan, Wang Yun, and Anthony L. J. Cole, 2003. Timber Press, Portland, Oregon.

Eyewitness Handbooks: Mushrooms by Thomas Laessoe and Gary Lincoff, 1998. DK Publishing, New York.

Mushrooms of North America by Roger Phillips, 1991. Little, Brown and Company, Boston.

Mushrooms of the World by Giuseppe Pace, Firefly Books, 1998. Willowdale, Ontario.

National Audubon Society Field Guide to North American Mushrooms by Gary Lincoff, 1991. Alfred A. Knopf, New York.

Psilocybin Mushrooms of the World by Paul Stamets, 1996. Ten Speed Press, Berkeley, California.

Eastern North America

Mushrooms of Cape Cod and the National Seashore by Arleen R. Bessette, Alan E. Bessette, and William J. Neill, 2001. Syracuse University Press, Syracuse, New York.

Mushrooms of Northeastern North America by Alan E. Bessette, Arleen R. Bessette, and David W. Fischer, 1997. Syracuse University, Syracuse, New York.

Mushrooms of Ontario and Eastern Canada by George Barron, 1999. Lone Pine Publishing, Edmonton.

Southern United States

A Field Guide to Southern Mushrooms by Nancy Smith-Weber and Alexander Smith, 1985. University of Michigan Press, Ann Arbor, Michigan.

Western United States

All That the Rain Promises and More by David Arora, 1991. Ten Speed Press, Berkeley, California.

Mushrooms of Colorado and the Southern Rocky Mountains by Vera Stucky Evenson, 1997. Denver Public Gardens, Denver, Colorado.

Mushrooms Demystified by David Arora, 1986, second edition. Ten Speed Press, Berkeley, California.

The New Savory Wild Mushroom by Margaret Mckenny and Daniel Stuntz, revised and enlarged by Joseph F. Ammirati, 1997. University of Washington Press, Seattle, Washington.

Subtropics to Tropics

Hong Kong Mushrooms by S. T. Chang, 1995. Chinese University of Hong Kong, Hong Kong, China.

Mushrooms of Hawaii: An Identification Guide by Don E. Hemmes and Dennis E. Desjardin, 2002. Ten Speed Press, Berkeley, California.

Recommended Cookbooks

A Cook's Book of Mushrooms by Jack Czarnecki, 1995. Artisan, New York.

Cooking with Mushrooms by John Pisto, 1997. Pisto's Kitchen, Pacific Grove, California.

The Complete Mushroom: The Quiet Hunt by A. Carluccio, 2003. Rizzoli International Publishers, New York.

Hope's Mushroom Cookbook by Hope Miller, 1993. Mad River Press, Eureka, California.

The Wild Vegetarian Cookbook by Steve Brill, 2002. Harvard Commons Press, Boston, Massachusetts.

Recommended Mycological Journals

Mycologia
The Mycological Society of America
c/o Allen Press
Box 1897
Lawrence, KS 66044
(800) 627-0629
www.mycologia.org

Mycologist
Cambridge University Press
100 Brook Hill Drive
West Nyack, NY 10994
(845) 353-7500
http://journals.cambridge.org

Mycological Research
Cambridge University Press
100 Brook Hill Drive
West Nyack, NY 10994
(845) 353-7500
http://journals.cambridge.org

Mycotaxon
Mycotaxon, Ltd.
Box 254
Ithaca, NY 14851
(607) 273-0508
www.mycotaxon.com

McIlvainea: Journal of American Amateur Mycology
North American Mycological Association
6615 Tudor Ct.
Gladstone, OR 97027
(503) 657-7358
www.namyco.org/publications/mcillvainea.html

Recommended Sources for Mushroom Spawn

I encourage supporting small spawn laboratories servicing local communities. Please send me your recommendations of spawn laboratories, preferably certified organic, whose mission seems compatible with the philosophy stated in this book. I will select the best and add them over time. If my business (www.fungi.com) or others listed below do not have the spawn that you need, see www.fungi.net.

North America

Fungi Perfecti, LLC
Box 7634
Olympia, WA 98507
(800) 780-9126
www.fungi.com

Europe

Mycelia
Jean Bethunestraat 9
9040 Gent, Belgium
+32 9 228 70 90
www.mycelia.be

Asia

The Arunyik Mushroom Center
Box 1
Bangkok 10162 Thailand

New Zealand

Mushroom Gourmet
82 McEntee Road
Waitakere Auckland 1007
http://homepages.ihug.co.nz/~mushspor/index.htm

BIBLIOGRAPHY

Abrams, D. 2002. Personal communication.

Abrams, D. 2004. Antihyperlipidemic effects of oyster mushrooms. Clinical trials. National Center for Complementary and Alternative Medicine (NCCAM), National Institutes of Health. (http://www.clinicaltrials.gov/ct/show/NCT00069524?amp;order=9)

Adachi, Y., N. Ohno, M. Ohsawa, S. Oikawa, and T. Yadomae. 1990. Change of biological activities of (1‡3)-beta-D-glucan from *Grifola frondosa* upon molecular weight reduction by heat treatment. *Chemical & Pharmaceutical Bulletin* 38(2): 477–481.

Adachi, Y., N. Ohno, and T. Yadomae. 1998. Activation of murine kupffer cells with gel-forming (1‡3)-beta-D-glucan from *Grifola frondosa*. *Biological & Pharmaceutical Bulletin* 21(3): 278–283.

Adachi, Y., M. Okazaki, N. Ohno, and T. Yadomae. 1994. Enhancement of cytokine production by macrophages stimulated with (1‡3)-beta-D-glucan, grifolan (GRN), isolated from *Grifola frondosa*. *Biological & Pharmaceutical Bulletin* 17(12): 1554–1560.

Adaskaveg, J. E., and R. L. Gilbertson. 1986. Cultural studies and genetics of sexuality of *Ganoderma lucidum* and *G. tsugae* in relation to the taxonomy of the *G. lucidum* complex. *Mycologia* 78(5): 694–705.

Adaskaveg, J. E., and R. L. Gilbertson. 1987. Vegetative incompatibility between intraspecific dikaryotic pairings of *Ganoderma lucidum* and *G. tsugae*. *Mycologia* 79(4): 603–613.

Adl, S., A. Simpson, M. Farmer, R. Andersen, O. Anderson, J. Barta, S. Bowser, G. Brugerolle, R. Fensome, S. Fredericq, T. James, S. Karpov, P. Kugrens, J. Krug, C. Lane, L. Lewis, J. Lodge, D. Lynn, D. Mann, R. McCourt, L. Mendoza, Ø. Moestrup, S. Mozley-Standridge, T. Nerad, C. Shearer, A. Smirnov, F. Spiegel, and M. Taylor. 2005. The new higher level classification of eukaryotes with emphasis on the taxonomy of protists. *Journal of Eukaryotic Microbiology* 52(5): 399–451.

Alexander, S. J., D. Pilz, N. S. Weber, E. Brown, and V. A. Rockwell. 2002. Mushrooms, trees and money: Value estimates of commercial mushrooms and timber in the Pacific Northwest. *Environmental Management* 30(1): 129–141.

Alfthan, G. V. 2000. Selenium and mercury in wild mushrooms—Temporal and geographical variation and nutritional aspects. *Selenium 2000 Conference*, Venice, October 1–5.

Allison, C., and D. Tait. 2000. The application of decision analysis to forest road deactivation problems—An example in coastal British Columbia. *Streamline: Watershed Restoration Technical Bulletin* 5(2): 1–11.

Amaranthus, M. P., D. Page-Dumroese, A. E. Harvey, and E. Cazares. 1996. *Soil Compaction and organic matter affect conifer seedling and nonmycorrhizal and ectomycorrhizal abundance*. Portland, Ore.: USDA Forest Service, Pacific Northwest Research Station. PNW-RP-494.

Amaranthus, M., and J. Trappe. 1993. Effects of erosion on ecto- and VA-mycorrhizal inoculum potential of soil following forest fire in southwest Oregon. *Plant and Soil* 150: 41–49.

Ammirati, J. 2004. Personal communication.

Andreacchi, A. 1995. Characterization of AA 567, a coronary vasodilator and Ca-channel antagonist produced by a basidiomycete species YL8006. Ph.D. dissertation, University of Rochester, Department of Chemical Engineering.

Andreacchi, A., T. Wang, and J. H. Wu. 1997. Cardio-vascular effects of the fungal extract of basidiomycetes sp. YL8006. *Life Sciences* 60(22): 1987–1994.

Aoki, M., T. Motomu, A. Fukushima, T. Hieda, S. Kubo, M. Takabayashi, K. Ono, and Y. Mikami. 1993. Antiviral substances with systemic effects produced by Basidiomycetes such as *Fomes fomentarius. Bioscience, Biotechnology, and Biochemistry* 57(2): 278–293.

Aoki, T., Y. Usuda, H. Miyakoshi, K. Tamura, and R. B. Herberman. 1987. Low natural killer syndrome: Clinical and immunologic features. *Natural Immunity and Cell Growth Regulation* 6: 116–128.

Arica, M. Y., C. Arpa, B. Kaya, S. Bektas, A. Denizli, and O. Genc. 2003. Comparative biosorption of mercuric ions from aquatic systems by immobilized live and heat-inactivated *Trametes versicolor* and *Pleurotus sajor-caju. Bioresource Technology* 89(2): 145–154.

Arnebrant, K., H. Ek, R. D. Finlay, and B. Sodersom. 1993. Nitrogen translocation between *Alnus glutinosa* (L.) Gaertn. seedlings inoculated with *Frankia* sp. and *Pinus contorta Doug.ex Loud* seedlings connected by a common ectomycorrhizal mycelium. *New Phytologist* 124: 213–242.

Arnold, A. E., and E. A. Herre. 2003. Canopy cover and leaf age affect colonization by tropical fungal endo-phytes: Ecological pattern and process in *Theobroma cacao* (Malvaceae). *Mycologia* 95(3): 388–398.

Arnold, A. E., L. C. Meija, D. Kyllo, E. I. Rojas, Z. Maynard, N. Robbins, and E. A. Herre. 2003. Fungal endophytes limit pathogen damage in a tropical tree. *Proceedings of the National Academy of Sciences* 100(23): 15649–15654.

Arora, D. 1986. *Mushrooms demystified.* Berkeley, Calif: Ten Speed Press.

Arora, D. 1991. *All that the rain promises and more.* Berkeley, Calif: Ten Speed Press.

Babitskaya, V., N. Bisko, N. Y. Mitropolskaya, and N. V. Ikonnikova. 2002. Melanin complex from medicinal mushroom *Inonotus obliquus* (Pers.: Fr.) Pilat (Chaga) (Aphyllophoromycetideae). *International Journal of Medicinal Mushrooms* 4(2).

Badham, E. R. 1982. Tropisms in the mushroom *Psilocybe cubensis. Mycologia* 74(2): 275–279.

Bagley, S. 1999. Desert road removal: Creative restora-tion techniques. *The Road-RIPorter* 4:4.

Baldrian, P., and J. Gabriel. 2003. Lignocellulose degra-dation by *Pleurotus ostreatus* in the presence of cad-mium. *FEMS Microbiology Letters* 220(2): 235–240.

Baldrian, P., C. in der Wiesche, J. Gabriel, F. Nerud, and F. Zadrazil. 2000. Influence of cadmium and mercury on activities of lignolytic enzymes and degradation of polycyclic aromatic hydrocarbons by *Pleurotus ostreatus* in soil. *Applied Environmental Microbiology* 66(6): 2471–2478.

Bao, X. F., Y. Zhen, L. Ruan, and J. N. Fang. 2002. Purification, characterization and modification of T. lymphocyte-stimulating polysaccharide from spores of *Ganoderma lucidum. Chemical & Pharmaceutical Bulletin* 50(5): 623–629.

Baum, S., N. Thomas, F. Sieber, W. M. R. Schwarze, and S. Fink. 2003. Latent infections of *Fomes fomentarius* in the xylem of European beech (*Fagus sylvatica*). *Mycological Progress* 2(2): 141–148.

Bebber, D. P., Hynes, J., Darrah, P. R., Boddy, L., and M. D. Fricker. 2007. Biological solutions to transport network design. *Proceedings of the Royal Society* 274: 2307–2315.

Beelman, R. 2003. Executive summary. Nutritional Research Advisory Panel Meeting, September 17. Washington, D.C.: Mushroom Council, American Mushroom Institute.

Bellaby, M. D. 2004. In long shadow of Chernobyl. *Philadelphia Inquirer.* November 25.

Berch, S. M., and W. Cocksedge. 2003. Commercially important wild mushrooms and fungi of of British Columbia: What the buyers are buying. British Columbia Ministry of Forest Science Program. Abstract of Technical Report 006.

Bessette, A. E., A. R. Bessctte, and D. W. Fischer. 1997. *Mushrooms of northeastern North America.* Syracuse, N.Y.: Syracuse University Press.

Bhatt, M., T. Cajthaml, and V. Sasek. 2002. Myco-remediation of PAH-contaminated soil. *Folia Micro-biologica* 47(3): 255–258.

Black, A. K., and C. A. Reddy. 1991. Cloning and char-acterization of a lignin peroxidase gene from the white rot fungus *Trametes versicolor. Biochemical and Bio-physical Research Communications* 179(1): 428–435.

Boa, E. 2004. *Wild edible fungi: A global overview of their use and importance to people.* Rome, Italy: Food and Agriculture Organization of the United Nations. FAO Technical Paper: Non-Wood Forest Products 17.

Bobek, P., and S. Galbavy. 2001. Effect of pleuran (beta-glucan from *Pleurotus ostreatus*) on the anti-oxidant status of the organism and on dimehtylhydrazine-induced pre-cancerous lesions in rat colon. *British Journal of Biomedical Science* 58(3): 164–168.

Bobek, P., E. Ginter, M. Jurcovicova, and L. Kuniak. 1999. Reviews for selected medicinal properties of mushrooms. Cholesterol reducing effect of Pleurotus species (Agaricomycetideae). *International Journal of Medicinal Mushrooms* 1: 371–380.

Bobek, P., L. Ozdin, and S. Galbavy. 1998. Dose- and time-dependent hypocholesterolemic effect of oyster mushroom *(Pleurotus ostreatus)* in rats. *Nutrition* 14(3): 282–286.

Bobek, P., O. Ozdin, and M. Mikus. 1995. Dietary oyster mushroom *(Pleurotus ostreatus)* accelerates plasma cholesterol turnover in hypercholesterolaemic rat. *Physiological Research* 44(5): 287–291.

Boyce, C.K., C. Hotton, M. Fogel, G. Cody, R. Hazen, and F. Hueber. 2007. Devonian landscape heterogeneity recorded by a giant fungus. *Geology* 35: 399–402.

Brandt, C. R., and F. Piraino. 2000. Mushroom antivirals. *Recent Research Developments for Antimicrobial Agents and Chemotherapy* 4:11–26.

Bresinski, A., M. Fischer, B. Meixner, and W. Paulus. 1987. Speciation in *Pleurotus*. *Mycologia* 79: 234–245.

Bressa, G., L. Cima, and P. Costa. 1988. Bioaccumulation of Hg in the mushroom *Pleurotus ostreatus*. *Ecotoxicology and Environmental Safety* 16(2): 85–89.

Burczyk, J., A. Gawron, M. Slotwinska, B. Smietana, and K. Terminska. 1996. Antimitotic activity of aqueous extracts of *Inonotus obliquus*. *Bollettino chimico farmaceutico* 135(5): 306–309.

Burford, E. P., M. Kierans, and G. M. Gadd. 2003. Geomycology: Fungi in mineral substrata. *Mycologist* 17: 98–107.

Caeser-TonThat, T. C. 2002. Soil binding properties of mucilage produced by a basidiomycete fungus in a model system. *Mycological Research* 106(8): 930–937.

Caeser-TonThat, T. C., and V. L. Cochran. 2000. Role of a saprophytic basidiomycete soil fungus in aggregate stabilization. In *Sustaining the global farm—Selected papers from the 10th International Soil Conservation Organization, May 24–29, 1999*, ed. D. E. Stott, R. H. Mohtar, and G. C. Steinhardt, pp. 575–579. West Lafayette, Ind.: Purdue University.

Cajthaml, T., M. Bhatt, V. Sasek, and V. Mateju. 2002. Bioremediation of PAH-contaminated soil by composting: A case study. *Folia Microbiologica* 47(6): 696–700.

Cajthaml, T., M. Moder, P. Kacer, V. Sasek, and P. Popp. 2002. Study of fungal degradation products of polycyclic aromatic hydrocarbons using gas chromatography with ion trap mass spectrometry detection. *Journal of Chromatography* A, 974: 213–222.

Carlile, M. J., S. Watkinson, and G. W. Gooday. 2001. *The fungi*. San Diego, Calif.: Elsevier Science & Technology.

Cao, L. Z., and Z. B. Lin. 2002. Regulation on maturation and function of dendritic cells by *Ganoderma lucidum* polysaccharides. *Immunology Letters* 83(3): 163–169.

Capasso, L. 1998. 5300 years ago, the Ice Man used natural laxatives and antibiotics. *Lancet* 352:1864.

Chairul, S. M., and Y. Hayashi. 1994. Lanostanoid triterpenes from *Ganoderma applanatum*. *Phytochemistry* 35(5): 1305–1308.

Chang, H. M., and P. P. But. 1986. *Pharmacology and applications of Chinese Materia Medica*. Vol.1. Singapore: World Scientific.

Chang, S. T., and W. A. Hayes. 1978. *The biology and cultivation of edible mushrooms*. New York: Academic Press.

Chapman, B., G. Xiao, and S. Meyers. 2004. Early results from field trails using *Hypholoma fasciculare* to reduce *Armillaria ostoyae* root disease. *Canadian Journal of Botany* 82: 962–969.

Chen, A., and P. Miles. 1996. Biomedical research and the application of mushroom nutriceuticals from *Ganoderma lucidum*. In *Mushroom biology and mushroom products*. University Park: Penn State University.

Chen, K., and W. Zhang. 1987. Advances in anti-aging herbal medicines in China. *Abstracts of Chinese Medicines* 1: 309–330.

Chen, S., Y. C. Kao, and C. A. Laughton. 1997. Binding characteristics of aromatase inhibitors and phytoestrogens to human aromatase. *The Journal of Steroid Biochemistry and Molecular Biology* 61: 107–115.

Chen, W. C., D. M. Hau, C. C. Wang, I. H. Lin, and S. S. Lee. 1995. Effects of *Ganoderma lucidum* and krestin on subset T-cell in spleen of gamma-irradiated mice. *American Journal of Chinese Medicine* 23(3-4): 289–298.

Chiu, S. W., M. L. Ching, K. L. Fong, and D. Moore. 1998. Spent oyster mushroom substrate performs better than many mushroom mycelia in removing the biocide pentachlorophenol. *Mycological Research* 102: 1553–1562

Chou, S. 2004. Minimum risk levels (MRLs) for hazardous substances. Agency for Toxic Substances and Disease Registry, Centers for Disease Control. Updated May 11, 2004. www.atsdr.cdc.gov/mrls.html.

Clark, L. C., G. F. Combs, B. W. Turnbull, E. H. Slate, D. K. Chalker, J. Chow, L. S. Davis, R. A. Glover, G. F. Graham, E. G. Gross, A. Konrad, J. L. Lesher, H. K. Park, B. B. Sanders, C. L. Smith, and J. R. Taylor. 1996. Effects of selenium supplementation for cancer prevention in patients with carcinoma of the skin. *Journal of the American Medical Association* 276: 1957–1963.

Cobb, A. R., N. M. Nadkarni, G. A. Ramsey, and A. J. Svoboda. 2001. Recolonization of bigleaf maple branches by epiphytic bryophytes following experimental disturbance. *Canadian Journal of Botany* 79(1): 1–8.

Coffan, R., D. Southworth, and J. Frank. 2008, in press.

Collins, R. A., and T. B. Ng. 1997. Polysaccharopeptide from *Coriolus versicolor* has potential for use against human immunodeficiency virus type 1 infection. *Life Sciences* 60(25): PL383–387.

Crisan, E., and A. Sands. 1978. Nutritional value of edible mushrooms. In *The biology and cultivation of edible mushrooms*, ed. S. T. Chang and W. A. Hayes, pp. 137–168. New York: Academic Press.

Currie, C. R., B. Wong, A. E. Stuart, T. R. Schultz, S. A. Rehner, U. G. Mueller, G.-H. Sung, J. W. Spatafora, and N. A. Straus. 2003. Ancient tripartite coevolution in the attine ant–microbe Symbiosis. *Science* 299(5605): 386–388.

Dadachova, E., R. A. Bryan, X. Huang, T. Moadel, A. D. Schweitzer, P. Aisen, J. D. Nosanchuk, and A. Casadevall. 2007. Ionizing radiation changes the electronic properties of melanin and enhances the growth of melanized fungi. *PLoS ONE* 2(5):e457.

Dahlstrom, J. L., J. E. Smith, and N. S. Weber. 2000. Mycorrhiza-like interaction by *Morchella* with species of the Pinaceae in pure culture synthesis. *Mycorrhiza* 9: 279–285.

Dahncke, R. M. 1993. *1200 Pilze*. Aarau, Switzerland: AT Verlag.

De Siqueira, M. F., A. Grainger, L. Hannah, L. Hughes, B. Huntley, A. V. Jaarsveld, G. F. Midgley, L. Miles, M. A. Ortega-Huerta, A. T. Peterson, O. L. Phillips, and S. E. Williams. 2004. Extinction risk from climate change. *Nature* 427(6970): 145–148.

Dong, Y., C. Y. Kwan, S. N. Chen, and M. Yang. 1996. Antitumor effects of a refined polysaccharide peptide fraction isolated from *Coriolus versicolor*: In vitro and in vivo studies. *Research Communications in Molecular Pathology and Pharmacology* 92(2): 140–147.

Dong, Y., M. M. Yang, and C. Y. Kwan. 1997. In vitro inhibition of proliferation of HL-60 cells by tetrandrine and *Coriolus versicolor* peptide derived from Chinese medicinal herbs. *Life Sciences* 60(8): PL135–140.

Duke, J. 1999. Personal communication.

Duncan, C. J. G., N. Pugh, D. S. Pasco, and S. A. Ross. 2002. Isolation of a galactomannan that enhances macrophage activation from the edible fungus *Morchella esculenta*. *Journal of Agricultural and Food Chemistry* 50: 5683–5685.

Eberhart, J. L., D. L. Luoma, D. Pilz, M. P. Amaranthus, R. Abbott, and D. Segotta. 1999. Effects of harvest techniques on American Matsutake *(Tricholoma magnivelare)* production. *Proceedings from the IXth International Congress of Mycology*. Sydney, Australia. http://www.matsiman.com/formalpubs /harvestmethodposter/harmethposter.htm.

Eggen, T., and V. Sasek. 2002. Use of edible and medicinal oyster mushroom [*Pleurotus ostreatus* (Jacq.:Fr.) Kimm.] spent compost in remediation of chemically polluted soils. *International Journal of Medicinal Mushrooms* 4: 225–261.

Eo, S. K., Y. S. Kim, C. K. Lee, and S. S. Han. 1999. Antiviral activities of various water and methanol soluble substances isolated from *Ganoderma lucidum*. *Journal of Ethnopharmacology* 68(1-3): 129–136.

Eo, S. K., Y. S. Kim, C. K. Lee, and S. S. Han. 2000. Possible mode of antiviral activity of acidic protein bound polysaccharide isolated from *Ganoderma lucidum* on herpes simplex viruses. *Journal of Ethnopharmacology* 72(3): 475–481.

Epik, O., and G. Yaprak. 2003. The mushrooms as bioindicators of radioscesium in forest ecosystem. *Proceedings of the Fifth General Conference of the Balkan Physical Union*, Vrnjaāka Banja, Serbia and Montenegro, August 25–29.

Faeth, S. H. 2002. Are endophytic fungi defensive plant mutualists? *Oikos* 98: 25–36.

Favero, N., G. Bressa, and P. Costa. 1990. Response of *Pleurotus ostreatus* to cadmium exposure. *Ecotoxicology and Environmental Safety* 20(1): 1–6.

Ferguson, B. A., T. A. Dreisbach, C. G. Parks, G. M. Filip, and C. L. Schmitt. 1998. Coarse-scale population structure of pathogenic *Armillaria* species in a mixed-conifer forest in the Blue Mountains of northeast Oregon. *Canadian Journal of Forest Research* 33(4): 612–623.

Field, J. A., E. de Jong, G. Costa Feijoo, and J. A. de Bont. 1992. Biodegradation of polycyclic aromatic hydrocarbons by new isolates of white rot fungi. *Journal of Applied and Environmental Microbiology* 589(7): 2219–2226.

Fielitz, V. U. 2001. Radioökologie: Abschlussbericht zum Forschungsvorhaben StSch 4206 im Auftrag des Bundesministeriums für Umwelt, Naturschutz und Reaktorsicherheit, Überprüfung von Ökosystemen nach Tschernobyl hinsichtlich der Strahlenbelastung der Bevölkerung.

Fisher, M., J. Jun, H. J. Wang, J. Chevrier, and L. X. Yang. 2003. In vitro cytotoxic and pro-apoptotic effects of *Coriolus versicolor* preparation on human carcinoma cell lines. San Franscisco, California: Radiology Laboratory, St. Mary's Medical Center, California Pacific Medical Research Institute.

Fujimiya, Y., Y. Suzuki, K. Oshiman, H. Kobori, K. Moriguchi, H. Nakashima, Y. Matumoto, S. Takahara, T. Ebina, and R. Katakura. 1998. Selective tumoricidal effect of soluble proteoglucan extracted from the basidiomycete, *Agaricus blazei* Murrill, mediated via natural killer cell activation and apoptosis. *Cancer Immunology and Immunotherapy* 46: 147–159.

Fujimoto, T. 1989. High speed year-round shiitake cultivation. *Shiitake News* 5(2): 108.

Fukuoka, M. 1978. *The One-Straw Revolution.* Emmaus, Penn.: Rodale Press.

Fullerton, S. A., A. A. Samadi, D. G. Tortorelis, C. Mallouh, H. Tazaki, and S. Kunno. 2000. Apoptosis in prostatic cancer cells with maitake mushroom extract: Potential alternative therapy. *Molecular Urology* 4(1): 7–13.

Gadd, G. M.1993. Interactions of fungi with toxic metals. *New Phytologist* 124: 25–60.

Gadd, G. M., ed. 2001. *Fungi in bioremediation.* Cambridge, UK: Cambridge University Press.

Gadd, G. M. 2004. Mycotransformation of organic and inorganic substrates. *Mycologist* 18(2): 60–70.

Gao, J. J., B. S. Min, E. M. Ahn, N. Nakamura, H. K. Lee, and M. Hattori. 2002. New triterpene aldehydes, lucialaldehydes A-C, from *Ganoderma lucidum* and their cytotoxicity against murine and human tumor cells. *Chemical & Pharmaceutical Bulletin* 50(6): 837–840.

Gao, X. X., X. F. Fei, B. X. Wang, J. Zhang, Y. J. Gong, M. Minami, T. Nagata, and T. Ikejima. 2000. Effects of polysaccharides (F10-b) from mycelium of *Ganoderma tsugae* on inflammatory cytokine production by THP-1 cells and human PBMC (I). *Acta Pharmacologica Sinica* 21(12): 1179–1185.

Garaudee, S., M. Elhabiri, D. Kalny, C. Robiolle, J. M. Trendel, R. Hueber, A. Van Dorsselaer, P. Albrecht, A. M. Albrecht-Gary. 2002. Allosteric effects in norbadione A: A clue for the accumulation process of 137Cs in mushrooms? *Chemical Communications* 9: 944–945.

Garcia, M. A., J. Alonso, M. I. Fernandez, and M. J. Melgar. 1998. Lead concentration in edible wild mushrooms in northwest Spain as indicator of environmental contamination. *Archives of Environmental Contamination and Toxicology* 34(4): 330–335.

Garcia-Lora, A., S. Pedrinaci, and F. Garrido. 2001. Protein-bound polysaccharide K and interleukin-2 regulate different nuclear transcription factors in the NKL human natural killer cell line. *Cancer Immunology and Immunotherapy* 50(4): 191–198.

Garrity, M. T. 1995. Economic impact of the Wildland Recovery System of the Northern Rockies Ecosystem Protection Act. Missoula, Mont.: Alliance for the Wild Rockies. Special Report 7.

Gau, J. P., C. K. Lin, S. S. Lee, and S. R. Lang. 1990. The lack of antiplatelet effect of crude extracts from *Ganoderma lucidum* on HIV-positive hemophiliacs. *American Journal of Chinese Medicine* 18(3-4): 175–179.

Ghoneum, M. 1995. Immunomodulatory and anticancer properties of (MGN-3), a modified xylose from rice bran, in 5 patients with breast cancer (abstract). Presented at the American Association for Cancer Research (AACR) special conference: The Interference between Basic and Applied Research. Baltimore, Md., November 5–8.

Ghoneum, M. 1998. Enhancement of human natural killer cell activity by modified arabinoxylane from rice bran (MGN-3). *International Journal of Immunotherapy* 14(2): 89–99.

Gilbertson, R. L., and L. Ryvarden. 1986–87. *North American Polypores.* 2 vols. Oslo, Norway: Fungiflora.

Gilbertson, R. L., and L. Ryvarden. 1993–94. *European Polypores.* 2 vols. Oslo, Norway: Fungiflora.

Gordon, M., B. Bihari, E. Goosby, R. Gorter, M. Greco, M. Guralnik, T. Mimura, V. Rudinicki, R. Wong, and Y. Kaneko. 1998. A placebo-controlled trial of the immune modulator, lentinan, in HIV-positive patients: A phase I/II trial. *Journal of Medicine* 29(5-6): 305–330.

Gorman, J., 2003. Microbial materials: Scientists co-opt viruses, bacteria and fungi to build new structures. *Science News* 164(1): 7–9.

Grob, C. 2004. Stage IV cancer patients and psilocybin. The Los Angeles Biomedical Research Institute at Harbor-UCLA Medical Center (LA BioMed) (http://www.medicalnewstoday.com/medicalnews.php?newsid=14202).

Gu, Y. H., and M. A. Belury. 2005. Selective induction of apoptosis in murine skin carcinoma cells (CH72) by an ethanol extract of *Lentinula edodes. Cancer Letters* 220(1): 21–28.

Gunde-Cimerman, N. 1999. Medicinal value of the genus *Pleurotus* (Fr.) P. Kast. (Agaricales s.l., Basidiomycetes). *International Journal of Medicinal Mushrooms* 1(1): 69–80.

Gunde-Cimerman, N. G., and A. Cimerman, 1995. *Pleurotus* fruiting bodies contain the inhibitor of 3-hydroxy-3-methylglutaryl coenzyme A reductase-lovastatin. *Experimental Mycology* 19(1): 1–6.

Gunde-Cimerman, N., and A. Plemenitas. 2002. *Pleurotus sporocarps:* A hypocholesterolemic nutraceutical. In *Proceedings of the 7th International Mycological Congress.* Oslo, Norway, August 11–17, p. 97.

Guzman, G. 1983. *The genus Psilocybe.* Lichtenstein, Germany: J. Cramer.

Hall, I., S. Stephenson, P. Buchanan, W. Yun, and T. Cole. 2003. *Edible and poisonous mushrooms of the world.* Portland, Ore.: Timber Press.

Han, M. D., E. S. Lee, and Y. K. Kim. 1998. Production of nitric oxide in raw 264.7 macrophages treated with ganoderan, the beta glucan of *Ganoderma lucidum. Korean Journal of Mycology* 26: 246–255.

Harris, B. 1986. *Growing shiitake commercially.* Madison, Wis.: Science Tech Publishers.

Hatvani, N. 2001. Antibacterial effect of the culture fluid of *Lentinus edodes* mycelium grown in submerged liquid culture. *International Journal of Antimicrobial Agents* 17(1): 71–74.

Hatvani, N., and L. Mecs. 2003. Effects of certain heavy metals on the growth, dye decolorization and enzyme activity of *Lentinula edodes. Ecotoxicology and Environmental Safety* 55(2): 199–203.

Hawken, P., A. Lovins, and L. H. Lovins. 1999. *Natural capitalism.* Boston: Little, Brown.

Hawkins, E. 2001. Arabinoxylane: Immune support from mushrooms. *Natural Pharmacy* 21: 26.

Henson, J. M., K. B. Sheehan, R. J. Rodriguez, and R. S. Redman. 2004. Use of endophytic fungi to treat plants. U.S. Patent Application 20,040,082,474. April 29, 2004.

Hickenbottom, J. A. S. 2000. *Comparison of sediment generation from existing and recontoured Forest Service roads.* Missoula, Mont.: USDA Forest Service, Lolo National Forest.

Hilborn, M. I. 1942. The biology of *Fomes fomentarius. Bulletin of the Maine Agricultural Experimental Station* 409.

Hirasawa, M., N. Shouji, T. Neta, K. Fukushima, and K. Takada. 1999. Three kinds of antibacterial substances from *Lentinus edodes* (Berk.) Sing. (shiitake, an edible mushroom). *International Journal of Antimicrobial Agents* 11(2): 151–157.

Hirotani M., and T. Furuya. 1986. Ganoderic acid derivatives, highly oxygenated lanostane-type triterpenoids, from *Ganoderma lucidum. Phytochemistry* 25: 1189–1193.

Hobbs, C. 1995. *Medicinal mushrooms: An exploration of tradition, healing and culture.* Lake Oswego, Ore.: Culinary Arts Ltd.

Hobbs, C. 2003. *Medicinal Mushrooms.* Summertown, Tenn.: Book Publishing Company.

Hong, S. G., and H. S. Jung. 2004. Phylogenetic analysis of Ganoderma based on nearly complete mitochondrial small-subunit ribosomal DNA sequences. *Mycologia* 96(4): 742–755.

Horner, W. E., E. Levetin, and S. B. Lehrer. 1993. Basidiospore allergen release: Elution from intact spores. *Journal of Allergy and Clinical Immunology* 92(2): 306–312.

Horwitz, W., ed. 2000. *Official methods of analysis of AOAC International.* Gaithersburg, Md.: AOAC International.

Hosford, D., D. Pilz, R. Molina, and M. Amaranthus. 1997. *Ecology and management of the commercially harvested American matsutake mushroom.* Portland, Ore.: USDA Forest Service, Pacific Northwest Research Station. PNW-GTR-412.

Hoshino, T., M. Kiriaki, S. Tsuda, S. Ohgiya, H. Kondo, Y. Yokata, and I. Yumoto. 2003. Antifreeze proteins from basidiomycetes. U.S. Patent Application 20,030,180,884. March 12, 2003.

Hseu, R. S., and H. H. Wang. 1991. A new system for identifying cultures of *Ganoderma* species. In *Science and cultivation of edible fungi,* ed. M. J. Maher, vol. 1, pp. 51–56. Rotterdam: Aa Balkema.

Hsieh, T. C., and J. M. Wu. 2001. Cell growth and gene modulatory activities of Yunzhi (Windsor Wunxi) from mushroom *Trametes versicolor* and in androgen-dependent and androgen-insensitive human prostate cancer cells. *International Journal of Oncology* 18(1): 81–88.

Humar, M., M. Bokan, S. A. Amartey, M. Sentjurc, P. Kalan, and F. Pohleven. 2004. Fungal bioremediation of copper, chromium and boron treated wood as studied by electron paramagnetic resonance. *International Biodeterioration & Biodegradation* 53: 25–32.

Iimura, Y., and K. Tatsumi. 1997. Isolation of mRNAs induced by a hazardous chemical in white-rot fungus, *Coriolus versicolor,* by differential display. *FEBS Letters* 12(2): 370–374.

Ikekawa, T. 2001. Beneficial effects of edible and medicinal mushrooms on health care. *International Journal of Medicinal Mushrooms* 3: 291–298.

Ikekawa, T. 2003. Personal communications.

Ikekawa, T., H. Maruyama, T. Miyano, A. Okura, Y. Sawaskai, K. Naito, K. Kawamura, and K. Shiratori. 1985. Proflamin, a new antitumor agent: Preparation, physicochemical properties and antitumor activity. *Japanese Journal of Cancer Research* 76: 142–148.

Ikekawa, T., N. Uehara, Y. Maeda, M. Nakanishi, and F. Fukuoka. 1969. Antitumor activity of aqueous extracts of edible mushrooms. *Cancer Research* 29: 734–735.

Imazeki, R. 1937. Reishi and *Ganoderma lucidum* that grow in Europe and America: Their differences. *Botany & Zoology* 5: 5.

Ito, H., K. Shimura, H. Itoh, and M. Kawade. 1997. Antitumor effects of a new polysaccharide-protein complex (ATOM) prepared from *Agaricus blazei* (Iwade strain 101) "Himematsutake" and its mechanisms in tumor-bearing mice. *Anticancer Research* 17(1A): 277–284.

Ito, H., M. Sugiura, and T. Miyazaki. 1976. Antitumor polysaccharide fraction from the culture filtrate of *Fomes fomentarius. Chemical & Pharmaceutical Bulletin* 24(10): 2575.

Itoh, H., H. Ito, H. Amano, and H. Noda. 1994. Inhibitory action of a (1‡6)-beta-D-glucan-protein complex (FIII-2-b) isolated from *Agaricus blazei* Murrill ("Himematsutake") on Meth A fibrosarcoma-bearing mice and its antitumor mechanism. *Japanese Journal of Pharmacology* 66(2): 265–271.

James, T. Y., and R. Vilgalys. 2001. Abundance and diversity of *Schizophyllum commune* spore clouds in the Caribbean detected by selective sampling. *Molecular Ecology* 10: 471–479.

Jane's Defence Weekly. 1999. Fungi could combat chemical weapons. *Jane's Defence Weekly* 32(7): 37.

Japans Times. 2001. Radiation detected in mushrooms. *The Japan Times Online,* Sunday, November 10. www.japantimes.co.jp/cgi-bin/makeprfy.pl5 ?nn20011110b8.htm.

Johannes, C., A. Majacherczyk, and A. Huttermann. 1996. Degradation of anthracene by laccase of *Trametes versicolor* in the presence of different mediator compounds. *Applied Microbiology and Biotechnology* 46(3): 313–317.

Kahlos, K., and R. Hiltunen. 1987. Gas chromatographic mass spectrometric study of some sterols and lupines from *Inonotus obliquus. Acta Pharmaceutica Fennica* 96(2): 85–89.

Kahlos, K., and R. Hiltunen. 1988. Gas chromatographic mass spectrometric identification of some lanostanes from *Inonotus obliquus. Acta Pharmaceutica Fennica* 97: 45–49.

Kahlos, K., R. Hiltunen, and T. Vares. 1990. Optimization of pH level and effect of pH on secondary metabolites of two strains of *Inonotus obliquus. Planta Medica* 56: 627.

Kahlos, K., L. Kangas, and R. Hiltunen. 1987. Antitumour activity of some compounds and fractions from an n-hexane extract of *Inonotus obliquus. Acta Pharmaceutica Fennica* 96: 33–40.

Kahlos, K., A. Lesnau, W. Lange, and U. Lindequist. 1996. Preliminary tests of antiviral activity of two *Inonotus obliquus* strains. *Fitopterapia* 6(4) 344–347.

Kahlos, K., M. V. Schantz, and R. Hiltunen. 1984. 3 ß-hydroxy-lanosta-8, 24-dien-21, a new triterpene from *Inonotus obliquus. Acta Pharmaceutica Fennica* 92: 197–198.

Kalac, P., J. Burda, and I. Staskov. 1991. Concentrations of lead, cadmium, mercury and copper in the vicinity of a lead smelter. *The Science of the Total Environment* 105: 109–119.

Kamm, Y. J., H. T. Folgering, and H. G. van den Bogart. 1991. Provocation tests in extrinsic allergic alveolitis in mushroom workers. *Netherlands Journal of Medicine* 38(1-2): 59–64.

Kamo, T., M. Asanoma, H. Shibata, and M. Hirota. 2003. Anti-inflammatory lansotane-type triterpene acids from *Piptoporus betulinus. Journal of Natural Products* 66(8): 1104–1106.

Kanamoto, T., Y. Kashiwada, K. Kanbara, K. Gotoh, and M. Yoshimori. 2001. Anti-human immunodeficiency virus activity of YI-FH 312 (a betulinic acid derivative), a novel compound blocking viral maturation. *Antimicrobial Agents and Chemotherapy* 45(4): 1225–1230.

Kariya, K., K. Nakamura, K. Nomoto, S. Matama, and K. Saigneji. 1992. Mimicking of superoxide dismutase activity by protein-bound polysaccharide of *Coriolus versicolor* QUEL., and oxidative stress relief for cancer patients. *Molecular Biotherapy* 4(1): 40–46.

Kasinath, A., C. Novotny, K. Svobodova, K. C. Patel, and V. Sasek. 2003. Decolorization of synthetic dyes by *Irpex lacteus* in liquid cultures and packed-bed bioreactor. *Enzyme and Microbial Technology* 32: 167–173.

Katagiri, N., Y. Tsutsumi, and T. Nishida. 1995. Correlation of brightening with cumulative enzyme activity related to lignin biodergradation during biobleaching of kraft pulp by white rot fungi in the solid-state fermentation system. *Journal of Applied and Environmental Biology* 61(2): 617–622.

Kawagishi, A., R. Shimada, R. Shirai, K. Okamoto, F. Ojima, H. Sakamoto, Y. Ishiguro, and S. Furukawa. 1991. Hericenones C, D and E, stimulators of nerve growth factor (NGF)-synthesis from the mushroom *Hericium erinaceum. Tetrahedron Letters* 32(35): 4561–4564.

Kawagishi, H. 2002. The inducer of the synthesis of nerve growth factor from lion's mane (*Hericium erinaceus*). *Explore!* 11(4): 4–51.

Kawagishi, H., K. Hamajima, and Y. Inoue. 2002. Novel hydroquinone as a matrix metallo-proteinase inhibitor from the mushroom, *Piptoporus betulinus. Bioscience, Biotechnology, and Biochemistry* 66(12): 2748–2750

Kawagishi, H., A. Nomura, T. Mizuno, A. Kimura, and S. Chiba. 1990. Isolation and characterization of a lectin from *Grifola frondosa* fruiting bodies. *Biochimica et Biophysica Acta* 1034(3): 247–252.

Kawagishi, H., A. Shimada, R. Shirai, K. Okamoto, F. Ojima, H. Sakamoto, Y. Ishiguro, and S. Furukawa. 1994. Erinacines A, B, C, strong stimulators of nerve growth factor synthesis, from the mycelia of *Hericium erinaceum. Tetrahedron Letters* 35(10): 1569–1572.

Kawakami, S., K. Minato, T. Hashimoto, H. Ashida, and M. Mizuno. 2002. TNF-A and NO production from macrophages is enhanced through up-regulation of NF-kB polysaccharides purified from *Agaricus blazei* Murrill. *Proceedings of the 7th International Mycological Congress.* Oslo, Norway, August 11–17.

Kenmoku, H., T. Shimai, T. Toyomasu, N. Kato, and T. Sassa. 2002. Erinacine Q, a new erinacine from *Hericium erinaceum*, and its biosynthetic route to erinacine C in the basidiomycete. *Bioscience, Biotechnology, and Biochemistry* 66(3): 571–575.

Kersten, P. J., B. Kalyanaraman, K. E. Hammel, B. Reinhammar, and T. K. Kirk. 1990. Comparison of lignin peroxidases, horseradish peroxidases and laccase in the odixation of methoxybenzenes. *Journal of Biochemistry* 268(2): 475–480.

Kiho, T., S. Sobue, and S. Ukai. 1994. Structural features and hypoglycemic activities of two polysaccharides from a hot-water extract of *Agrocybe cylindracea. Carbohydrate Research* 251: 81–87.

Kim, B. K., H. W. Kim, and E. C. Choi. 1994. Anti-HIV effects of *Ganoderma lucidum*. In *Ganoderma: Systematics, phytopathology and pharmacology: Proceedings of contributed symposium* 59 A,B. *5th International Mycological Congress.* Vancouver, August 14–21.

Kim, D. H., S. B. Shim, N. J. Kim, and I. S. Jang. 1999. Beta-glucuronidase-inhibitory activity and hepatoprotective effect of *Ganoderma lucidum. Biological & Pharmaceutical Bulletin* 22(2): 162–164.

Kim, H. S., S. Kacew, and B. M. Lee. 1999. In vitro chemopreventive effects of plant polysaccharides (*Aloe barbadensis* Miller, *Lentinus edodes, Ganoderma lucidum* and *Coriolus versicolor*). *Carcinogenesis* 20(8): 1637–1640.

Kim, W. G., I. K. Lee, J. P. Kim, I. J. Ryoo, H. Koshino, and I. D. Yoo. 1997. New indole derivatives with free radical scavenging activity from *Agrocybe cylindracea. Journal of Natural Products* 60(7): 721–723.

Kirby, A. 2002. Insect species "fewer than thought." *BBC News Online:* Sci/Tech, April 25. http://news.bbc.co.uk/1/low/sci/tech/1949109.stm

Knapp, J. S., E. J. Vantoch-Wood, F. Zhang. 2001. Use of wood-rotting fungi for the decolorization of dyes and industrial effluents. In *Fungi in Bioremediation*, ed. G. M. Gadd, pp. 242–304. Cambridge, UK: Cambridge University Press.

Kobayashi, H., K. Matsunaga, and Y. Oguchi. 1995. Antimetastatic effects of PSK (Krestin), a protein-bound polysaccharide obtained from basidiomycetes: An overview. *Cancer Epidemiology, Biomarkers & Prevention* 4(3): 275–281.

Kobayashi, Y., K. Kariya, K. Saigenji, and K. Nakamura. 1994a. Enhancement of anti-cancer activity of cisdiaminedicholoroplatinum by the protein-bound polysaccharide of *Coriolus versicolor* QUEL (PS-K) in vitro. *Cancer Biotherapy* 9(4): 351–358.

Kobayashi, Y., K. Kariya, K. Saigenji, and K. Nakamura. 1994b. Suppression of cancer cell growth in vitro by the protein-bound polysaccharide of *Coriolus versicolor* QUEL (PS-K) with SOD mimicking activity. *Cancer Biotherapy* 9(1): 63–69.

Kobayashi, H., K. Matsunaga, and M. Fujii. 1993. PSK as a chemopreventive agent. *Cancer Epidemiology, Biomarkers & Prevention* 2(3): 271–276.

Kodama, N., K. Komuta, and H. Nanba. 2002. Can maitake MD fraction aid cancer patients? *Alternative Medicine Review* 7(3): 236–239.

Kolotushkina, E. V., M. G. Moldavan, K. Y. Voronin, and G. G. Skibo. 2003. The influence of *Hericium erinaceus* extract on myelination process *in vitro. Fiziologicheskii zhurnal* 49(1): 38–45.

Komatsu, M., Y. Nozaki, A. Inoue, and M. Miyauchi. 1980. Correlation between temporal changes in moisture contents of the wood after felling and mycelial growth of *Lentinus edodes* (Berk.) Sing. *Report of the Tottori Mycological Institute* 18: 169–187.

Komatsu, N., H. Terakawa, K. Nakanishi, and Y. Watanabe. 1963. Flammulin, a basic protein protein of *Flammulina velutipes* with antitumor activities. *Journal of Antibiotics* (Tokyo) 16(3): 139–143.

Kondo, R., K. Sakai, and K. Wakao 2003. White rot fungi and method for decomposing dioxins using them. Bio Remediation Technologie, Inc. (JP); U.S. Patent 6,653,119. November 25, 2003.

Konno, S., D. G. Tortorelis, S. A. Fullerton, A. A. Samadi, J. Hettiarcarchchi, and H. Tazaki. 2001. A possible hypoglycaemic effect of maitake mushroom on Type 2 diabetic patients. *Diabetes Medicine* 18(12): 1010.

Kubo, K., H. Aoki, and H. Nanba. 1994. Anti-diabetic activity present in the fruit body *Grifola frondosa* (maitake). *Biological & Pharmaceutical Bulletin* 17(8): 1106–1110.

Kuo, D. D., and M. H. Kuo. 1983. *How to grow forest mushroom (shiitake)*. Naperville, Ill.: Mushroom Technology Corp.

Kuo, M. 2004. North American morels in the MDCP. Retrieved from the MushroomExpert.Com website: http://www.bluewillowpages.com/mushroomexpert /morels/mdcp_legend.html.

Kuo, M. 2005. *Morels*. Ann Arbor: University of Michigan Press.

Kurashige, S., Y. Akuzawa, and F. Endo. 1997. Effects of *Lentinus edodes, Grifola frondosa* and *Pleurotus ostreatus* administration on cancer outbreak, and activities of macrophages and lymphocytes in mice treated with a carcinogen, N-butyl-N-butanolnitrosoamine. *Journal of Immunopharmacology and Immunotoxicology* 19(2): 175–183.

Lal, R., M. Griffin, J. Apt, L. Lave, and M. Granger Morgan. 2004. Managing Soil Carbon. *Science* 304: 393.

Leatham, G. 1982. Cultivation of shiitake, the Japanese forest mushroom, on logs: A potential industry for the United States. *Forest Products Journal* 332: 29–35.

Lee, J. M., H. Kwon, H. Jeong, J. W. Lee, S. Y. Lee, S. J. Baek, and Y. J. Surh. 2001. Inhibition of lipid peroxidation and oxidative DNA damage by *Ganoderma lucidum*. *Phytotherapy Research* 15(3): 245–249.

Lelley, J. I., and J. Vetter. 2004. Orthomolecular medicine and mushroom consumption: An attractive aspect for promoting production. In *Science and Cultivation of Edible and Medicinal Fungi*, ed. C. P. Romaine, C. B. Keil, D. L. Rinker, and D. J. Royse, pp. 637–643. Penn State.

Li, M. C., L. S. Lei, D. S. Liang, Z. M. X, J. H. Yuan, S. Q. Yang, and L. S. Sun. 2000. Effect of *Ganoderma lucidum* polysaccharides on oxygen free radicals in murine peritoneal macrophages. *Zhongguo Yaolixue Yu Dulixue Zazhi Chi: Journal of Pharmacology and Toxicology* 14: 65–68.

Lieu, C. W., S. S. Lee, and S. Y. Wang. 1992. The effect of *Ganoderma lucidum* on induction of differentiation in leukemic U937 cells. *Anticancer Research* 12(4): 1211–1215.

Lin, I. H., D. M. Hau, and Y. H. Chang. 1996. Restorative effect of *Coriolus versicolor* polysaccharides against gamma-irradiation-induced spleen injury in mice. *Acta Pharmacologica Sinica* 17(2):102–104.

Lin, J. M., C. C. Lin, M. F. Chen, T. Ujiie, and A. Takada. 1995. Radical scavenger and antihepatoxic activity of *Ganoderma formosanum, Ganoderma lucidum* and *Ganoderma neo-japonicum*. *Journal of Ethnopharmacology* 47(1): 33–41.

Lin, Y., C. C. Lin, H. F. Chiu, J. J. Yang, and S. G. Lee. 1993. Evaluation of the anti-inflammatory and liver-protective effects of *Anoectochilus formosanus, Ganoderma lucidum* and *Gynostermma pentaphyllum* in rats. *American Journal of Chinese Medicine* 21: 59–69.

Lincoff, G. 1981. *The National Audubon Society field guide to North American mushrooms*. New York: Alfred A. Knopf.

Liu, X., J. P. Yuan, and X. J. Chen. 2002. Antitumor activity of the sporoderm-broken germinating spores of *Ganoderma lucidum*. *Cancer Letters* 182: 155–161.

Lovy, A., B. Knowles, R. Labbe, and L. Nolan. 1999. Activity of edible mushrooms against the growth of human T4 leukemia cancer cells, and *Plasmodium falciparum*. *Journal of Herbs, Spices & Medicinal Plants* 6(4): 49–57.

Madej, M. A., B. Barr, T. Curren, A. Bloom, and G. Gibbs. 2001. *Effectiveness of road restoration in reducing sediment loads*. Arcata, Calif.: U.S. Geological Survey, Redwood Field Station. www.werc.usgs.gov/redwood/project-doc.pdf.

Mader, P., A. Fleissbach, D. Dubois, L. Gunst, P. Fried, and U. Niggli. 2002. Soil fertility and biodiversity in organic farming. *Science* 296(5573): 1694–1697.

Manez, S., M. C. Recio, R. M. Giner, and J. L. Rios. 1997. Effect of selected triterpenoids on chronic dermal inflammation. *European Journal of Pharmacology* 334(1): 103–105.

Manohar, V., N. A. Talpur, B. W. Echard, S. Liberman, and H. G. Preuss. 2002. Effects of a water-soluble extract of maitake mushroom on circulating glucose/insulin concentrations in KK mice. *Diabetes, Obesity & Metabolism* 4(1): 43–48.

Maret, S. 1991. Fungi in Khanty folk medicine. *Journal of Ethnopharmacology* 31: 175–179.

Mau, J. L., P. R. Chen, and J. H. Yang. 1998. Ultraviolet irradiation increased vitamin D2 content in edible mushrooms. *Journal of Agricultural Food Chemistry* 46: 5269–5272.

Mayer, J., and J. Drews. 1980. The effect of a protein-bound polysaccharide from *Coriolus versicolor* on immunological parameters and experimental infections in mice. *Infection* 8(1): 13–21.

McClure, R. 2002. Extensive arsenic and lead pollution revealed: Report finds smelter fallout at many sites in King County. *Seattle Post Intelligencer*. Thursday, April 4, 2002.

Medsafe. 2000. Selenium. Prescriber Update Articles. New Zealand Medicines and Medical Devices Safety Authority: A Business Unit of the Ministry of Health. http://www.medsafe.govt.nz/Profs/PUarticles/Sel.htm.

Miller, K. 2004. Bacterial integrated circuits: By interfacing bacteria to silicon chips, NASA-supported researchers have created a device that can sense almost anything. Science@Nasa. June 10, 2004. http://science.nasa.gov/headlines/y2004/10jun _bbics.htm.

Milstein, O., R. Gersonde, A. Hutterman, M. J. Chen, and J. J. Meister. 1992. Fungal biodegradation of lignopolystyrene graft polymers. *Journal of Applied and Environmental Microbiology* 58(10): 3225–3232.

Mimura, S. 1904. Notes on shiitake culture. *Journal of the Forestry Society of Japan* 4.

Mimura, S. 1915. Notes on shiitake (*Cortinellus shiitake* Schrot) culture. *Bulletin of the Forest Experiment Station*, Bureau of Forestry, Department of Agriculture and Commerce, Meguro, Tokyo.

Min, B. S., J. J. Gao, N. Nakamura, and M. Hattori. 2000. Triterpenes from the spores of *Ganoderma lucidum* and their cytotoxicity against Meth-A and LLC tumor cells. *Chemical Pharmacology Bulletin* 48(7): 1026–1033.

Ming, D., J. Chilton, F. Fogarty, and G. H. Towers. 2002. Chemical constituents of *Ganoderma applanatum* of British Columbia forests. *Fitoterapia* 73(2): 147–152.

Minussi, R. C., S. G. de Moraes, G. M. Pastore, and N. Duran. 2001. Biodecolorization screening of synthetic dyes by four white-rot fungi in a solid medium: Possible role of siderophores. *Letters of Applied Microbiology* 33(1): 21–25.

Mizuno, T., ed. 1995. Mushrooms: The versatile fungus—Food and medicinal properties. *Food Reviews International* 11(1): 1–235.

Mizuno, T., and B. K. Kim. 1996. *A medicinal mushroom*, Ganoderma lucidum. Seoul, Korea: Il-Yang Pharmacy Co.

Mizuno, T., and C. Zhuang. 1995. Maitake, *Grifola frondosa*, pharmacological effects. *Food Reviews International* 111: 135–149.

Mizuno, T., C. Zhuang, K. Abe, H. Okamoto, T. Kiho, S. Ukai, S. Leclerc, and L. Meijer. 1996. Studies on the host-mediated antitumor polysaccharides, part XXVII. *Mushroom Science and Biotechnology* 3(2): 53–60.

Mizuno, T., C. Zhuang, K. Abe, H. Okamoto, T. Kiho, S. Ukai, S. Leclerc, and L. Meijer. 1999. Antitumor and hypoglycemic activities of polysaccharides from the sclerotia and mycelia of *Inonotus obliquus* (Pers.:Fr.) Pil. (Aphyllophoromycetideae). *International Journal of Medicinal Mushrooms* 1: 301–316.

Moder, M., T. Cajthaml, S. Schrader, and V. Sasek. 2002. Solid phase microextraction (SPME) used for direct fruit-body sampling in comparison to describe PAH partitioning in mushroom cultures. *Fresnius Environmental Bulletin* 11(6): 284–288.

Molina, R., D. Pilz, J. Smith, S. Dunham, T. Driesbach, T. O'Dell, and M. Castellano. 1997. Conservation and management of forest fungi in the Pacific Northwestern United States: An integrated ecosystem approach. In *Fungal conservation: Issues and solutions (British Mycological Society Symposia)*, ed. D. Moore, N. N. Nauta, S. E. Evans, and M. Rotheroe, pp. 19–63. Cambridge, UK: Cambridge University Press.

Mollison, B. 1990. *Permaculture: A practical guide for a sustainable future*. Washington, D.C.: Island Press.

Mondoa, E., and M. Kitei. 2001. *Sugars that heal: The new healing science of glyconutrients*. New York: Ballantine Publishing Group.

Money, N. P. 1998. More g's than the Space Shuttle: Ballistospore discharge. *Mycologia* 90(4): 547–558.

Money, N. P. 2004. The fungal dining habit: A biochemical perspective. *Mycologist* 18: 71–76.

Moore, D. 1998. *Fungal morphogenesis*. Cambridge, UK: Cambridge University Press.

Moreno, F. A., and P. L. Delgado. 1997. Hallucinogen-induced relief of obsessions and compulsions. *Journal of American Psychiatry* 154(7): 1037–1038.

Mori, S., K. Nakagawa-Yoshida, H. Tschuhasi, Y. Koreeda, M. Kawabata, Y. Nishura, M. Ando, and M. Osame. 1998. Mushroom worker's lung resulting from indoor cultivation of *Pleurotus ostreatus*. *Occupational Medicine* 48(7): 465–468.

Moyers, B. 2001. Trade Secrets: A Moyers Report. Public Affairs Television, Thirteen/WNET, New York.

Mueller, J. C., J. R. Gawley, and W. A. Hayes. 1985. Cultivation of the shaggy mane mushroom *(Coprinus comatus)* on cellulosic residues from pulp mills. *Mushroom Newsletter for the Tropics* (6)1: 15–20.

Murasugi, A., S. Tanaka, N. Komiyama, N. Iwata, K. Kino, H. Tsunoo, and S. Sakuma. 1991. Molecular cloning of a cDNA and a gene encoding an immunomodulatory protein, Ling Zhi-8, from a fungus, *Ganoderma lucidum*. *Journal of Biological Chemistry* 266(4): 2486–2493.

Mushworld. 2004. *Oyster mushroom cultivation: Mushroom grower's handbook 1*. Seoul, Korea: Mushworld.

Mushworld. 2005. *Shiitake mushroom cultivation*. Seoul, Korea: Mushworld.

Nakagaki, T., H. Yamada, and A. Toth. 2000. Maze-solving by an amoeboid organism. *Nature* 407: 470.

Nakazato, H., A. Koike, S. Saji, N. Ogawa, and J. Sakamoto. 1994. Efficacy of immunotherapy as adjuvant treatment after curative resection of gastric cancer. *The Lancet* 343(8906): 1122–1126.

Nanba, H. 1992. Immunostimulant activity in-vivo and anti-HIV activity in-vitro of 3 branched ß-1-6 glucans extracted from maitake mushroom *(Grifola frondosa)*. *Proceedings of the 8th International Conference on AIDS and the 3rd STD World Congress*. Amsterdam.

Nanba, H. 1993. Antitumor activity of orally administered D-fraction from maitake mushroom. *Journal of Naturopathic Medicine* 41: 10–15.

Nanba, H. 1995. Activity of maitake D-fraction to inhibit carcinogenesis and metastasis. *Annals of the New York Academy of Sciences* 768: 243–245.

Nanba, H. 1997. Maitake D-fraction: Healing and preventative potential for cancer. *Journal of Orthomolecular Medicine* 12: 43–49.

National Cancer Institute. 2004. Selenium and Vitamin E Cancer Prevention Trial (SELECT): Questions and answers. *Cancer Facts.* National Cancer Institute. http://cis.nci.nih.gov/fact/4_20.htm.

Ng, M. L., and A. T. Yap. 2002. Inhibition of human colon carcinoma development by lentinan from shiitake mushrooms (*Lentinus edodes*). *Journal of Alternative and Complementary Medicine* 8(5): 581–589.

Ng, T. B. 1998. A review of research on the protein-bound polysaccharide (polysaccharopeptide, PSP) from the mushroom *Coriolus versicolor. General Pharmacology* 30(1): 1–4.

Novotny, C., B. Rawal, M. Bhatt, M. Patel, V. Sasek, and H. P. Molitoris. 2001. Capacity of *Irpex lacteus* and *Pleurotus ostreatus* for decolorization of chemically different dyes. *Journal of Biotechnology* 89: 113–122.

Novotny, C., B. Rawal, M. Bhatt, M. Patel, V. Sasek, and H. P. Molitoris. 2003. Screening of fungal strains for remediation of water and soil contaminated with synthetic dyes. In *The utilization of bioremediation to reduce soil contamination: Problems and solutions,* ed. V. Sasek, J. A. Glaser, and P. Baveye, pp. 247–266. Dordrecht, The Netherlands: Kluwer Academic Publishers.

Obst, J., W. Coedy, and R. Bromley. 2001. Heavy metal analyses of wild edible mushrooms in the North Great Slave Lake region, Northwest Territories, Canada. Yellowknife, Northwest Territories: Report for the Government of Canada, Department of Indian and Northern Affairs.

Odani, S., K. Tominaga, S. Kondou, H. Hori, T. Koide, S. Hara, M. Isemura, and S. Tsunasawa. 1999. The inhibitory properties and primary structure of a novel serine proteinase inhibitor from the fruiting body of the basidiomycete, *Lentinus edodes. European Journal of Biochemistry* 262(3): 915–923.

O'Donnell, K., N. Weber, S. Rehner, and G. Mills. 2003. Phylogeny and biogeography of *Morchella. Fungal Genetics Conference Proceedings.* ARS/USDA.

Ohno, N., N. Asada, Y. Adachi, and T. Yadomae. 1995. Enhancement of LPS triggered TNF-alpha (tumor necrosis factor-alpha) production by 1‡3)-beta-D-glucans in mice." *Biological & Pharmaceutical Bulletin* 18(1): 126–133.

Ohno, N., T. Harada, S. Masuzawa, N. N. Miura, Y. Adachi, M. Nakajima, and T. Yadomae. 2002. Antitumor activity and hematopoietic response of a ß-glucan extracted from an edible and medicinal mushroom *Sparassis crispa* Wulf.:Fr. (Aphyllophoromycetideae). *International Journal of Medicinal Mushrooms* 4: 13–26.

Ohno, N., K. Iino, T. Takeyama, I. Suzuki, K. Sato, S. Oikawa, T. Miyazaki, and T. Yadomae. 1985. Structural characterization and antitumor activity of the extracts from matted mycelium of cultured *Grifola frondosa. Chemical & Pharmaceutical Bulletin* 33(8): 3395–3401.

Ohno, N., N. N. Miura, M. Nakajima, and T. Yadomae. 2000. Antitumor 1,3-beta-glucan from cultured fruit body of *Sparassis crispa. Biological & Pharmaceutical Bulletin* 23(7): 866–872.

Ohno, N., N. Sachiko, H. Toshie, M. N. Noriko, A. Yoshiyuki, N. Mitsuhiro, Y. Kenshi, Y. Hitoji, and Y. Toshiro. 2003. Immunomodulating activity of a b-glucan preparation, SCG, extracted from a culinary-medicinal mushroom, *Sparassis crispa* Wulf.: Fr. (Aphyllophoromycetideae), and application to cancer patients. *International Journal of Medicinal Mushrooms* 5(4): 359–368.

Ohno, N., I. Suzuki, S. Okawa, K. Sato, T. Miyazaki, and T. Yadomae. 1984. Antitumor activity and structural characterization of glucans extracted from cultured fruitbodies of *Grifola frondosa. Chemical & Pharmaceutical Bulletin* 32(3): 1142–1151.

Ohnogi, H., K. Sugiyama, H. Sagawa, and I. Kato. 2004. Remedies. U.S. Patent Application 20,040,175,396. Filed September 9, 2004.

Ohtomo, M. 2001. In vivo and in vitro test study: Physiological activity in immune response system of representative basidiomycetes. Unpublished research report provided to Fungi Perfecti from Tamagawa University, Japan.

Ooi, L. S., V. E. Ooi, and M. C. Fung. 2002. Induction of gene expression of the immunomodulatory cytokines in the mouse by a polysaccharide from *Ganoderma lucidum* (Curt.; Fr.) P. Karst. (Aphyllophoromycetideae). *International Journal of Medicinal Mushrooms* 4: 27–35.

Overholts, L. O. 1953. *The polyporaceae of the United States, Alaska and Canada.* Ann Arbor: University of Michigan Press.

Pagony, H. 1973. Publication details not available.

Pearce, M. H., and N. Malajczuk. 1990. Inoculation of Eucalyptus diversicolor thinning stumps with wood decay fungi for control of Armillaria luteobubalina. Mycological Research 94: 32–37.

Perera, C. O., V. J. Jasinghe, F. L. Ng, and A. S. Mujumdar. 2003. The effect of moisture content on the conversion of ergosterol to vitamin D in shiitake mushrooms. *Drying Technology* 21(6): 1091–1099.

Perry, D. 1994. *Forest ecosystems.* Baltimore, Md.: John Hopkins Press.

Petersen, R. H., and I. Krisai-Greilhuber. 1996. An epitype specimen for *Pleurotus ostreatus. Mycological Research* 100: 229–235.

Pilz, D., R. Molina, and L. H. Liegel. 1998. Biological productivity of chanterelle mushrooms in and near the Olympic Peninsula Biosphere Reserve. *Ambio– A Journal of the Human Environment.* Special Report Number 9, September.

Pilz, D., L. Norvell, E. Danell, and R. Molina. 2003. *Ecology and management of commercially harvested chanterelle mushrooms.* Portland, Ore.: USDA Forest Service, Pacific Northwest Research Station. PNW-GTR-576.

Piraino, F., and C. Brandt. 1999. Isolation and partial characterization of an antiviral, RC-183, from the edible mushroom, *Rozites caperata. Antiviral Research* 43: 67–68.

Pisha, E., H. Chai, I. S. Lee, T. E. Chagwedera, N. R. Farnsworth, G. A. Cordell, C. W. Beecher, H. H. Fong, A. D. Kinghorn, and D. M. Brown. 1995. Discovery of betulinic acid as a selective inhibitor of human melanoma that functions by induction of apoptosis. *Nature Medicine* 1(10): 1046–1051.

Prescott, C. 2001. Rehabilitation of forest roads and landings with wood waste. Vancouver: University of British Columbia, Science Council of British Columbia.

Price, M. S., J. J. Classen, and G. A. Payne. 2001. *Aspergillus niger* absorbs copper and zinc from swine wastewater. *Bioresource Technology* 77(1): 41–49.

Protiva, J., H. Skorkovska, J. Urban, and A. Vystrcil. 1980. Triterpenes and steroids from *Ganoderma applanatum. Collection of Czechoslovak Chemical Contributions* 45(10): 2710–2713.

Przybylowicz, P., and J. Donoghue. 1988. *Shiitake growers handbook: The art and science of mushroom cultivation.* Dubuque, Iowa: Kendall Hunt.

Purkayastha, R. P., A. K. Mitra, and B. Bhattacharyya. 1994. Uptake and toxicological effects of some heavy metals on *Pleurotus sajor-caju* (Fr.) Singer. *Ecotoxicology and Environmental Safety* 27(1): 7–13.

Qingtian, Z., et al. 1991. Antitumor activity of *Flammulina velutipes* polysaccharide (FVP). *Edible Fungi of China* 10:2: 11–15.

Qiu, X., and M. J. McFarland. 1991. Bound residue formation in PAH contaminated soil composting using *Phanerochaete chrysosporium. Hazardous Waste & Hazardous Materials* 8(2): 115–126.

Redberg, G., D. S. Hibbett, J. F. Ammirati, and R. Rodriguez. 2003. Phylogeny and genetic diversity of *Bridgeoporus nobilissimus* inferred using mitochondrial and nuclear rDNA sequences. *Mycologia* (95): 836–845.

Reshef, A., I. Moulalem, and P. Weiner. 1984. Acute and long-term effect of exposure to basidiomycetes spores to mushroom growers. *The Journal of Allergy and Clinical Immunology* 81(1): 275.

Ritz, K., and I. M. Young. 2004. Interactions between soil structure and fungi. *Mycologist* 18: 52–59.

Rodriguez, E., F. J. Ruiz-Dueñas, A. T. Martínez, and M. J. Martínez. 2004. Cloning and characterization of a new laccase from the white-rot fungus *Pleurotus eryngii.* 7th European Conference on Fungal Genetics Copenhagen, April 17–20.

Rosecke, J., and W. A. Konig. 2000. Constituents of various wood-rotting basidiomycetes. *Phytochemistry* 54(6): 603–610.

Rossman, A. 1994. A strategy for an all-taxa inventory of fungal biodiversity. In *Biodiversity and terrestrial ecosystems,* ed. C.-I. Peng and C. H. Chou, pp. 169–194. Taipei: Institute of Botany. Academia Sinica Monograph Series 14.

Ruttman-Johnson, C., D. Cullen, and R. T. Lamar. 1994. Manganese peroxidases of the white rot fungus *Phanerochaete sordida. Applied and Environmental Microbiology* 60(2): 599–605.

Rzymowska, J. 1998. The effect of aqueous extracts from *Inonotus obliquus* on the mitotic index and enzyme activities. *Bollettino Chimico Farmaceutico* 137(1): 13–15.

Sack, U., and T. Gunther. 1993. Metabolism of PAH by fungi and correlation with extracellular enzymatic activities. *Journal of Basic Microbiology* 33(4): 269–277.

Sakagami, H., T. Aoki, A. Simpson, and S. I. Tanuma. 1991. Induction of immunopotentiation activity by a protein-bound polysaccharide, PSK. *Anticancer Research* 11: 993–1000.

Sakagami, H., K. Sugaya, A. Utsumi, S. Fujinaga, T. Sato, and M. Takeda. 1993. Stimulation by PSK of interleukin-1 production by human peripheral blood mononuclear cells. *Anticancer Research* 13(3): 671–675.

Sakagami, H., and M. Takeda. 1993. Diverse biological activity of PSK (Krestin): A protein-bound polysaccharide from *Coriolous versicolor* (Fr.) Quel. In *Mushroom Biology and Mushroom Products*, ed. S.-T. Chang, J. A. Buswell, and S.-W. Chiu, pp. 237–245. Hong Kong: The Chinese University Press.

Samajpati, N. 1979. Nutritive value of some Indian edible mushrooms. In *Mushroom Science X: Proceedings of the Tenth International Congress on the Science and Cultivation of Edible Fungi*, ed. J. Delmas, part 2, pp. 695–703.

Sano, M., K. Yoshino, T. Matsuzawa, and T. Ikekawa. 2002. Inhibitory effects of edible higher basidiomycetes mushroom extracts on mouse type IV allergy. *International Journal of Medicinal Mushrooms* 4(1): 37–41.

Sanzen, I., N. Imanishi, N. Takamatsu, S. Konosu, N. Mantani, K. Terasaw, K. Tazaw, Y. Odaira, M. Watanabe, M. Takeyama, and H. Ochiai. 2001. Nitric oxide–mediated antitumor activity induced by the extract of *Grifola frondosa* (maitake mushroom) in a macrophage cell line, RAW264.7. *Journal of Experimental Clinical Cancer Research* 20(4): 591–597.

Sargar, N. 1992. The occurrence of *Macrolepiota rachodes* on wood ant nests in England and on the ground in Oregon. *Transactions of the Mycological Society of Japan* 33: 487–496.

Sarkar, S., J. Koga, R. J. Whitley, and S. Chatterjee. 1993. Antiviral effect of the extract of culture medium of *Lentinus edodes* mycelia on the replication of herpes simplex virus 1. *Antiviral Research* 20(4): 293–303.

Sasaki, T., Y. Arai, T. Ikekawa, G. Chihara, and F. Fukuoka. 1971. Antitumor polysaccharides from some Polyporaceae, *Ganoderma applanatum* (Pers.) Pat and *Phellinus linteus* (Berk. et Curt.) Aoshima. *Chemical & Pharmacological Bulletin* 19(4): 821–826.

Sasek, V. 2003. Why mycoremediations have not yet come into practice. In *The utilization of bioremediation to reduce soil contamination: Problems and solutions*, ed. V. Sasek, J. A. Glaser, and P. Baveye, pp. 247–266. Dordrecht, The Netherlands: Kluwer Academic Publishers.

Sasek, V., M. Bhatt, T. Cajthaml, K. Malachova, and D. Lednicka. 2003. Compost-mediated removal of polycyclic aromatic hydrocarbons from contaminated soil. *Archives of Environmental Contamination and Toxicology* 44: 336–342.

Savage, D., N. Neal, and R. Villard. 2003. Hubble helps confirm oldest known planet. NASA News, Release 03-234, July 10.

Schlegel, B., U. Luhmann, A. Hartl, and U. Grafe. 2000. Piptamine, a new antibiotic produced by *Piptoporus betulinus* Lu 9-1. *Journal of Antibiotics* 53(9): 973–974.

Schliephake, K., W. L. Baker, and G. T. Longergan. 2003. Decolorization of industrial wastes and degradation of dye water. In *Fungal Biotechnology in Agricultural, Food and Environmental Applications*, ed. D. Arora, P. D. Bridge, and D. Bhatnagar. New York: Marcel Dekker.

Schwartz, P., and D. Randall. 2003. *An abrupt climate change scenario and its implacations for United States national security.* Report prepared for the Global Business Network by the Department of Defense. http://www.ems.org/climate/pentagon_climatechange.pdf.

Shin, Y., Y. Tamai, and M. Terazawa. 2000. Chemical constituents of *Inonotus obliquus* I. *Eurasian Journal of Forestry Research* 1:43–50.

Sia, G. M., and J. K. Candlish. 1999. Effect of shiitake (*Lentinus edodes*) extract on human neutrophils and the U937 monocytic cell line. *Phytotherapy Research* 13(2): 133–137.

Silliman, B., and S. Y. Newell. 2003. Fungal farming by a snail. *Proceedings of the National Academy of Sciences* 100(26): 15643–15648.

Simard, S. W., D. A. Perry, M. D. Jones, D. D. Myrold, D. M. Durall, and R. Molina. 1997. Net transfer of carbon between ectomycorrhizal tree species in the field. *Nature* 388: 579–582.

Singer, R. 1986. *The Agaricales in modern taxonomy.* 4th ed. Königstein, Germany: Koeltz Scientific Books.

Slejkovec, Z., A. R. Byrne, T. Stijve, W. Goessler, and K. J. Irgolic. 1997. Arsenic compounds in higher fungi. *Applied Organometallic Chemistry* 11: 673–682.

Slivova, V., T. Valachovicova, J. Jiang, and D. Silva. 2004. *Ganoderma lucidum* inhibits invasiveness of breast cancer cells. *Journal of Cancer Integrative Medicine* 2(1): 25–30.

Small, E. J., M. W. Frohlich, and R. Bok. 2000. Prospective trial of the herbal supplement PC-SPES in patients with progressive prostate cancer. *Journal of Clinical Oncology* 18: 3595–3603.

Smania, A., F. Delle Monache, E. F. A. Smania, and R. S. Cuneo. 1999. Antimicrobial activity of steroidal compounds isolated from *Ganoderma applanatum* (Pers.) Pat. (Aphyllophoromycetideae) fruitbody. *International Journal Of Medicinal Mushrooms* 1: 325–330.

Smania, A., F. D. Monache, C. Loguericio-Leite, E. F. A. Smania, A. L. Gerber. 2001. Antimicrobial activity of basidiomycetes. *International Journal of Medicinal Mushrooms* 3: 87.

Smith, A. H. 1949. *Mushrooms in their natural habitats.* New York: Hafner Press.

Smith, A. H., and L. R. Hesler. 1968. *The North American species of* Pholiota. New York: Hafner Publishing.

Smith-Weber, N., and A. H. Smith. 1985. *A field guide to southern mushrooms.* Ann Arbor: University of Michigan Press.

Spolar, M. R., R. B. Beelman, D. J. Royse, E. Schaffer, and J. A. Milner. 1998. Selenium enrichment of fresh mushrooms *(Agaricus bisporus)*. *Proceedings of the Sixth International Symposium on the Uses of Selenium and Tellurium,* pp. 65–73.

Stamets, P. 1978. *Psilocybe mushrooms and their allies.* Seattle, Wash.: Homestead Book Co.

Stamets, P.1999. *Psilocybin mushrooms of the world.* Berkeley, Calif.: Ten Speed Press.

Stamets, P. 2000a. *Growing gourmet and medicinal mushrooms.* Berkeley, Calif.: Ten Speed Press.

Stamets, P. 2000b. Techniques for the cultivation of the medicinal mushroom Royal Sun Agaricus—*Agaricus blazei* Murr. (Agaricomycetideae). *International Journal of Medicinal Mushrooms* 2: 151–160.

Stamets, P. 2001. New anti-viral compounds from mushrooms. *Herbalgram* 51: 24, 27.

Stamets, P. 2002a. The noble polypore *(Bridgeporus nobilissimus)*: An ancient mushroom of many mysteries. *International Journal of Medicinal Mushrooms* 4(4): 355–358.

Stamets, P. 2002b. Novel antimicrobials from mushrooms. *Herbalgram* 54: 29–33.

Stamets, P. 2003. Potentiation of cell-mediated host defense using fruitbodies and mycelia of medicinal mushrooms. *International Journal of Medicinal Mushroom* 5(2): 179–192.

Stamets, P. 2005. Antimicrobial activity from medicinal mushrooms *(Fomitopsis)*. U.S. Patent Application Serial No. 11/029,861. Filed January 4, 2005.

Stamets, P., and J. Chilton. 1983. *The mushroom cultivator.* Olympia, Wash.: Agarikon Press.

Stamets, P., and D. Yao. 2002. *MycoMedicinals: An informational treatise on the medicinal properties of mushrooms.* Olympia, Wash.: Mycomedia Productions, Fungi Perfecti.

Standish, L. J. K. Lawson, D. Yee, D. McKenna, C. Werner, K. Rinn, P. Stamets, G. Ostroff, and M. Haberman. 2004. Approved, in vitro tests on-going. J. S. McCune, "Project 3: Phase I/II clinical trial of *Trametes versicolor* in women with breast cancer" University of Minnesota/Bastyr University Developmental CAM Research Center Proposal. http://nccam.nih.gov/research/extramural/awards /2004/.

Stavinoha, W. B. 1997. Status of *Ganoderma lucidum* in United States: *Ganoderma lucidum* as an anti-inflammatory agent. *Proceedings of the 1st International Symposium on Ganoderma lucidum.* Tokyo, November 17–18, pp. 99–103.

Stavinoha, W. B., J. T. Slama, and S. T. Weintraub. 1996. The anti-inflammatory activity of *Ganoderma lucidum* 6.1. In *Ganoderma lucidum*, a Medicinal Mushroom, *Ganoderma lucidum* Ganoderma, Polyporaceae and others. Oriental Tradition, Cultivation, Breeding, Chemistry, Biochemistry and Utilization of *Ganoderma lucidum*, ed. T. Mizuno, pp.193–196. Seoul, Korea: Il-Yang Pharm Co.

Stavinoha, W. B., S. Weintraub, T. Opham, A. Colorado, R. Opieda, and J. Slama. 1990. Study of the anti-inflammatory activity of *Ganoderma lucidum*. *Proceedings from the Academic/Industry Conference (AIJC)*, Sapporo, Japan, August 18–20.

Stierle, A.,G. Strobel, D. Stierle, P. Grothaus, and G. Bignami. 1995. The search for a taxol-producing microorganism among the endophytic fungi of the Pacific yew, *Taxus brevifolia*. *Journal of Natural Products* 58(9): 1315–1324.

Stijve, T., D. Andrey, G.-F. Luchini, and W. Goessler. 2001. Simultaneous uptake of rare earth elements, aluminium, iron, and calcium by various macromycetes. *Australasian Mycologist* 20: 92–98.

Stijve, T., A. Pittet, D. Andrey, M. A. L. de Almeida Amazonas, and W. Goessler. 2003. Potential toxic constituents of *Agaricus brasiliensis* (*A. blazei ss.* Heinem.), as compared other cultivated and wild-growing edible mushrooms. *Deutsche Lebensmittel-Rundschau* 99: 475–481.

Straight, R. 2001. Getting buffers on the ground: A group effort. *Inside Agroforestry* 3: 1–7.

STUK (Radiation and Nuclear Safety Authority). 2003. EC recommendation sets new limits to cesium concentrations in wild food products on the market. July 8, 2003. www.stuk.fi/english/news2/news_12.html.

Suay, I., F. Arenal, F. Asenio, A. Basilio, M. Cabello, M. T. Diez, J. B. Garcia, A. Gonzalez del Val, J. Gorrochategui, P. Hernandez, F. Pelaez, and M. F. Vicente. 2000. Screening of basidiomycetes for antimicrobial activities. *Antonie van Leeuwenhoek* 78: 129–139.

Sugimachi, K., Y. Maehara, M. Ogawa, T. Kakegawa, and M. Tomita. 1997. Dose intensity of uracil and tegafur in postoperative chemotherapy for patients with poorly differentiated gastric cancer. *Cancer Chemotherapy and Pharmacology* 40(3): 233–238.

Sumerlin, D. 2004. Personal communication.

Sun, H., C. G. Zhao, X. Tong, and Y. P. Qip. 2003. A lectin with mycelia differentiation and antiphytovirus activities from the edible mushroom *Agrocybe aegerita*. *Journal of Biochemistry and Molecular Biology* 36(2): 214–222.

Sundareshwar, P. V., J. Morris, and E. Koepfler. 2003. Phosphorus limitation of coastal ecosystem processes. *Science* 299: 563–565

Suzuki, H., K. Iiyama, O. Yoshida, S. Yamazaki, N. Yamamoto, and S. Toda. 1990. Structural characterization of immunoactive and antiviral water-solubilized lignin in an extract of the culture medium of *Lentinus edodes* mycelia (LEM). *Agricultural and Biological Chemistry* 54(2): 479–487.

Suzuki, I., T. Itani, N. Ohno, S. Oikawa, K. Sato, T. Miyazaki, and T. Yadomae. 1984. Antitumor activity of a polysaccharide fraction extracted from cultured fruitbodies of *Grifola frondosa*. *Journal of Pharmacobio-dynamics* 7(7): 492–500.

Takeyama, T., I. Suzuki, N. Ohno, S. Oikawa, K. Sato, M. Ohsawa, and T. Yadomae. 1987. Host-mediated antitumor effect of Grifolan NMF-5N, a polysaccharide obtained from *Grifola frondosa*. *Journal of Pharmacobio-dynamics* 10(11): 644–651.

Tao, J., and K. Y. Feng. 1990. Experimental and clinical studies on inhibitory effect of *Ganoderma lucidum* on platelet aggregation. *Journal of Tongji Medical University* 10(4): 240–243.

Thomas, S. A. 2000. Personal communication.

Thomas, S. A., P. Becker, M. R. Pinza, J. Q. Word, and P. Stamets. 1999. Mycoremediation: A method for test-to-pilot scale application. In *Phytoremediation and innovative strategies for specialized remedial applications*, ed. A. Leeson and B. C. Alleman. Columbus, Ohio: Battelle Press.

Tokuyama, T., Y. Hayashi, M. Nishizawa, H. Toruda, S. M. Chairul, and Y. Hayashi. 1991. Applanoxidic acids A, B, C and D, biologically active tetracyclic triterpenes from *Ganoderma applanatum*. *Phytochemistry* 30(12): 4105–4109.

Tornberg, K., E. Baath, and S. Olsson. 2003. Fungal growth and effects of different wood decomposing fungi on the indigenous bacterial community of polluted and unpolluted soils. *Biology and Fertility of Soils* 37: 190–197.

Trudell, S. A., P. T. Rygiewicz, and R. L. Edmonds. 2003. Nitrogen and carbon stable isotope abundances support the myco-heterotrophic nature of host-specificity of certain achlorophyllus plants. *New Phytologist* 160: 391–401.

Tsujinaka, T., M. Yokota, J. Kambayashi, M. C. Ou, Y. Kido, and T. Mori. 1990. Modification of septic processes by beta-glucan administration. *European Surgical Research* 22(6): 340–346.

Tsukagoshi, S., Y. Hashimoto, G. Fujii, H. Kobayashi, K. Nomoto, and K. Orita. 1984. Krestin (PSK). *Cancer Treatment Review* 11: 131–155.

Turner, N. J., R. Bouchard, and D. I. D. Kennedy. 1980. Ethnobotany of the Okanagan-Colville Indians of British Columbia and Washington. *Occasional Papers of the British Columbia Provincial Museum* 21: 16–17.

Ukai, S., T. Kiho, C. Hira, I. Kuruma, and Y. Tanaka. 1983. Polysaccharides in fungi. XIV. Anti-inflammatory effect of the polysaccharides from the fruitbodies of several fungi. *Journal of Pharmacobiodynamics* 6(12): 983–990.

Urbanelli, S., C. Fanelli, A. A. Fabbri, V. Della Rosa, L. Maddau, F. Marras, and M. Reverberi. 2002. Molecular genetic analysis of two taxa of the *Pleurotus eryngii* complex: *P. eryngii* (DC.Fr.) Quèl. var. *eryngii* and *P. eryngii* (DC.Fr.) Quel. var. *ferulae. Biological Journal of the Linnean Society* 75(1): 125–136.

USDA. 1993. *Forest ecosystem management: An ecological, economic, and social assessment.* Report of the Forest Ecosystem Management Assessment Team, Northwest Forest Plan Documents, USDA Forest Service.

Van de Bogart, F. 1976–1979. The genus *Coprinus* in western North America I: *Mycotaxon* 4: 233–275; II: *Mycotaxon* 8:243–291; *Mycotaxon* 10: 154–174.

Varma, A., S. Verma, Sudha, N. Sahay, B. Butehorn, and P. Franken. 1999. *Piriformospora indica*, a cultivatable plant-growing root endophyte. *Applied and Environmental Microbiology* 65(6): 2741–2744

Vellinga, E. C. 2003. Type studies in Agaricaceae — *Chlorophyllum rachodes* and allies. *Mycotaxon* 85: 259–270.

Vellinga, E. C., P. J. de K. Rogier, and T. D. Burns. 2003. Phylogeny and taxonomy of Macrolepiota (Agaricaceae). *Mycologia* 95(3): 442–456.

Venturella, G. 2002. On the real identity of *Pleurotus nebrodensis* in Spain. *Mycotaxon* 84: 445–446.

Vilgalys, R., and B. L. Sun. 1994. Ancient and recent patterns of geographic speciation in the oyster mushroom *Pleurotus* revealed by phylogenetic analysis of ribosomal DNA sequences. *Proceedings of the National Academy of Sciences* 91: 4599–4603.

Vole, J., N. P. Denisova, F. Nerud, and V. Musilek. 1985. Glucose-2-oxidase activity in mycelial cultures of basidiomycetes. *Folia Microbiologica* (Prague) 30(2): 141–147.

Wang, H. X., and T. B. Ng. 2000. Isolation of a novel ubiquitin-like protein from *Pleurotus ostreatus* mushroom with anti-human immunodeficiency virus, translation-inhibitory, and ribonuclease activities. *Biochemical and Biophysical Research Communications* 276(2): 587–593.

Wang, J., J. J. Zhang, and W. W. Chen. 1985. Study of the action of *Ganoderma lucidum* on scavenging hydroxyl radicals from plasma. *Journal of Traditional Chinese Medicine* 5(1): 44–60.

Wang, S. Y., M. L. Hsu, H. C. Hsu, C. H. Tzeng, S. S. Le, M. S. Shiao, and C. K. Ho. 1997. The anti-tumor effect of *Ganoderma lucidum* is mediated by cytokines released from activated macrophages and T lymphocytes. *International Journal of Cancer* 70(6): 669–705.

Wasser, S., M. Berreck, and K. Haselwandter. 2003. Radiocesium contamination of wild-growing mushrooms in Ukraine. *International Journal of Medicinal Mushrooms* 5: 61–86.

Wasser, S. P., M. Y. Didukh, A. de Meijer, M. L. A. de A. Amazonas, E. Nevo, and A. F. da Eira. 2002. Is a widely cultivated culinary-medicinal Royal Sun Agaricus (the Himematsutake mushroom) indeed *Agaricus blazei* Murrill? *International Journal of Medicinal Mushrooms* 4: 267–290.

Wasser, S. P., and A. L. Weis. 1999. Medicinal properties of substances occurring in higher basidiomycetes mushrooms: Current perspectives (Review). *International Journal of Medicinal Mushrooms* 1(1): 31–62.

Watanabe, Y., K. Nakanishi, N. Komatsu, T. Sakabe, and H. Terakawa. 1964. Flammulin, antitumor substance. *Bulletin of the Chemical Society of Japan* 37: 747–750.

Watling, R. 1998. Larger fungi and some of earth's major catastrophes. In *Fungi and ecological disturbance: Proceedings of the Royal Society of Edinburgh*, ed. L. Boddy, R. Watling, and A. J. Hyon, series B, pp. 49–60.

Weil, A. 1963. The strange case of the Harvard drug scandal. *Look Magazine* 27(22): 46.

Weil, A. 1998. *Natural health, natural medicine.* Boston: Houghton Mifflin.

Weil, A. 2004. Ask Dr. Weil: Fish oil; Selenium and breast cancer. *Self Healing* February, p. 6.

Wines, M. 2002. Radioactive Produce Season in Moscow. *New York Times.* Thursday, September 12.

Woodward, S., H. Y. Sultan, D. K. Barrett, and R. B. Pearce. 1993. Two new antifungal metabolites produced by *Sparassis crispa* in culture and in decayed trees. *Journal of General Microbiology* 139: 153–159.

Word, J., S. A. Thomas, P. Becker, M. Huesemann, T. E. Divine, F. Roberto, and P. Stamets. 1997. Adaptation of mycofiltration phenomena for wide-area and point-source decontamination of CW/BW agents: Proof of concept. LDRD Battelle Marine Sciences Laboratory report, July 14.

Yamada, Y., H. Nanba, and H. Kuroda. 1990. Antitumor effect of orally administered extracts from fruit body of *Grifola frondosa* (maitake). *Chemotherapy* (Tokyo) 38(8): 790–796.

Yamamoto Y., H. Shirono, K. Kono, and Y. Ohashi. 1997. Immunopotentiating activity of the water-soluble lignin rich fraction prepared from LEM— the extract of the solid culture medium of *Lentinus edodes* mycelia. *Bioscience, Biotechnology, and Biochemistry* 61(11): 1909–1912.

Yang, M. M., Z. Chen, and J. S. Kwok. 1992. The anti-tumor effect of a small polypeptide from *Coriolus versicolor* (SPCV). *American Journal of Chinese Medicine* 20(3-4): 221–232.

Yang, Q. Y., and M. Wang. 1994. The effect of *Ganoderma lucidum* extract against fatigue and endurance in the absence of oxygen. *Proceedings of Contributed Symposia 59 A,B. 5th International Congress.* Vancouver, August 14–21, pp. 101–104.

Yang, W., T. Hu, and D. Win. 1992. The experiments on the *Ganoderma lucidum* extract for its anti-aging and invigoration effects. The 4th International Symposium on Ganoderma lucidum. Seoul, Korea.

Yaoita Y., Y. Yoshihara, R. Kakuda, K. Machida, and M. Kikuchi. 2002. New sterols from two edible mushrooms, *Pleurotus eryngii* and *Panellus serotinus*. *Chemical & Pharmaceutical Bulletin* 50(4): 551–553.

Ying, J. 1987. *Icons of medicinal fungi.* Beijing: Science Press.

Yoshida, I., T. Kiho, S. Usui, M. Sakushima, and S. Ukai. 1996. Polysaccharides in fungi. XXXVII. Immuno-modulatory activities of carboxymethylated derivatives of linear (1↓3)-alpha-D-glucans extracted from the fruiting bodies of *Agrocybe cylindracea* and *Amanita muscaria Biological & Pharmaceutical Bulletin* 19(1): 114–121.

Zadrazil, F. 1977. The conversion of starch into feed by basidiomycetes. *European Journal of Applied Microbiology* 4: 273.

Zeng, Q., J. Zhao, and Z. Deng, 1990. The anti-tumor activity of *Flammulina velutipes* polysaccharide (FVP). *Edible Fungi of China* 10: 2.

Zervakis, G. I., G. Venturella, and K. Papdopoulou. 2001. Genetic polymorphism and taxonomic infra-structure of the *Pleurotus eryngii* species-complex as determined by RAPD analysis, isozyme profiles and ecomorphological characters. *Microbiology* 147(Pt11): 3183–3194.

Zhang, H., F. Gong, Y. Feng, and C. Zhang. 1999. Flammulin purified from the fruitbodies of *Flammulina velutipes* (Curt.:Fr.) P. Karst. *International Journal of Medicinal Mushrooms* 1: 89–92.

Zhang, J., Q. Tang, M. Zimmerman-Kordmann, W. Reutter, and H. Fan. 2002. Activation of B lymphocytes by GLIS, a bioactive proteoglycan from *Ganoderma lucidum*. *Life Science* 71(6): 623–638.

Zhang, Y., G. L. Mills, and M. G. Nair. 2003. Cyclo-oxygenase inhibitory and antioxidant compounds from the fruiting body of an edible mushroom, *Agrocybe aegerita*. *Phytomedicine* 10(5): 386–390.

Zhao, J. D. 1989. *The Ganodermataceae in China.* Berlin and Stuttgart.: J. Cramer.

Zhou, S., and Y. Gao. 2002. The immunomodulating effects of *Ganoderma lucidum* (Curt.: Fr) P. Karst. (Ling Zhi, reishi mushroom) (Aphyllophoromyce-tideae). *International Journal of Medicinal Mushrooms* 4(1): 1–12.

Zhu, H. S., X. L. Yang, L. B. Wang, D. X. Zhao, and L. Chen. 2000. Effects of extracts from sporoderm-broken spores of *Ganoderma lucidum* on HeLa cells. *Cell Biology and Toxicology* 16(3): 201–206.

Zhu, M., Q. Chang, L. K. Wong. F. S. Chong, and R. C. Li. 1999. Triterpene antioxidants from *Ganoderma lucidum*. *Phytotherapy Research* 13: 529–531.

Zusman, I., R. Reifen, O. Livini, P. Smirnoff, P. Gurevich, B. Sandler, A. Nyska, R. Gal, Y. Tendler, and Z. Madar. 1997. Role of apoptosis, proliferating cell nuclear antigen and p53 protein in chemically induced colon cancer in rats fed corncob fiber treated with the fungus *Pleurotus ostreatus*. *Anti-cancer Research* 17(3C): 2105–2113.

PHOTOGRAPHY AND ARTWORK CREDITS

Unless otherwise indicated below, photographs are by Paul Stamets.

Steve Allen: Figure 246
Mike Amaranthus Group: Figures 42, 81, and 227
Anonymous: Figures 340 and 341
David Arora: Figures 44 and 298
Michael Beug: Figures 110 and 354
Shiuan Chen: Figures 229 and 230
Bill Cheswick and Hal Burch: Figure 4
Ann Drum: Figures 58, 104, and 255–257
Jonathan Frank: Figure F
Lawrence Gilbert: Figure 115
Richard Gaines: Figure 114
Roger Gold Group: Figure 120
James Gouin: Figure 50
Linda Greer: Figure 198
Ann Gunter: Figures 48, 116, and 141
Ed Handja: Figure 80
Marie Heerkens: Figure 262
Omon Isikhuemhen: Figure 273
Jmol: an open-source Java viewer for chemical structures in 3D: Figure B
Bala Kottapalli: Figures G and H
Michael Kuo: Figure 300
Diana Leemon: Figure C
Charles Lefevre: Figures 40 and 41
Andrew Lenzer: Figure 226 and frontispiece
Taylor Lockwood: Figures 243, 247, 322, 326, 327, and 349
Barrett Lyon: Page vi
Charles Meissner: Figure D

Neal Mercado: Figure 119
Hank Morgan: Figure 3
Thomas Newmark: Figure 54
Toshuyiki Nakagaki: Figure 5
NASA: Figure 11
George Osgood: Figure 105
Mary Parrish: Figure E
John Plischke: Figure 319
Reinhold Poeder: Figure 254
David Price: Figure 361
Zeger Reyers: Figure 328
Steve Rooke: Figures 1, 12, 18, and 19 (enhanced)
Catharine Scates: Figures 201 and 336
Ethan Schaffer: Figures 205 and 206
Elinoar Shavit: Figure 274
Luiz Amaro Pachoa de Silva: Figure 303
Mitchel Sogin: Figure A (adapted by Andrew Lenzer)
Azureus Stamets: Figure 66
David Sumerlin: Figure 113
Satit Thaithatgoon: Figure 43
Susan Thomas: Figures 100–102
Tim Thornewell: Figures 355 and 356
Co du Trong: Figures 271 and 281
U. S. Forest Service: Figure 9
Thomas Volk: Figures 357 and 358
Solomon Wasser: Figure 324
Rebecca and William Webb: Figure 187
Dusty Yao: Figure 29

INDEX